Audel™
Basic Electronics

Paul Rosenberg

Wiley Publishing, Inc.

Vice President and Executive Group Publisher: Richard Swadley
Vice President and Publisher: Joseph B. Wikart
Executive Editor: Carol Long
Senior Production Editor: Pamela Hanley
Development Editor: Kevin Shafer
Text Design & Composition: TechBooks

Published simultaneously in Canada

ISBN: 0-7645-7900-2
ISBN 13: 978-0-7645-7900-9

Library of Congress Cataloging-in-Publication Data:

ISBN: 0-7645-7900-2

Manufactured in the United States of America

10 9 8 7 6 5 4 3 2 1

Contents

Introduction

Nothing changed life on Earth during the twentieth century more than electronic technologies, and nothing in the past can compare to it. The development of electronics has been a unique benefit to humankind.

This book explains electronics from the ground up, covering the core essentials of this ever-improving field in understandable terms and following (more or less) the chronology of the real developments in the field.

Chapters 1 and 2 of this book explain what electricity is, where it comes from, and why it is so incredibly important.

Chapters 3 through 5 explain all the basics of electrical and electronic circuits, power, voltage, current, resistance, impedance, resonance, magnetism, and more. These chapters lay the necessary foundation for everything that follows.

Chapters 6 and 7 cover semiconductors and semiconductor devices. These are the components that more or less define electronics today, and they are explained in considerable detail.

Chapter 8 explains optoelectronics—the intersection of electronics and optics. Chapter 9 switches gears momentarily and covers all of the components (such as wires, solder, and fuses) that are necessary for constructing electronic circuits and systems.

Chapters 10, 11, and 12 cover amplifiers, oscillators, and filters, which are some of the core electronic technologies. These specialized circuits are used in hundreds (perhaps thousands) of applications.

Chapter 13 turns the reader's attention to the most important of new developments in electronics—digital electronics. This is what underlies all computer and microprocessor technologies.

Chapters 14 through 19 cover all of the great electronic technologies: radio, television, fiber optics, audio, radar, and computers.

Also included are a full glossary and appendixes that contain electronics symbols, abbreviations, and tips on a number of electronic circuits.

No doubt *Basic Electronics* will find broad use in technical schools, as well as for self-instruction. It should apply equally well to either. Complex mathematics is avoided, and a separate appendix is devoted to the mathematic skills that are required.

Finally, review questions and exercises are included with each chapter. This allows students to test themselves, to reinforce what they have learned, and provides instructors with extra learning tools.

I wrote this book with one primary thought in mind: to explain the most complex and difficult electronic technologies in the simplest and clearest ways possible. I hope that after reading it you will agree that I've accomplished my objective.

Paul Rosenberg

About the Author

Paul Rosenberg is a contributing editor at *Power Outlet* magazine, a past president of the Fiber Optic Association, and teaches engineering courses for Iowa State University. He has written approximately 40 books and hundreds of articles for the electrical and electronics industries. Paul frequently lectures at industry events and works as an expert witness.

Chapter 1

What is Electricity?

Among all the discoveries mankind has made, almost nothing can compare to the use of electricity. Perhaps learning how to use fire was as big an event in its day, but that day was long, long ago.

I won't bother trying to list the ways electricity contributes to all our lives; the list is endless. As I used to tell my students, "without electricity, we all go back to the farm." (Out of simple decorum I passed over an explanation of what farm life was really like before electricity revolutionized it. It wasn't pretty.)

I should add that *electricity* and *electronics* are really the same thing. We separate them for some convenience, but these are rather artificial divisions. In general, we use the word electricity for higher-powered applications and electronics for more intricate applications. However, electronic devices such as semiconductors are commonly used for very high levels of power, and electrical devices are used in very complex circuitry. In this chapter, I am using the term electricity to apply to everything elecronic as well as to power transformers and the like.

The Invisible Force

Of course, the great problem with understanding electricity is that it is invisible. We may see its effects, but we never see the electricity itself.

For ages, men saw the lightning (explosions from the skies) and even the mysterious effects of electricity's twin, magnetism. But they could never see the cause of these effects—electricity itself. And so, it remained a deep and frightening mystery for a very long time.

The truth is that even today, we do not see electricity itself. We have come to understand electricity by application of the scientific method. We have experimented, hypothesized, and verified enough times that we are now 100 percent certain of how electricity behaves.

What is certain about electricity is that an electrical current consists of electrical charges moving through a conductor. But the explanations of this certainty have changed over time. Many texts describe electrical current as a flow of electrons. I have described electricity that way, and some time-honored electrical formulas still use that terminology. But recent research indicates that it is not a flow of electrons per se, but a flow of charges, that makes an electrical current.

Not many years ago, electrical current was thought to flow from positive to negative. Texts containing the positive-to-negative theory remained into the 1970s. However, we now have much better information that indicates a negative-to-positive flow.

The interesting thing is that, on a daily basis, an explanation of what exactly electricity might be does not matter. We know with certainty how it works, and that is more than enough. A perfect understanding of what electricity is would be very nice, but it isn't necessary for making profligate use of the stuff.

Amazing Usefulness

Electricity is unique among any form of energy known to man. It can be used at supremely small or large levels. So far, we've been able to do this with no other form of energy. None of us for example, has ever seen a nuclear-powered car, much less a nuclear-powered clock radio. As impressive and as useful as nuclear technology can be at large scale, we've never been able to use it at smaller scales.

Electricity can be used at minute and finely controlled levels in the smallest microcircuitry. Billions of complex circuits are combined in one computer chip smaller than a postage stamp. At the same time, however, it can be used to power entire cities. High-voltage lines operate at hundreds of thousands of volts, and a single circuit of this type can power huge factories, or even cities. To put it into half-technical illustrative terms, a single wire can carry as much power as could be provided by 10,000 horses.

Added to the amazing range of electricity is the fact that it can be easily delivered to wherever we want it. Small, cheap copper wires are easily used to provide large amounts of power to almost any location imaginable. In fact, the wiring in your house makes the power of three horses available at every receptacle in your living room, dining room, bedroom, and so on. (And for you city folks: Horses are *really* strong.) Your home's central air conditioner is provided with more power than nine horses!

Consider this: There is a real historical reason why we use the term *horsepower*. Before electricity, people really did have to use horses for power. When electricity came in, people had to equate its power with something they were familiar with, hence horsepower. Just for fun, try imagining the actual process of using horses to provide power for the things you do every day. It's pretty entertaining.

Now, beyond just being able to use bizarrely large or small amounts of electricity, we can also do a lot of other things to it.

We can turn electricity on and off very easily. In fact, we can turn it on and off hundreds or thousands of times in one second. (Think about that horse again.)

We can change the intensity level of electricity. We can easily, cheaply, and instantly make it a lot stronger or weaker. It can be strong enough to literally blow things up or be made so mild as to be undetectable by humans, animals, or even to microbes. That, by itself, is more than amazing.

We can make electricity change directions. Moreover, we can make it change direction as slowly or as quickly as we like. We can use electricity in one direction only (called *direct current*, or *DC*), or we can make it alternate its directions, going back and forth (*alternating current*, or *AC*). And, if we want to use an alternating current, we can make it change directions with whatever *frequency* we like. If we want it to reverse 10 times a second, we can. If we want to increase the frequency of the reversals to a million times per second, we can. The really intriguing thing about this is that electricity acts differently at these different frequencies of reversal. It's astonishing, really.

There is so much to be said on this subject that deciding where to stop is a real judgment call. I will conclude by saying that there are almost innumerable tricks that we can do with electricity. We can speed it up, slow it down, make it larger, make it smaller, and change its characteristics in a dozen different ways. When we begin to combine these tricks, we find that their number increases all the more. Amazing, amazing stuff.

The Invisible Twin

At around the time of Ben Franklin's famous kite experiment, mankind was beginning to learn how to control electricity. Immeditely following these discoveries, they stumbled across something that astonished them all: Electricity and magnetism are twin phenomena.

Every time an electrical current flows through a conductor, it creates a magnetic field around that conductor. There are no exceptions, and this phenomenon cannot be eliminated. Electricity and magnetism go together, or they go not at all.

This process goes in reverse as well. Anytime you pass a conductor through a magnetic field, a current is forced through the conductor.

Electricity causes magnetism. Magnetism causes electricity. They cannot be separated.

We do have a fairly good idea of how magnetism works, in that it involves the alignment of tiny *magnetic domains* (think of these as micromagnets) in magnetic materials. (Iron is the only really useable magnetic material.) Somehow (and I will not attempt to explain this here), the passing of a current through a wire and the alignment of these magnetic domains causes the same phenomenon.

The critical thing is that electrical energy can be turned into magnetic energy, and magnetic energy can be turned into electrical energy. This can be done instantly and easily, and the characteristics of the electricity and magnetism can be modified during the process. This is truly an amazingly useful fact.

When you understand that this transformation between electricity and magnetism lies at the core of radio transmission, power transformers, and lighting, you begin (but *only* begin) to see the incredible usefulness of the invisible twins: electricity and magnetism.

Devices That Do Intelligent Work

Other energy technologies provide power. Steam engines are very strong, for example. But no other technology can do intelligent work—electricity can.

By putting together two very basic electrical devices (resistors and capacitors), we can build timers. Furthermore, these timers are 100 percent reliable, and a beginning electronics student can build timers that will react at whatever length of time he or she wants.

We have devices that turn light into electricity and electricity into light. We have others that turn alternating current into direct current, that switch things on and off with no moving parts, that allow a small current to control a large current, and that filter electronic signals, even circuits that make decisions. All of this is done easily and with 100 percent reliability.

Note that when I say "done easily," I am referring to using the truths about electronics that we now hold. Making the discoveries themselves required a lot of very hard work by a scattered group of determined geniuses and eccentric rogues.

On our end of this chain of development, we have machines that do our mathematics for us, correct our spelling, and allow us to communicate, easily and at almost no expense, with people half a globe away. How much further we will go and how fast, no one knows, but every year sees several important new developments. It will be an interesting ride to be sure. And a few of us will rise to the level of determined geniuses and eccentric rogues and make new discoveries that open half a dozen new fields of opportunity—thus doing more to improve the daily lives of men and women than any politician could ever hope to.

Summary

This chapter provided background information on electricity. Chapter 2 discusses the three basic factors affecting all electricity and electronics.

Review Questions

1. How have we learned to predict how electricity will behave?
2. Explain, in your own words, what the text refers to as "the amazing range of electricity."
3. When we refer to the *frequency* of an electric current, what exactly do we mean?
4. Explain the connection between electricity and magnetism.
5. What is a *magnetic domain*, and in what kind of material would you find one?
6. Describe a few of the things that electrical devices can do.

Chapter 2

The Primary Factors

There are only three basic factors regarding the operation of all electrical and electronic circuits. Yes, there are many other exotic and useful factors that are used, but there are only three basics:

- Force
- Current
- Resistance

This chapter explains these three factors in some detail, as well as some introductory coverage of the other factors that are derived from them. It is very important that you understand this material completely. This is the foundation, and everything subsequent will be built upon this base.

These three basic factors are not difficult to understand, but you will have to go through these explanations as slowly as necessary to grasp them completely. Pay close attention in this chapter (and in all the chapters, really) to understanding *why* these things operate as they do. It is not enough to know only what they do; you must also understand the reasons why they behave as they do. If you can understand why, everything in electronics will become much easier for you, and you'll be much better in whatever area of electronics you work.

Once we have covered the three primary factors related to electrical current in circuits, we will move on to magnetism, insulators, and associated subjects. These are all basics that will underly everything you see or do in electronics. It is important that you understand them fully.

The Three Primaries

The three primary factors in the operation of electricity are (in common language) voltage, current flow, and resistance. These are the fundamental things that control every electrical circuit everywhere. This section introduces these three factors, and uses a comparison to the flow of water to provide a comparative illustration of how electricity operates.

Voltage is the force that pushes the current through electrical circuits. The scientific name for voltage is *electromotive force*, and is represented in formulas with the capital letter *E*. (You may occasionally see electromotive force represented as *V* in a formula, but *E* is the correct symbol to use.) It is measured in volts. The scientific definition of *volts* is the electromotive force necessary to force 1 ampere of current to flow through a resistance of 1 ohm.

In comparing electrical systems to water systems, voltage is comparable to water pressure. The more pressure there is, the faster the water will flow through the system. Likewise, with electricity, the higher the voltage (electrical pressure), the more current will flow through any electrical system.

Current (which is measured in *amperes*, or simply *amps*) is the rate of flow of electrical current. The scientific description for current is intensity of current flow, and it is represented in formulas with the capital letter I. The scientific definition of ampere is a flow of 6.25×10^{23} electrons (called one *coulomb*) per second.

I compares with the rate of flow in a water system, which is typically measured in gallons per minute. In simple terms, electricity is thought to be the flow of electrons through a conductor. Therefore, a circuit that has 12 amps flowing through it will have three times as many electrons (or electrical charges) flowing through it as a circuit that has a current of 4 amps.

Resistance is opposition to the flow of electricity. Some materials, such as glass, offer a very high resistance to the flow of electricity. Other materials, such as metals, offer little resistance to the flow of electricity. To continue with the water illustration, a drinking straw would have a high resistance to water flow. Even at a relatively high pressure, getting a barrel of water through a drinking straw would take a long time. A 36-inch sewer pipe, on the other hand, would have a very low resistance to the flow of water. You could easily run a barrel of water through this pipe in less than a second, even at a very low pressure.

Resistance is the more traditional and more common term for resistance to current flow. You will see it used in formulas, represented by the symbol R. Resistance is measured in ohms, and ohms are represented by the Greek capital letter omega (Ω).

A more modern (and better) term for the opposition to the flow of electrical current is *impedance*. Impedance is the total opposition to the flow of electricity. Like resistance, impedance is measured in ohms, but is represented in formulas by the letter Z.

The scientific definition of an ohm is the amount of resistance that will restrict 1 volt of potential to a current flow of 1 ampere.

It is important to differentiate between impedance and resistance. Resistance is the more traditional term. Unfortunately, it is the less accurate term. Impedance is better. Resistance is a fine term for a direct-current circuit, but not for an alternating-current circuit.

Impedance can be used for either direct- or alternating-current circuits.

Chapter 4 explains the differences between resistance and impedance. Until then, let's generally use direct current and the term *resistance*. This is a simpler way of explaining the operation of electricity. However, once the distinctions between resistance and impedance have been properly made, the terms will be used precisely.

Origins Inside of Atoms

Electricity begins in atoms. This book is not a physics text, but it is necessary for you to understand a few terms.

Key Terms

Atoms are regarded as the smallest particles that retain the properties of an element. Figure 2-1 shows the simplest atom, hydrogen.

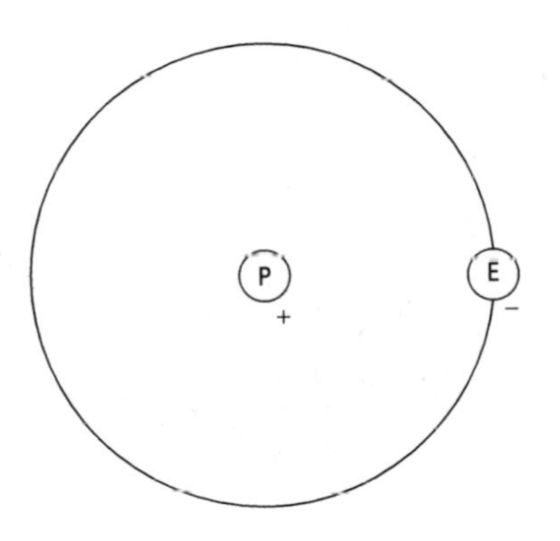

Figure 2-1 Hydrogen atom with one proton and one electron.

Elements are substances that can't be changed, decomposed by ordinary types of chemical change, or made by chemical union. There are more than 100 known elements, distinguishable by their chemical and physical differences. Some common elements are copper, silver, gold, oxygen, hydrogen, sulphur, zinc, lead, iron, and nitrogen. Table 2-1 provides a list of the known elements and their characteristics.

A *molecule* is the smallest unit quantity of matter that can exist by itself and retain all the properties of the original substance. It consists of one or more atoms. A common molecule, water (H_20), contains two atoms of hydrogen, one atom of oxygen, as shown in Figure 2-2. Note also that a molecule can contain one of more elements of the same type. An example of this, a molecule of nitrogen, appears in Figure 2-3.

Table 2-1 The Elements and Their Characteristics

Element	Symbol	Atomic No.	Atomic wt.	Specific Gravity	Melting Point °C	Boiling Point °C
Actinium	Ac	89	227	10.07	1051	3198
Aluminum	Al	13	26.981538	2.6989	660.32	2519
Americium	Am	95	243	13.67	1176	2011
Antimony	Sb	51	121.76	6.61	630.63	1587
Argon	Ar	18	39.948	1.7837	−189.35	−185.85
Arsenic	As	33	74.9216	5.73	817	603
Astatine	At	85	210	—	302	—
Barium	Ba	56	137.327	3.5	727	1897
Berkelium	Bk	97	247	14.00	1050	—
Beryllium	Be	4	9.012182	1.848	1287	2471
Bismuth	Bi	83	208.98038	9.747	271.40	1564
Bohrium	Bh	107	264	—	—	—
Boron	B	5	10.811	2.37	2075	4000
Bromine	Br	35	79.904	3.12	−7.2	58.8
Cadmium	Cd	48	112.411	8.65	321.07	767
Calcium	Ca	20	40.078	1.55	842	1484
Californium	Cf	98	251	—	900	—
Carbon	C	6	12.0107	1.8–3.5	4492	3825
Cerium	Ce	58	140.116	6.771	798	3443
Cesium	Cs	55	132.90545	1.873	28.5	671
Chlorine	Cl	17	35.453	1.56	−101.5	−34.04
Chromium	Cr	24	51.9961	7.18–7.20	1907	2671

Cobalt	Co	27	58.9332	8.9	1495	2927
Copper	Cu	29	63.546	8.96	1084.62	2562
Curium	Cm	96	247	13.51	1345	3100
Darmstadtium	Ds	110	281	—	—	—
Dubnium	Db	105	262	—	—	—
Dysprosium	Dy	66	162.5	8.540	1412	2567
Einsteinium	Es	99	252	—	860	—
Erbium	Er	68	167.259	9.045	1529	2868
Europium	Eu	63	151.964	5.283	822	1529
Fermium	Fm	100	257	—	1527	—
Fluorine	F	9	18.9984032	1.108	−219.67	−188.12
Francium	Fr	87	223	—	27	—
Gadolinium	Gd	64	157.25	7.898	1313	3273
Gallium	Ga	31	69.723	5.904	29.76	2204
Germanium	Ge	32	72.64	5.323	938.25	2833
Gold	Au	79	196.96655	19.32	1064.18	2856
Hafnium	Hf	72	178.49	13.31	2233	4603
Hassium	Hs	108	277	—	—	—
Helium	He	2	4.002602	0.1785	−272.2	−268.934
Holmium	Ho	67	164.93032	8.781	1474	2700
Hydrogen	H	1	1.00794	0.070	−259.34	−252.87
Indium	In	49	114.818	7.31	156.60	2072
Iodine	I	53	126.90447	4.93	113.7	184.4
Iridium	Ir	77	192.217	22.42	2446	4428

(continued)

Table 2-1 (continued)

Element	Symbol	Atomic No.	Atomic wt.	Specific Gravity	Melting Point °C	Boiling Point °C
Iron	Fe	26	55.845	7.894	1538	2861
Krypton	Kr	36	83.8	3.733	–157.38	–153.22
Lanthanum	La	57	138.9055	6.166	918	3464
Lawrencium	Lr	103	262	—	1627	—
Lead	Pb	82	207.2	11.35	327.46	1749
Lithium	Li	3	6.941	0.534	180.50	1342
Lutetium	Lu	71	174.967	9.835	1663	3402
Magnesium	Mg	12	24.305	1.738	650	1090
Manganese	Mn	25	54.938049	7.21–7.44	1246	2061
Meitnerium	Mt	109	268	—	—	—
Mendelevium	Md	101	258	—	827	—
Mercury	Hg	80	200.59	13.546	–38.83	356.73
Molybdenum	Mo	42	95.94	10.22	2623	4639
Neodymium	Nd	60	144.24	6.80 & 7.004	1021	3074
Neon	Ne	10	20.1797	0.89990	–248.59	–246.08
Neptunium	Np	93	237	20.25	644	
Nickel	Ni	28	58.6934	8.902	1455	2913
Niobium (Columbium)	Nb	41	92.90638	8.57	2477	4744
Nitrogen	N	7	14.0067	0.808	–210.00	–195.79
Nobelium	No	102	259	—	827	—
Osmium	Os	76	190.23	22.57	3033	5012

Oxygen	O	8	15.9994	1.14	–218.79	–182.95
Palladium	Pd	46	106.42	12.02	1554.9	2963
Phosphorous	P	15	30.973761	1.82	44.15	280.5
Platinum	Pt	78	195.078	21.45	1768.4	3825
Plutonium	Pu	94	244	19.84	640	3228
Polonium	Po	84	209	9.32	254	962
Potassium	K	19	39.0983	0.862	63.5	759
Praseodymium	Pr	59	140.90765	6.772	931	3520
Promethium	Pm	61	145	—	1042	3000
Protactinium	Pa	91	231.03588	15.37	1572	—
Radium	Ra	88	226	5.0	700	—
Radon	Rn	86	222	4.4	–71	–61.7
Rhenium	Re	75	186.207	21.02	3186	5596
Rhodium	Rh	45	102.9055	12.41	1964	3695
Rubidium	Rb	37	85.4678	1.532	39.30	688
Ruthenium	Ru	44	101.07	12.44	2334	4150
Rutherfordium	Rf	104	261	—	—	—
Samarium	Sm	62	150.36	7.536	1074	1794
Scandium	Sc	21	44.95591	2.989	1541	2836
Seaborgium	Sg	106	266	—	—	—
Selenium	Se	34	78.96	4.79	220.5	685
Silicon	Si	14	28.0855	2.33	1414	3265
Silver	Ag	47	107.8682	10.5	961.78	2162
Sodium	Na	11	22.98977	0.971	97.80	883

(continued)

Table 2-1 (continued)

Element	*Symbol*	*Atomic No.*	*Atomic wt.*	*Specific Gravity*	*Melting Point °C*	*Boiling Point °C*
Strontium	Sr	38	87.62	2.54	777	1382
Sulfur	S	16	32.065	2.07	95.3	444.60
Tantalum	Ta	73	180.9479	16.654	3017	5458
Technetium	Tc	43	98	11.50	2157	4265
Tellurium	Te	52	127.60	6.24	449.51	988
Terbium	Tb	65	158.92534	8.234	1356	3230
Thallium	Tl	81	204.3833	11.85	304	1473
Thorium	Th	90	232.0381	11.72	1750	4788
Thulium	Tm	69	168.93421	9.314	1545	1950
Tin (white)	Sn	50	118.71	7.31	231.93	2602
Titanium	Ti	22	47.867	4.55	1668	3287
Tungsten	W	74	183.84	19.3	3422	5555
Uranium	U	92	238.02891	19.05	1135	4131
Vanadium	V	23	50.9415	6.11	1910	3407
Xenon	Xe	54	131.293	3.52	−111.79	−108.12
Ytterbium	Yb	70	173.04	6.972	819	1196
Yttrium	Y	39	88.90585	4.457	1522	3345
Zinc	Zn	30	65.39	7.133	419.5	907
Zirconium	Zr	40	91.224	6.506	1855	4409

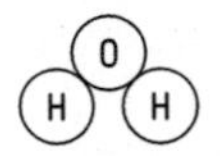

Figure 2-2 One molecule of water, containing one atom of oxygen and two of hydrogen.

Figure 2-3 One molucule of nitrogen, containing two nitrogen atoms.

The Basic Structure of an Atom

Atoms are composed of two main parts: a nucleus, and electrons that move around the nucleus. Figure 2-4 shows a very simple atom, helium. Note that there are two electrons moving around the nucleus in this case.

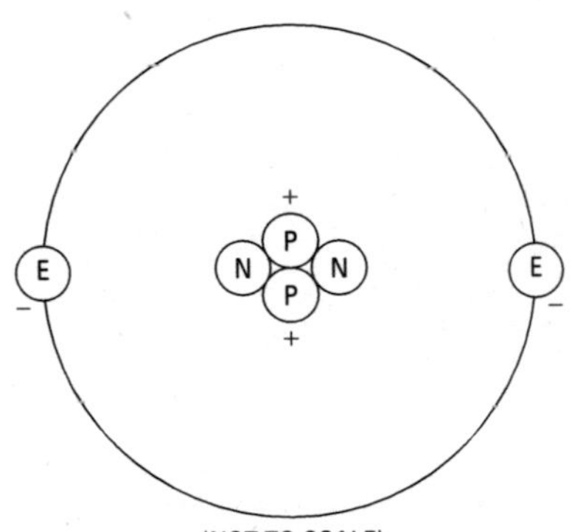

Figure 2-4 Helium atom with two protons, two neutrons, and two electrons.

There are three types of particles that make up atoms:

- *Protons* are positively charged particles.
- *Electrons* are negatively charged particles.
- *Neutrons* are particles that have no electrical charge at all.

Referring to Figure 2-4, notice that the nucleus contains both protons and neutrons. This is always the case. Protons and neutrons always form the nucleus of an atom, and electrons always move around the nucleus. Note that it is common for people to say that electrons orbit around the nucleus. That is not entirely true.

Electrons seem to travel in figure-eight types of paths rather than circular orbits.

An electron in a hydrogen atom is the same as an electron in a uranium atom. A proton or a neutron from one type of atom is the same as a proton or neutron in any other type of atom.

Protons, neutrons, and electrons all have mass. That is, they all have weight, although, obviously, a very small amount of weight because of their submicroscopic sizes. An electron has less than one-thousandth the mass of a proton or a neutron, which are approximately equal in mass.

Referring to Table 2-1, you can see that the *atomic number* of an atom is a measure of the number of protons in the nucleus. So, the atomic number for hydrogen is 1, for helium is 2, for lithium is 3, and so on, which means that hydrogen has 1 proton, helium has 2 protons, lithium has 3 protons, and so on.

The number of protons and electrons is equal in a stable atom. The positive charges and the negative charges balance themselves.

Electrons may be released from their atoms by various means. Some atoms of certain elements release their electrons more readily than atoms of other elements. If an atom has an equal number of electrons and protons, it is said to be *in balance*. If an atom has given up some of its electrons, the atom will then have a *positive charge*, and the matter which received the electrons from the atom have a *negative charge*. Some external force must be used to transfer the electrons.

Current

As mentioned earlier, an electrical current is the flow of electrons and/or charges through a conductor. For purposes of explanation and for simplicity, this discussion uses examples of electrons moving through a conductor.

Since conductors are composed from atoms, this means that these electrons must move from atom to atom, traveling from one end of a conductor toward the other end. Figure 2-5 shows the movement of atoms along a string of atoms. As a free electron enters the string of atoms from the left, it enters the electron orbit of the first atom, throwing the atom out of balance (giving it a negative charge, rather than being balanced). This has the effect of pushing another electron out of its orbit, and on to the next atom to the right. This process is repeated at supremely high speed, with the electron (or charge) traveling down the length of the conductor at near the speed of light (300,000 km/sec).

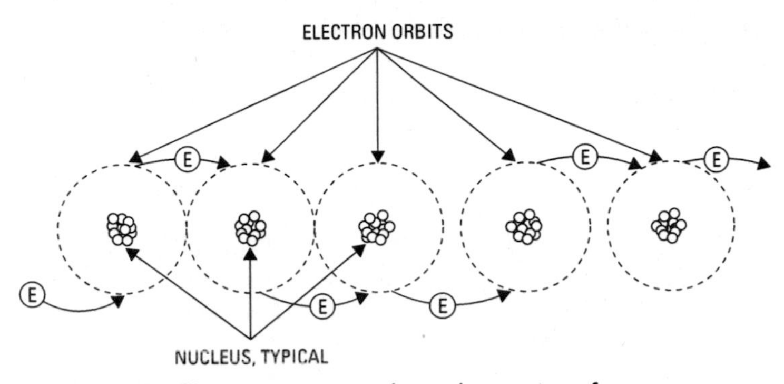

Figure 2-5 Electrons moving through a series of atoms.

The amount of current is measured in *amperes*. Like most electrical terms, this one commemorates one of the great discoverers of electricity. Andre-Marie Ampere was a French physicist of the early nineteenth century and did groundbreaking work on the relationships between electricity and magnetism.

One ampere represents a current flow of one *coulomb* of electrons (or charge) per second. A coulomb is 6.24×10^{18} electrons. So, 1 ampere of current requires the movement of 6,240,000,000,000,000,000 electrons or charges through a conductor. So, by comparing Figure 2-5 to this (huge) number, you can see that a tremendous number of electrons and atoms are involved in this process, traveling at very high speeds.

You may also hear people describe an ampere as the amount of current that will flow when 1 volt is applied to a resistance of 1 ohm. This is true, but it is not the best way to state the definition of an ampere. An ampere is 1 coulomb of electrical charge passed through one cross-sectional plane of a conductor in 1 second.

Voltage

As described earlier, voltage is properly termed electromotive force. As the name implies, voltage is the force that moves electrons and charges (refer to Figure 2-5). Voltage is the thing that forces the electrons in Figure 2-5 to move through the conductor. The conductor, left alone, would have no electrons moving through it. Even the most useful type of conductor—copper wire—has no current at all moving through it until a voltage is applied. Once a voltage is connected to the wire, a great deal of electricity can move through that wire, but without voltage it is as just as electrically dead as a rock.

There are several ways to obtain this electrical force. The critical thing to understand now, however, is that voltage is the invisible force that makes the electrons move.

The unit of voltage, the volt, is named after Alessandro Volta, the Italian scientist who invented the electric battery. The volt has a technical definition (1 joule of energy per coulomb of charge) that is neither necessary nor helpful to a first-level student. A volt is the amount of electromotive force required to push 1 ampere of current through a resistance on 1 ohm.

Resistance

As its name implies, resistance is the resistance of a material to the flow of electricity.

We measure resistance in *ohms*. One ohm is the amount of resistance that will allow one volt of electrical pressure to cause one ampere of current to flow. The symbol for ohm is the Greek letter omega: Ω. (Using the letter "O" would have been confusing—people would have continually mistaken it for a zero. The Greek omega roughly corresponds to the O in English.)

The ohm is named after Georg Simon Ohm, the German physisist who developed the first laws of electricity. (The fundamentals of these laws are discussed later in this chapter in the section "Ohm's Law.")

Some materials, such as copper wire, have a very low resistance to the flow of electricity. Others, such as glass or rubber, have a very high resistance to the flow of electricity. This, rather obviously, is why we use copper wires as electrical conductors, and why we use rubber and plastic to insulate those wires. We divide these materials into two general classifications: conductors and insulators. Copper is a conductor, and most plastics would be insulators.

As with voltage and current, resistance begins with atoms. Atoms may have several layers of electrons orbiting around their nucleus. These are usually called *electron shells*. In Figure 2-6, you can see that the Boron atom has two electron shells. In this case, the inner shell contains two electrons, and the outer shell contains three electrons. Also notice that the nucleus contains five protons, which is equal to the number of electrons in the atom. Looking at Table 2-1 you can see that Boron has an atomic number of 5, indicating that Boron has five protons and electrons.

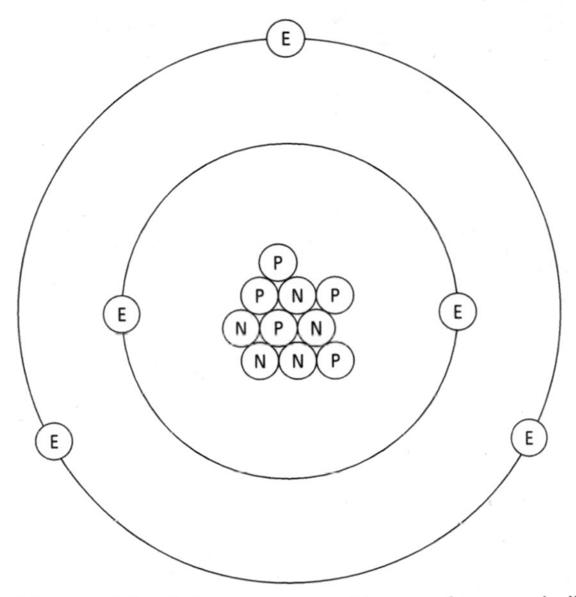

Figure 2-6 A boron atom with two electron shells.

The innermost electron shell of an atom can have only as many as two electrons. The outermost electron shell can have between one and eight electrons. Intermediate shells can have up to 18 electrons.

The only shell that really matters, as far as conducting electricity matters, is the outer electron shell. The outer shell encases the rest of the atom, and it is the part of the atom that interacts with all other atoms. We call electrons in this outer shell *valence* electrons. It should also be mentioned that chemical reactions also depend on the outer shell of electrons. The number of valence electrons determines how molecules form and how chemicals react.

Elements that are good electrical conductors have only one electron in their outer electron shells. Elements that are good insulators (poor conductors) have seven or eight valence electrons. It seems that it is much easier to get a lone electron to move from its place than to move one that is grouped with several other electrons.

Ohm's Law

From the explanations of the three primary factors of electricity, you may have noticed that voltage, current, and resistance have constantly been compared to each other. These relationships are calculated by using what is called *Ohm's Law*.

Ohm's Law explains the relationship between voltage, current, and resistance in an electrical circuit. This law is accurate and absolute. (When our discussion turns to alternating current in Chapter 4, we will find that it still applies, with a few other factors being involved.)

Ohm's Law states that current is directly proportional to voltage, and inversely proportional to resistance. Accordingly, the amount of voltage is equal to the amount of current multiplied by the amount of resistance. Ohm's Law goes on to say that current is equal to voltage divided by resistance, and that resistance is equal to voltage divided by current.

If there is any one electrical formula to be remembered, it is certainly Ohm's Law.

The fundamental statement of Ohm's Law is as follows:

The current in any electrical circuit is equal to the electromotive force (emf or voltage) imposed upon that circuit, divided by the entire resistance of the circuit.

The equation may be expressed as follows:

$$I = \frac{E}{R}$$

where the following is true:

I = intensity of current in amperes

E = emf in volts

R = resistance in ohms

If any two quantities of this equation are known, the third quantity may be found by transposing as follows:

$E = IR$

$R = E/I$

$I = E/R$

Analogy of Ohm's Law

A water system provides a workable analogy to illustrate electrical currents and the effect of friction. We all know from experience that water running through a hose encounters resistance. Thus with *X* pounds of pressure at the hydrant, we will get a considerable flow of water directly out of the open hydrant. If we use 50 feet of 3/4-inch hose, the flow will be cut down, and if we use 100 feet of 3/4-inch hose, the flow will be cut even more. The same will be true if we use 50 feet of 1/2-inch hose: We get less water with the same *X* pounds of pressure at the hydrant than we did with the 3/4-inch hose.

The *X* pounds of pressure may be compared to volts. The quantity of water delivered may be compared to current (amperes), and the friction (resistance) of the various hoses may be compared to the resistance of the conductor in ohms.

Illustrations of Ohm's Law

Now, if we deliver 100 volts to the circuit in Figure 2-7A, and the load resistance is 50 ohms, there will be a current of 2 amperes, as shown by the following equation:

$$I = \frac{E}{R} = \frac{100\ V}{50\ \Omega} = 2\,A$$

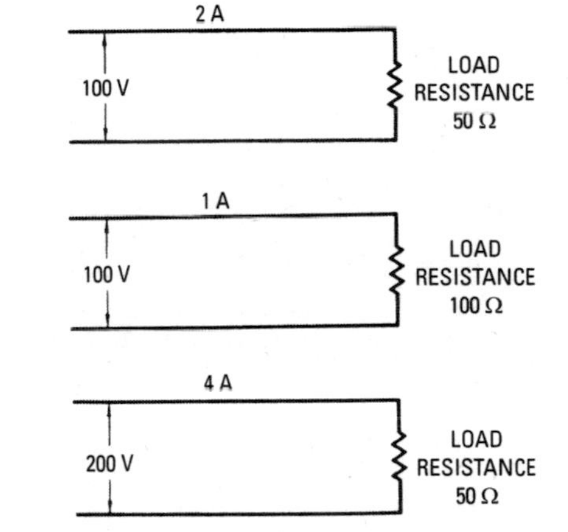

Figure 2-7 Examples of Ohm's Law.

In Figure 2-7B, we have the same voltage, but the load resistance is 100 ohms, so we find the following:

$$I = \frac{E}{R} = \frac{100\ V}{100\ \Omega} = 1\,A$$

In Figure 2-7C, we have doubled the voltage to 200 volts and have 50 ohms of load resistance, so the calculation is as follows:

$$I = \frac{E}{R} = \frac{200\ V}{50\ \Omega} = 4\,A$$

So, you can see that doubling the voltage on the same load as shown in Figure 2-7A will double the current from 2 to 4 amperes.

A couple of examples will help explain the principle.

Problem 1

An incandescent lamp is connected on a 110-volt system. The resistance (R) of the heated filament is 275 ohms. What amount of current will the lamp draw?

Answer:

$$I = \frac{E}{R} = \frac{110\ V}{275\ \Omega} = 0.4 A$$

Problem 2

Two electric heaters have resistances of 20 and 40 ohms, respectively, with 120 and 240 volts available. Calculate current amperes for the following combinations:

(a) 20 ohms on 120 volts:

$$I = \frac{E}{R} = \frac{120 V}{20\ \Omega} = 6 A$$

(b) 40 ohms on 120 volts:

$$I = \frac{E}{R} = \frac{120 V}{40\ \Omega} = 3 A$$

(c) 20 ohms on 240 volts:

$$I = \frac{E}{R} = \frac{240 V}{20\ \Omega} = 12 A$$

(d) 40 ohms on 240 volts:

$$I = \frac{E}{R} = \frac{240 V}{40\ \Omega} = 6 A$$

From these examples, you may readily see that doubling the resistance on the same voltage halves the current. Also, doubling the voltage with the same resistance doubles the current.

You may deduce that the current in any circuit will vary directly with the emf and that the current in an electrical circuit varies inversely with the resistance. If the current in an electrical circuit is to be kept constant, the resistance must be varied directly with the emf.

On the other hand, if the emf is to be kept constant, the resistance must be varied inversely with the current.

The Ohm's Law Circle

One of the easiest ways to remember, learn, and use Ohm's Law is with the circle diagram shown in Figure 2-8. This Ohm's Law circle can be used to obtain all three of these formulas easily.

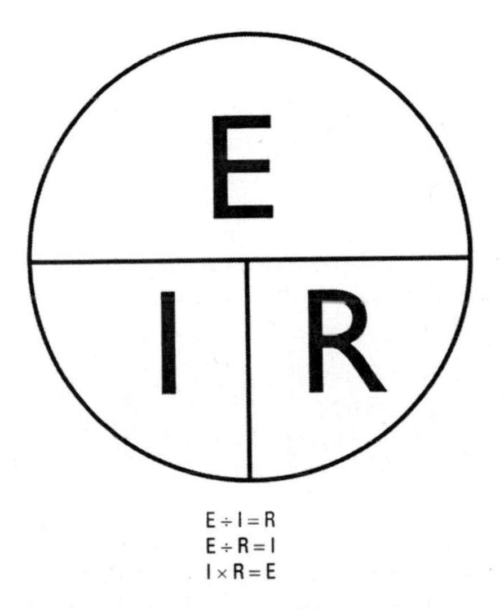

Figure 2-8 The Ohm's Law circle.

The method can be described as follows. Place your finger over the value that you want to find (*E* for voltage, *I* for current, or *R* for resistance), and then the other two will make up the formula.

For example, if you place your finger over the E in the circle, the remainder of the circle will show $I \times R$. If you then multiply the current by the resistance, you will get the value for voltage in the circuit.

If you wanted to find the value for current, you would put your finger over the *I* in the circle, and then the remainder of the circle will show *E* over *R*, or $E \div R$. So, to find current we divide voltage by resistance.

Lastly, if you place your finger over the R in the circle, the remaining part of the circle shows $E \div I$. These Ohm's Law formulas apply to any electrical circuit, no matter how simple or how complex.

Once you are comfortable with Ohm's Law, you will be able to continue onto the next important subject, which is power.

Power

Power is a measurement of work done per unit of time. So, if you were to move five bags of concrete from one place to another in 1 minute, and another person required 2 minutes to move the same concrete the same distance, you have twice as much power as the other person. You did the same amount of work in half the time. Power is work over time.

We measure power in many ways. One term you've heard used many times is *horsepower*, which has a real scientific definition. One horsepower equals 550 foot-pounds per second. This measurement was determined by James Watt (about whom you'll hear more in just a moment), who determined that a typical horse could lift 550 pounds at a rate of 1 foot per second. You can see from this how power is equal to work over time.

The term horsepower is used occasionally in the electrical field, but mostly as applies to the older inventions (for example, to electric motors, which are almost always measured in horsepower). However, a much more useful measurement of power is the *watt*.

The watt takes its name from the aforementioned James Watt, the English inventor who developed the first rotary steam engine. This is the machine that made the Industrial Revolution possible. The watt is represented by the letter *W* and is defined as the amount of power produced by a current of 1 ampere, under a pressure of 1 volt. For comparison, 1 horsepower equals 746 watts. One *kilowatt* (the measurement the power companies use on their bills) equals 1000 watts. Watt discovered the relationships between voltage, current, resistance, and power. We call this set of relationships Watt's Law, although Watt's discoveries are generally expressed as a set of formulas rather than as a specific statement of law.

A watt is the base unit of electrical power. The most commonly used formula for power (or watts) is voltage times current ($E \times I$). For example, if a certain circuit had a voltage of 40 volts with 4 amps of current flowing through the circuit, the wattage of that circuit would be 160 watts (40×4).

Figure 2-9 shows a circle for figuring power, voltage, and current, the same as the Ohm's Law circle that was used to calculate voltage, current, and resistance in Figure 2-8. For example, if you know that a certain appliance uses 200 watts and that it operates on 120 volts, you would find the formula $P \times E$ and calculate the current that flows through the appliance, which in this instance comes to 1.67 amps.

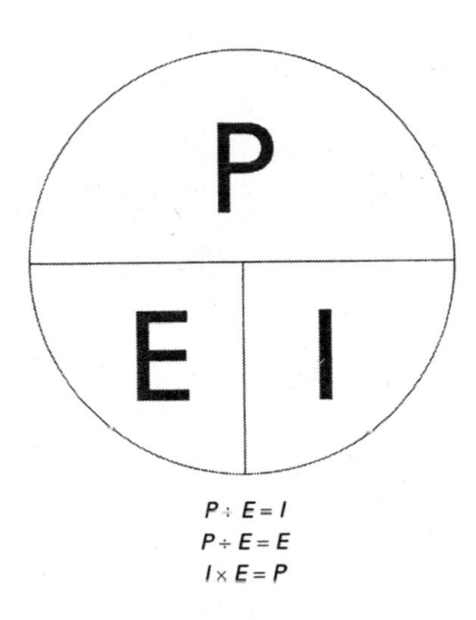

Figure 2-9 Watt's Law circle.

In all, there are 12 formulas that can be used by combining both Ohm's and Watt's Laws. These are shown in Figure 2-10. This is a very useful chart.

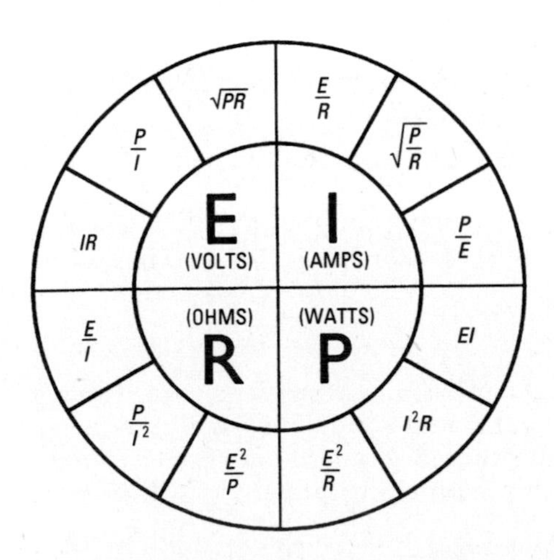

Figure 2-10 The 12 Watt's Law formulas.

The formulas in Figure 2-10 allow you to find a missing value for a wide variety of circuits. For example, you may be working with a circuit in which the voltage and resistance are known and need to find the amount of current flow. Again, you may know the voltage and power in the circuit and need to find the amount of current flow. Or, you may know the resistance and power in the circuit and need to find the amount of current flow.

Consider the following:

- You can find the voltage of a circuit if you know the current and resistance, or the current and the power, or the resistance and the power.
- You can find the resistance in a circuit if you know the voltage and current, or the current and power, or the voltage and power.
- You can find the power in a circuit if you know the voltage and current, or the current and resistance, or the voltage and resistance.

Review the previous list and compare the items to the formulas in Figure 2-10.

Following are a few examples of how the formulas work. Follow through them slowly and be sure you understand. Don't rush.

If $P = 100$ watts and $I = 4$ amperes, we find $E = P/I$, or 100 W/4 A = 25 V. Since $R = E/I$ and $E = P/I$, the following calculation can be made:

$$R = P/I^2, \quad I^2 = P/R \quad \text{and} \quad I = \sqrt{P/R}, \quad \text{and} \quad P = I^2R$$

Then, if $P = 100$ watts and $R = 4$ ohms, then $I^2 = 100\ \text{W}/4\ \Omega = 25\ \text{A}^2$, and the square root of 25 is 5, so $I = 5$ amperes. Further, since $P = EI$ and $I = E/R$, we have:

$$E^2/R = P \quad R = E^2/P \quad E^2 = RP \quad \text{and} \quad E = \sqrt{RP}$$

If $R = 4$ ohms and $P = 100$ watts, then $E^2 = PR = 100\text{W} \times 4\Omega = 400\text{V}^2$. The square root of 400 V^2 is 20 V. Thus, $E = 20$ volts.

Magnetism

As we stated in Chapter 1, magnetism is electricity's twin. Having covered some of the basics regarding circuits, we will now begin covering magnetism. As you proceed in your study of electronics, magnetism and its relationship with electrical current will become increasingly important.

In central Greece, and particularly in Magnesia, there is found a hard lead-colored mineral called *magnetite.* This is, in reality, iron oxide (Fe_3O_4), composed of three atoms of iron and four atoms of oxygen. In early times, people knew that this substance would attract small particles of iron. Later, it was discovered that a piece of magnetite would always turn to point in one direction. Some traditions states that Hoang-ti, a Chinese navigator, used a piece of magnetite floated on water to navigate a fleet of ships when out of sight of land. This was about 2400 B.C.

This substance was later found to exist in many places throughout the world and was called *lodestone* or *leading stone.*

In 1600 A.D., Dr. William Gilbert conducted extensive research and published an account of his magnetic discoveries. In these he found that the attractive force appeared at two regions on the lodestone, which he designated as the *poles*. The region between the two poles becomes less magnetic, and a point may be reached at which no magnetic forces exist. This may be illustrated by taking a magnetized bar and dipping it in iron filings (see Figure 2-11). The filings accumulate at the ends and not at the middle of the bar. This middle region on the bar was termed the *equator.*

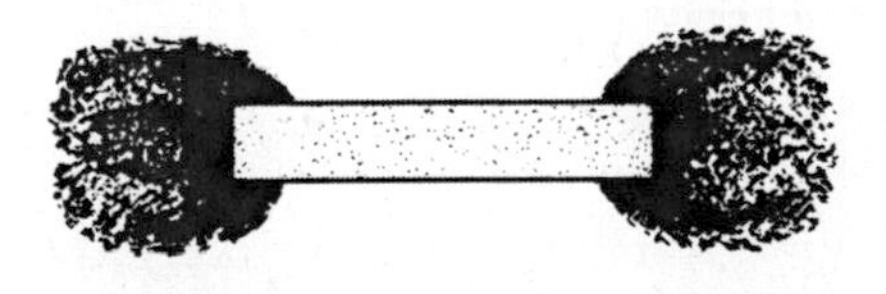

Figure 2-11 Magnetized iron bar and its attraction for iron filings.

In 1730, a scientist named Servington Savery discovered that hard steel retained magnetism better than soft iron.

Magnetic Poles

You may observe the effect of magnetism by using a magnetic needle and a bar magnet, as shown in Figure 2-12. You will observe that one end of the needle is attracted to one and only one end of the bar magnet and repelled by the other end of the bar magnet. For purposes of this discussion, let's characterize this difference in terms of poles, calling one end the *north pole* and the other end the *south pole.*

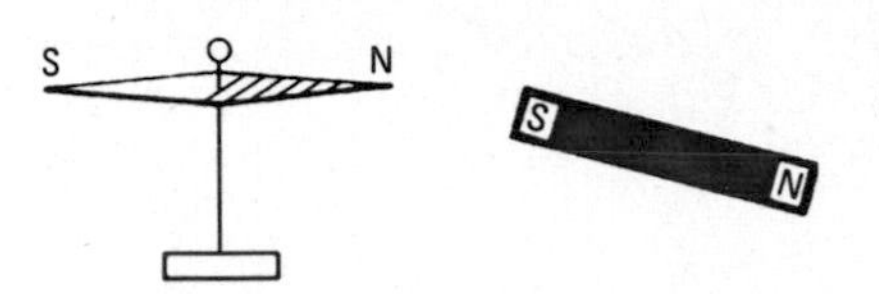

Figure 2-12 Magnetic needle and bar magnet.

This can be confusing, because the south pole of the magnet attracts the north pole of the needle, while the north pole of the magnet repels the north pole of the needle. From this you can deduce that like poles repel, and unlike poles attract. This confusion comes from the fact that the north pole of a compass always points to the north magnetic pole of the Earth. This tends to disrupt our attraction and repulsion observations. The French and Chinese call the north-pointing end of the compass the south pole. However it is customary in most Western countries to call the pole of the magnet that points north if suspended the "north pole," and the end that points to the south magnetic pole, the "south pole."

The first law of magnetism is very simple:

Like poles repel, and unlike poles attract.

Magnetic and Nonmagnetic Substances

A distinction must be made between magnets and magnetic substances. A *magnet* is a magnetic material in which polarity has been developed. Substances that are affected by magnets (that is, attracted by magnets) are *magnetic substances*. Some of these may retain magnetism, and some others may not. Iron and steel are the most highly magnetic substances. Nickel is a magnetic substance to a degree.

All substances may be classified magnetically as follows:

- *Magnetic substances*—Those substances that are attracted by a magnet.
- *Diamagnetic substances*—Those substances that are (slightly) repelled by a magnet.
- *Nonmagnetic substances*—Those substances that are not affected by a magnet in any way.

The Earth as a Magnet

The Earth itself is a great magnet, with its magnetic poles nearly coinciding with the geographic poles. For example, in Denver, the compass points about 14 degrees east of the true north-south line. In New York, the magnetic variation from the true north-south line is about 11 degrees west of the true line. These variations change from year to year, and this information is readily available.

If a steel needle is magnetized and balanced by a thread, it is free to move in any direction. It is free to align itself vertically with the declination of the Earth's magnetic field. The angle of dip increases the closer you get to the Earth's magnetic poles. Figure 2-13 illustrates a dipping needle.

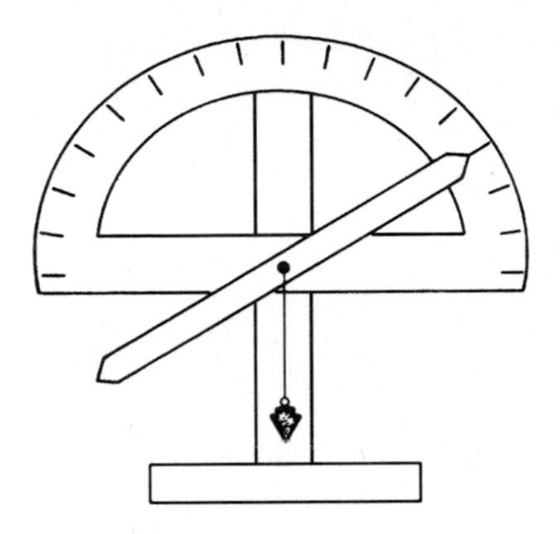

Figure 2-13 A dipping needle.

Magnetic Lines of Force

If an unmagnetized iron bar is approached by a pole of a magnet, the iron bar becomes magnetized, even though it is not touched by the magnet. This is called *magnetic induction.* If the amount of magnetism is comparatively high, the bar is said to possess *high permeability.*

Figure 2-14 illustrates the lines of force that radiate from a bar magnet. Since the lines of magnetic force may neither be seen or felt, a good method of proof that they exist is shown in Figure 2-15. A sheet of paper is placed over various magnets. The paper is then sprinkled with iron filings and tapped sharply. The filings align themselves with the magnetic lines of force.

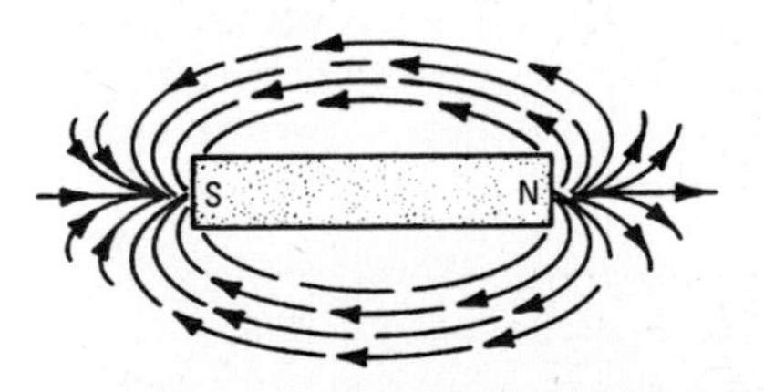

Figure 2-14 Direction of lines of magnetic force about a bar magnet.

There is no good insulator of magnetism. If, however, a hollow iron cylinder is placed in a magnetic field, the magnetic lines of force will follow a path through the iron cylinder and will be deflected away from the cylinder's interior as shown in Figure 2-16.

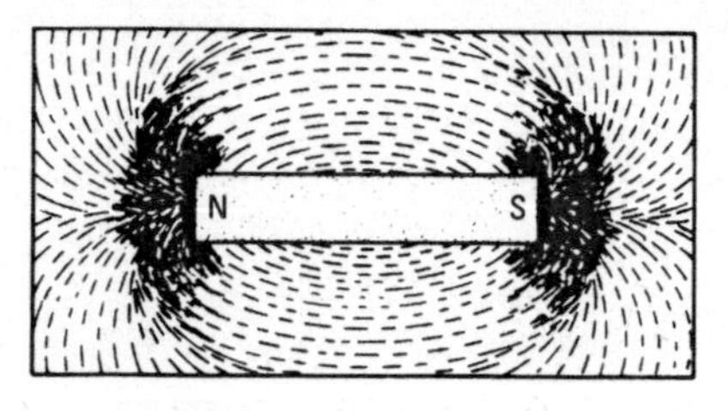

(A) Magnetic field between unlike poles.

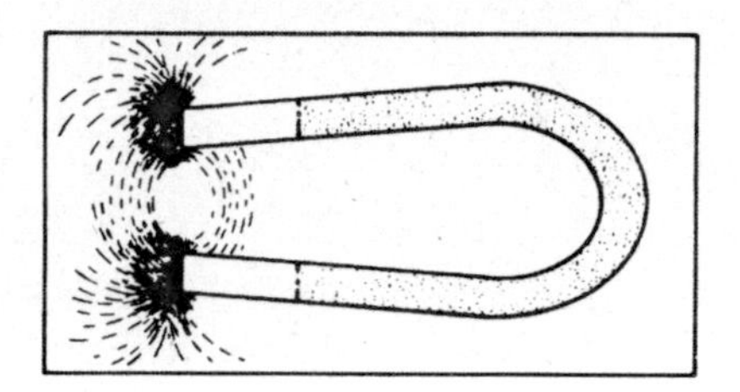

(B) Magnetic field around a horseshoe magnet.

Figure 2-15 Use of iron filings to illustrate magnetic field.

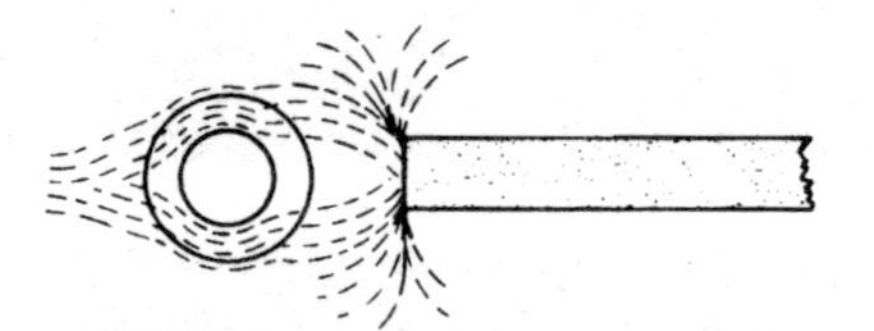

Figure 2-16 Deflection of magnetic lines of force.

The Molecular Origins of Magnetism

If a bar magnet is broken into pieces (as shown in Figure 2-17), each piece will become a magnet within itself, with both north and south poles. The broken bar magnet will produce as many magnets as there are pieces into which it is broken.

Figure 2-17 A bar magnet, when broken, produces as many magnets as there are pieces.

As the study of magnetism proceeded, it was found that an iron bar increases slightly in length when strongly magnetized. It was also discovered that still stronger magnetization causes the iron to subsequently shrink in length. If a bar is rapidly magnetized, first in one direction and then in the other, the bar will heat and a decided hum will be heard coming from the bar.

Experiments such as these have led to the theory that magnetism is an action affecting the arrangement of the molecules. Figure 2-18 shows the molecules of an unmagnetized bar of iron. It is theorized that when an iron bar is magnetized, its molecules are magnetized and line up, as shown in Figure 2-19. High-carbon steel was the original material used in making permanent magnets.

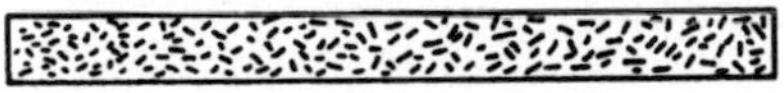

Figure 2-18 Chaotic arrangement of molecules in an unmagnetized iron bar.

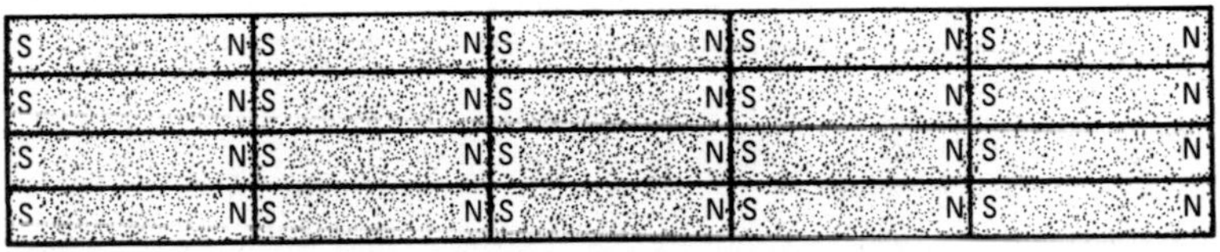

Figure 2-19 Arrangement of molecules in a magnet.

Magnets that are stronger and more permanent may be made from special alloys of steel containing tungsten, chromium, cobalt, aluminum, and nickel.

Table 2-2 illustrates some of the percentages of these alloys that are added to steel in the making of these stronger magnets.

The Strength of a Magnet

Soft iron will magnetize while under the influence of a magnetic field but will retain very little (if any) magnetism when the magnetic field is removed. We have also just seen that the alloy content of steel affects the magnetic strength.

The strength of a magnet may be determined by the magnetic force that it exhibits at a distance from other magnets. Thus, suppose there are two magnets acting upon a suspended needle. If at the same distance from the needle, the two magnets produce equal deflections, their strengths would be equal. In other words, the strength of a magnet may be defined as *the amount of free magnetism at its pole*.

The Lifting Power of a Magnet

A distinction must be drawn between the *strength* of a magnet and its *lifting power.*

Table 2-2 Alloys Added to Steel for Magnetization

Tungsten Steel		*Chromium Steel*		*Cobalt Steel*		*Alnico Steel*	
Tungsten	5 to 6 percent	Chromium	1 to 6 percent	Cobalt	3.5 to 8.0 percent	Aluminium	10 to 12 percent
Carbon	0.7 percent	Carbon	0.6 to 1.0 percent	Chromium	3 to 9 percent	Nickel	17 to 28 percent
Manganese	0.3 percent	Manganese	0.2 to 0.6 percent	Tungsten	1 to 9 percent	Cobalt	5 to 13 percent
Chromium	0.3 percent			Carbon	0.9 percent		
				Manganese	0.3 to 0.8 percent		

The lifting power of a magnet depends upon the shape of its pole and the number of lines of force passing through its pole. A horseshoe magnet with both poles connected by a soft iron keeper will lift three to four times as much as with one pole alone. A bar magnet will lift more or less depending upon the shape of its poles.

Suppose that there are three bar magnets of equal strength, as shown in Figure 2-20, and that there is the same amount of free magnetism at the poles of all three. If the end of bar *A* flares out, it will lift the least. If pole *B* is the same area as the bar, it will lift more. If the pole is chamfered off so that its face has a smaller cross section than the bar, as in *C*, it will lift the most.

Figure 2-20 Illustration of different pole shapes that give different lifting powers.

The reason for the various lifting powers is found in the law that governs the lifting power of a magnet. This law states:

The lifting power is proportional to the square of the number of magnetic lines of force per unit of cross section.

Thus, if the pole is chamfered off until the area is reduced by one-half, and the original amount of magnetism is crowded into this reduced area, the flux density of the magnetism would be doubled. This would cause the lifting power for the reduced area to be quadrupled. Since the cross section has been reduced by one-half, the lifting power is actually only doubled. As a practical matter, it would be impossible to concentrate the magnetism to such an extent as to bring about this result. Nevertheless, the lifting power may often be increased by diminishing the cross section of the pole.

Static Electricity

Probably the first type of electricity to be recognized was *static electricity*. The word *static* means "at rest." There are some applications where static electricity is put to use, but in other cases it is detrimental and must be avoided.

Lightning is a large discharge of static electricity, in an attempt to neutralize opposite charges. This is the same effect as experiencing a shock when touching a doorknob after rubbing your feet on a dry carpet. The voltages and distances are much larger, but the fundamentals are the same. (Note that people frequently use terms that impute an intelligence to electricity. Obviously this is not true, but it is sometimes useful and descriptive to say things like "electrical charges want to balance themselves.")

Static electricity occurs when electrons are moved from one substance to another. We often think of this as rubbing electrons off of

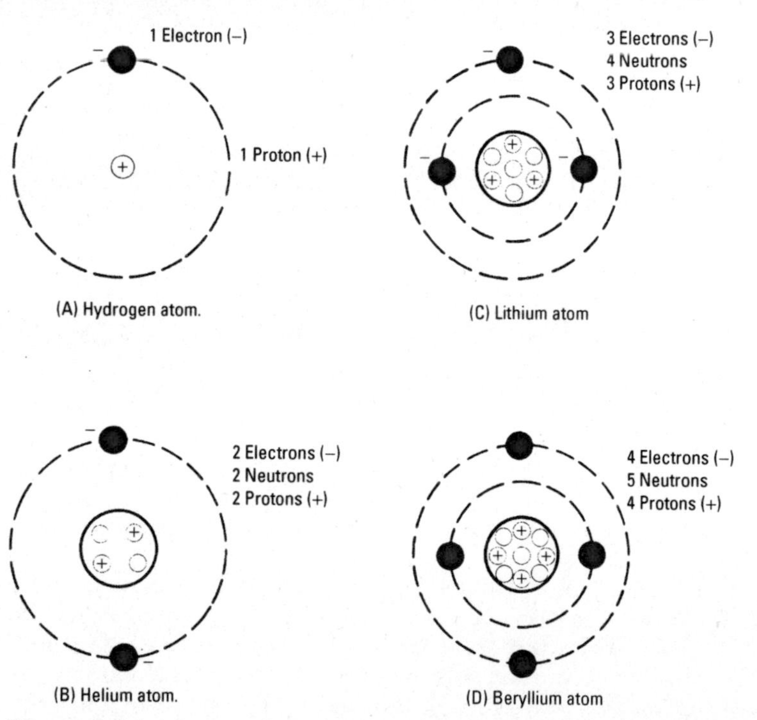

Figure 2-21 Atoms: electrons, neutrons, and protons. Electrons have a negative (−) charge, protons have a positive (+) charge, and neutrons are neutral.

one surface and collecting them on another. Please refer to Figure 2-21 and notice that rubbing the surface of an atom would affect the electrons, not anything in the nucleus. So, we are removing electrons from their atoms. When collected on the second (new) surface, a voltage forms because of the excess negative charge.

One method of moving electrons from their atoms is by the friction of rubbing a hard rubber rod with a piece of fur. The fur will give up some electrons to the hard rubber rod, leaving the fur with a positive charge, and the hard rubber rod will gain a negative charge. Then, again, a glass rod rubbed with silk will give up electrons to the silk, making the silk negatively charged and leaving the glass rod positively charged.

What actually transpires is that the intimate contact between the two surfaces results in the fur being robbed of some of its negative electrons, thereby leaving it positively charged, while the rubber rod acquires a surplus of negative electrons and is thereby negatively charged. It is important to note that this surplus of negative electrons doesn't come from the atomic structure of the fur itself. It is found that, in addition to the electrons involved in the structure of materials, there are also vast numbers of electrons at large. It is from this source that the rubber rod draws its negative charge of electrons.

If a hollow brass sphere is supported by a silk thread as shown in Figure 2-22 (silk is an insulator), and a hard rubber rod that has received a negative charge, as previously described, is touched to the brass sphere, the brass sphere will also be charged negatively by a transfer of electrons from the rod to the ball. The ball will remain negatively charged because it is supported by the insulating silk thread.

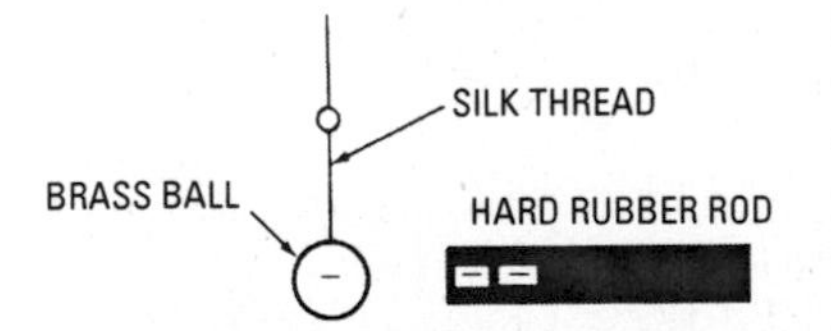

Figure 2-22 A negatively charged hard rubber rod touched to a hollow brass ball supported by a silk thread will negatively charge the brass ball.

Now, if the same experiment is tried with the hollow brass sphere supported from a metal plate by a wire, the rubber rod will transfer electrons to the ball, but the electrons will continue through the wire and metal plate and eventually to Earth, as shown in Figure 2-23.

When a body acquires an electrical charge as, for example, the hard rubber rod or the glass rod previously described, it is customary

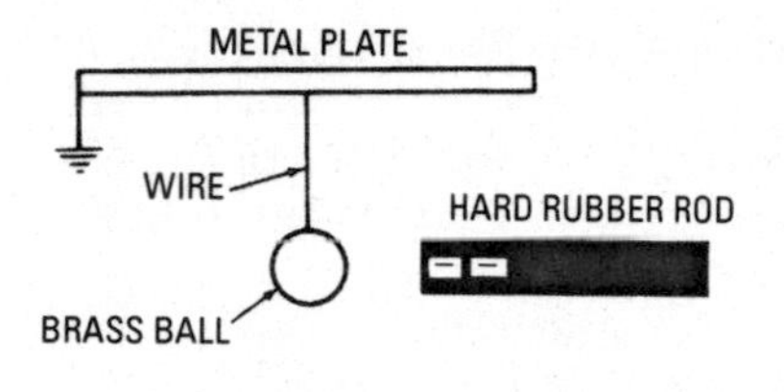

Figure 2-23 When a negatively charged hard rubber rod is touched to a hollow brass ball supported from a metal plate by a wire, the negative charge will move through the metal wire and on to Earth.

to say that the lines of force emanate from the surface of the electrified body. By definition, *a line of electrical force is an imaginary line in space along which electrical force acts.* The space occupied by these lines in the immediate vicinity of an electrified body is called an *electrostatic field of force* or an *electrostatic field.*

In Figure 2-22, the hollow ball was negatively charged and the lines of force emanated from it or converged on it in all directions (see Figure 2-24).

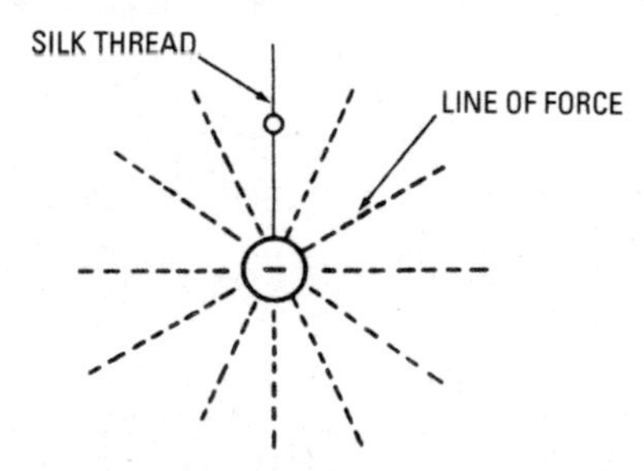

Figure 2-24 Lines of force from an electrically charged hollow ball emanate in, or converge from, all directions.

Static electrical charges may be detected by an *electroscope.* The simplest form of an electroscope is a light wooden needle mounted on a pivot so that it may turn about freely. A feather or a pith ball suspended by silk thread may also be employed for the purpose.

The gold-leaf electroscope most used was devised by a man named Abraham Bennett, and consists of a glass jar (see Figure 2-25) with the mouth of the jar closed by a cork. A metal rod with a metal ball on one end (outside the jar) and a stirrup on the other passes through the cork, and a piece of gold leaf is hung over the stirrup so that the ends drop down on both sides.

When an electrified rod is brought close to the hollow brass ball, the electrostatic field charges the ball. In Figure 2-25, the rod is negatively charged, so the electrons in the ball are repelled and the ball becomes positively charged. The electrons that were repelled from the ball go to the gold leaf, charging both halves of the gold leaf negatively, and the leaves fly apart, as illustrated in Figure 2-25.

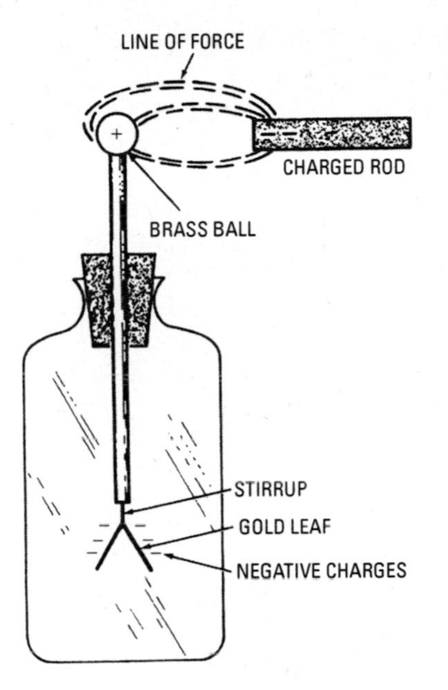

Figure 2-25 Gold-leaf electroscope.

Like charges repel each other, and unlike charges attract. Since both halves of the gold leaf are charged the same, they repel. Remember that we have not touched the rod to the ball in this experiment; the electrostatic charges are transmitted by *induction.*

If a positively electrified ball (*A* in Figure 2-26) mounted on an insulated support is brought near an uncharged insulated body (*B–C*), the positive charge on ball *A* will induce a negative charge at point *B* and a positive charge at point *C*. If pith balls are mounted on wire and suspended by cotton threads, as shown, the presence of these charges will be manifested. The pith ball (*D*), electrified by

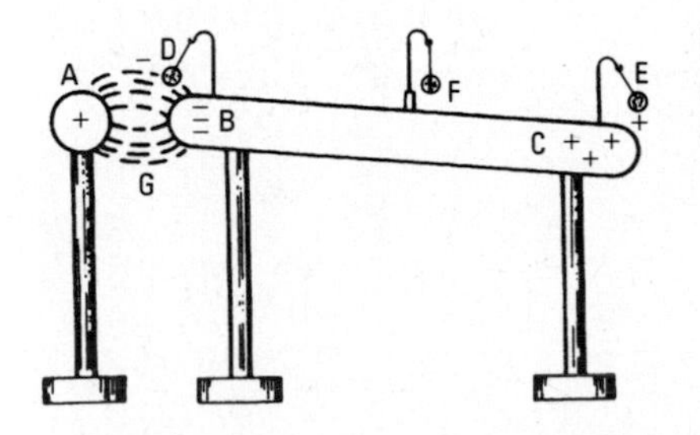

Figure 2-26 Illustration of charges produced by electrostatic induction.

contact with *B*, acquires a negative charge. It will be repelled by *B* and attracted toward *A* and stands off at some distance. The ball (*E*) is charged by contact positively and will be repelled from *C* a lesser distance because there is no opposite charge in the vicinity to attract it, while ball *F* at the center of the body will remain in its original position, indicating the absence of any charge at this point. This again shows electrostatic induction. The electric strain has been transmitted through the intervening air (*G*) between *A* and *B* and reappears at point *C*.

In Figure 2-26, the air in the space (*G*) between *A* and *B* is called a *dielectric*. The definition of a dielectric is *any substance that permits induction to take place through its mass*. All dielectrics are insulators, although the dielectric and insulating properties of a substance are not directly related. A dielectric is simply a transmitter of a strain.

When a dielectric is subjected to electrostatic charges, the charge tries to dislodge the electrons of the atoms of which the dielectric is composed. If the stress is great enough, the dielectric will break down and there will be an arc-over, that is a spark breaking through the dielectric.

Types of Current

There are basically three forms of electrical current:

- Direct current (DC)
- Pulsating direct current (pulsating DC)
- Alternating current (AC)

Figure 2-27 compares the flow of water to DC. Pump *A* may be compared to a battery or a generator driven by some external force,

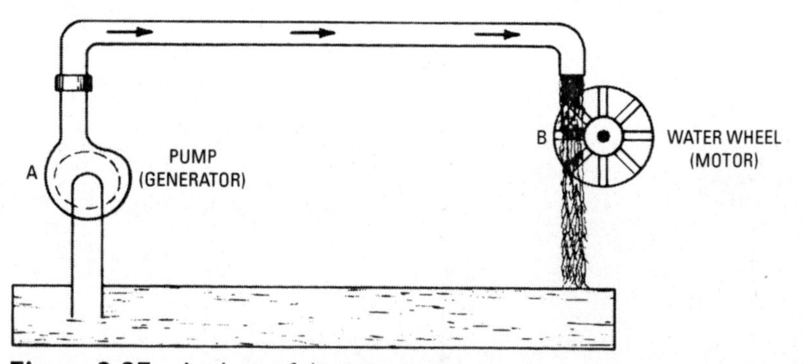

Figure 2-27 Analogy of direct current.

and wheel *B* may be compared to a DC motor, with the current flowing steadily in the direction represented by the arrows. This may also be represented as shown in Figure 2-28.

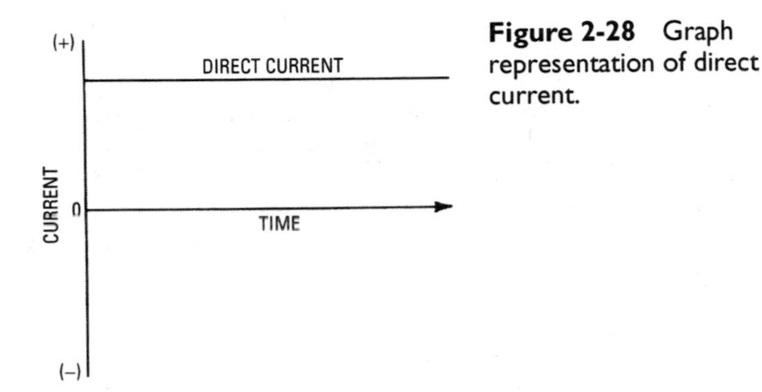

Figure 2-28 Graph representation of direct current.

Now, if generator *A* in Figure 2-27 were alternately slowed down and speeded up, the current would be under more pressure when the pump was speeded up and less pressure when the pump slowed down, so the water flow would pulsate in the same direction as represented in Figure 2-29. It would always be flowing in the same direction, but in different quantities.

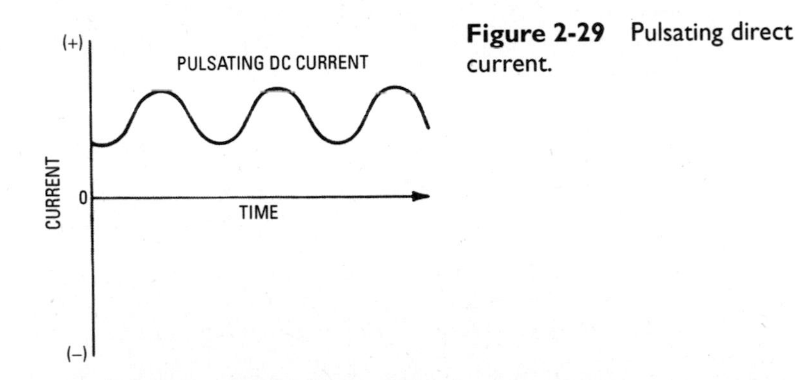

Figure 2-29 Pulsating direct current.

Now, in Figure 2-30 we find a piston pump (*A*) alternately stroking back and forth and thus driving piston *B* in both directions alternately. Thus, the water in pipes *C* and *D* flows first in one direction and then in the other. Figure 2-31 illustrates the flow of AC, which will be covered in detail in Chapter 4.

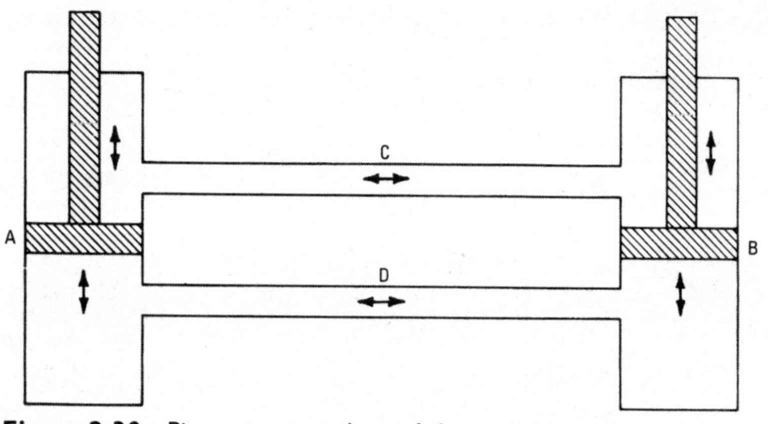

Figure 2-30 Piston pump analogy of alternating current.

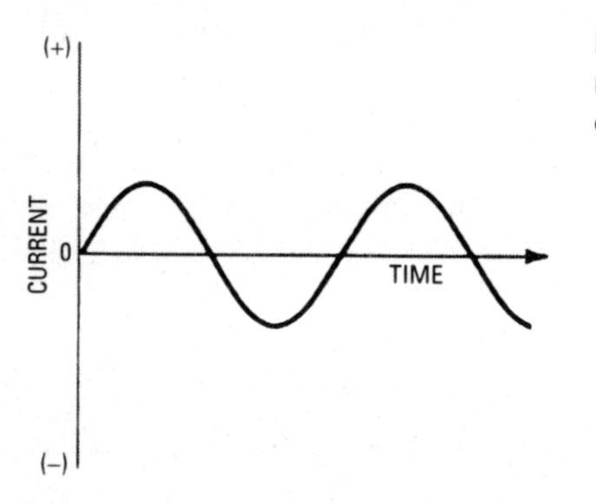

Figure 2-31 Graph representation of alternating current.

Insulators and Conductors

An insulator opposes the flow of electricity through it, while a conductor permits the flow of electricity through it. It is recognized that there is no perfect insulator. Pure water is an insulator, but the slightest impurities added to water make it into a conductor. Glass, mica, rubber, dry silk, and so on, are insulators, while metals are conductors.

Although silver is not exactly a 100 percent conductor of electricity, it is the best conductor known and is used as a basis for the comparison of the conducting properties of other metals, so it is commonly said to have 100 percent conductivity.

There are certain materials called *superconductors* that conduct electricity with zero loss. Typically, these are special ceramics that obtain zero resistance only at very low temperatures. These materials are useful only in very special conditions, and we will not cover them in this book. However, as new types of superconductors that operate at normal temperatures are developed, they could be extremely useful.

Table 2-3 shows some metals listed in the order of their conductivity.

Table 2-3 Metals and Their Conductivity

Metal	*Conductivity*
Silver	100 percent
Copper	98 percent
Gold	78 percent
Aluminum	61 percent
Zinc	30 percent
Iron	16 percent
Lead	15 percent
Tin	9 percent
Nickel	7 percent
Mercury	1 percent

Electromagnetism

Electromagnetism was first discovered by Hans Oersted, a Danish physicist. As he conducted experiments with electricity in 1819 or 1820, he noticed that a compass needle was affected by an early type of electrical battery called a *voltaic pile*. Upon further experimenting, he discovered that a compass needle that was placed immediately above or below a conductor carrying current, as in Figure 2-32, was deflected. This was a tremendously important discovery

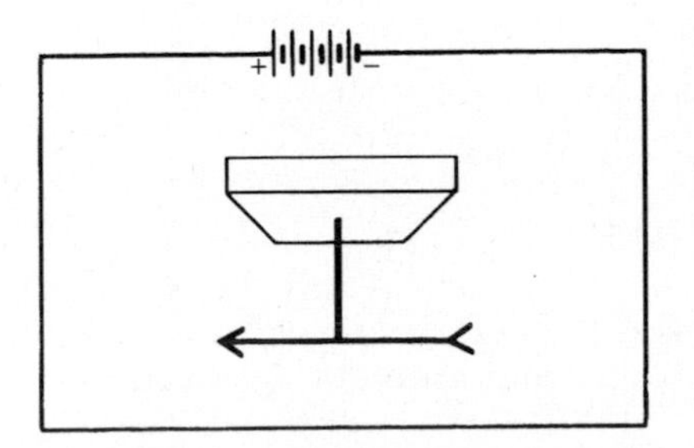

Figure 2-32 Compass needle is deflected by a current through a conductor.

and a complete surprise. Remember that electricity and magnetism were both being seriously explored for the first time. Both were intriguing, to be sure, but no one had any idea that they might be associated with one another.

The effect Oersted discovered can be demonstrated by passing a conductor through a paper on which iron filings have been sprinkled. When direct current is passed through the conductor, the iron filings will arrange in a configuration such as shown in Figure 2-33.

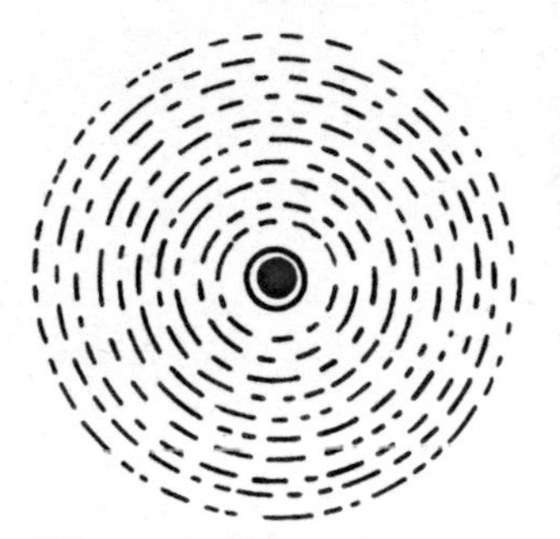

Figure 2-33 Iron filings around a current-carrying conductor.

In Figure 2-34, the lines of the magnetic field around a conductor are more clearly illustrated. The lines of force around a current-carrying conductor form in circular paths. To establish in which direction those lines of force travel, we will use Figures 2-35 and 2-36. In Figure 2-35, the current in the conductor is going away from us, and is represented by the end of the conductor having a plus in it. The lines of force (or flux) go around counterclockwise. In Figure 2-36, the current is coming toward us as represented by the dot in the circle. Here, the lines of flux are clockwise.

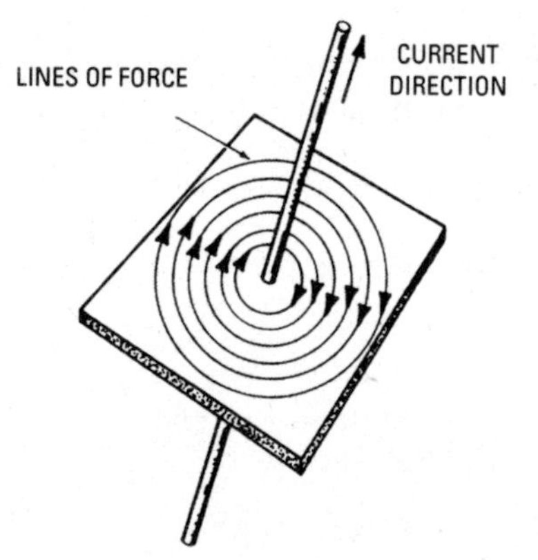

Figure 2-34 Magnetic field around a current-carrying conductor.

An easy way to remember this is by means of the *left-hand rule.* In your imagination, place your left hand around a conductor, as

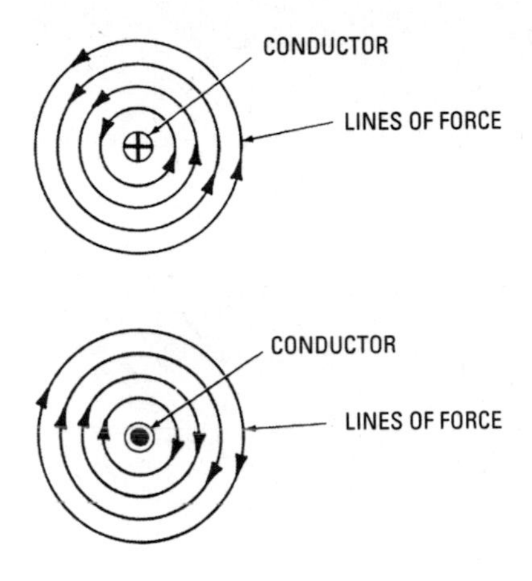

Figure 2-35 Magnetic field of a current moving away from us.

Figure 2-36 Magnetic field of a current moving toward us.

shown in Figure 2-37, so that your thumb points in the direction of the current (negative to positive). Then, your fingers will be pointing in the direction of the lines of force.

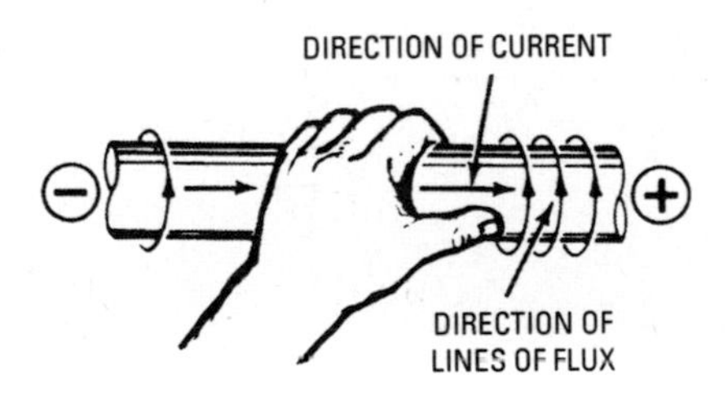

Figure 2-37 Current and direction of lines of flux.

Refer to Figure 2-38 to determine the direction in which a compass needle will be deflected.

Parallel conductors carrying currents in the same direction attract each other, as shown in Figure 2-39. Conductors *A* and *B* are carrying current away from us, so the lines of force are counterclockwise. Instead of circling *A* and *B* separately, as at *C* and *D*, they combine and encircle both conductors, as at *E* and *F*.

Wires carrying current in opposite directions repel each other, as shown in Figure 2-40. Conductor *A* is carrying current away from us and conductor *B* is carrying current to us. Lines of force around *A* are counterclockwise and clockwise around *B*. Since the lines of force are oriented in opposite directions, they will not combine as in Figure 2-39, but will tend to push the wires apart.

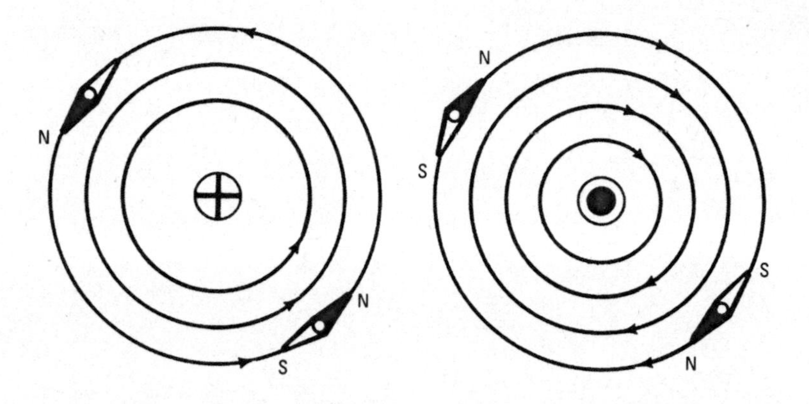

Figure 2-38 Direction of deflection of a compass needle.

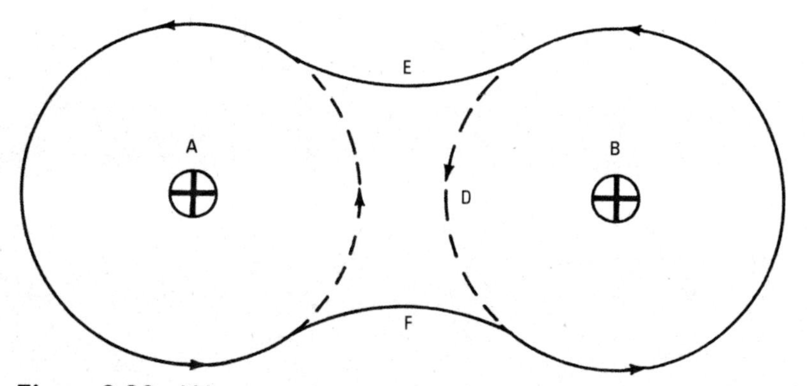

Figure 2-39 Wires carrying current in the same direction attract each other.

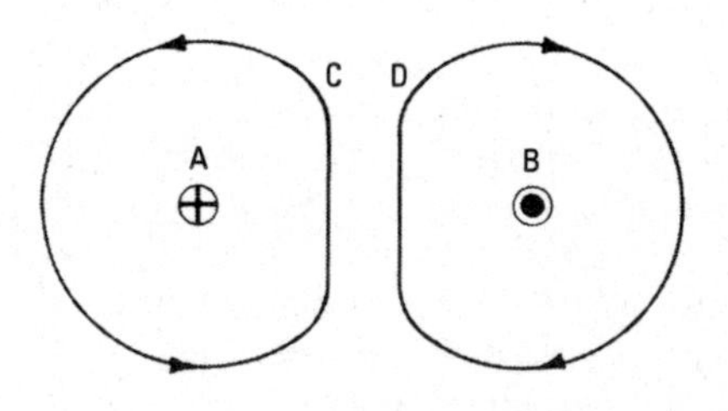

Figure 2-40 Wires carrying current in opposite directions repel each other.

James Clerk Maxwell was the British physicist who developed a unified theory of magnetism. *Maxwell's rule* states:

Every electrical circuit is acted upon by a force that urges it in such a direction as to cause it to include within its embrace the greatest possible number of lines of force.

Figure 2-41 can be used to explain Maxwell's rule. In Figure 2-41A, there is a circuit doubled back on itself. The lines of force around half the wire will oppose the lines of force around the other half, as was covered in the explanation of Figure 2-40. This means the wire will have a tendency to be pushed apart and, if free to move, would theoretically take the shape of the circuit shown in Figure 2-41B, which would be a circle.

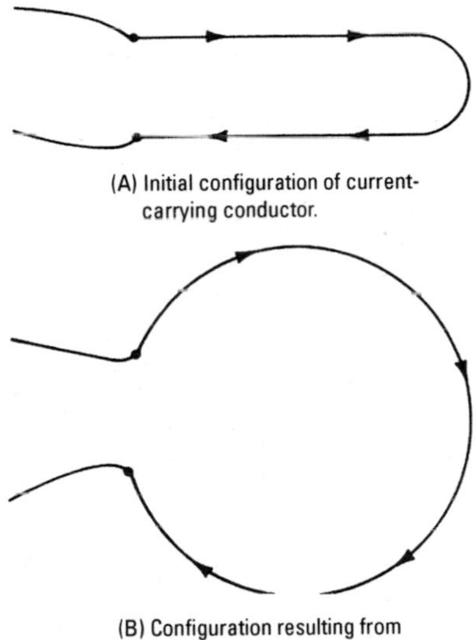

Figure 2-41 Illustration of Maxwell's rule.

A paraphrase of Maxwell's rule is:

Every electrical circuit tends to so alter its shape as to make the magnetic flux through it a maximum.

This rule explains motor action and also the action of many measuring instruments.

To illustrate, consider that every electric motor has a loop of wire carrying current. This loop is placed in such a position in a magnetic field that the lines of force pass parallel to, but not through, it. From Maxwell's rule, the loop tends to turn in such a direction so as to include within it the lines of force of the magnetic field. Take time to review and remeber this action, because it will have a far-reaching effect in later chapters.

On large-capacity circuits (which will have high currents available should the conductors of a circuit short together), Maxwell's rule must be prepared for before the short occurs. When the short occurs, the magnetic forces tending to cause the circuit to embrace the greatest possible number of lines of force will tend to throw the conductors apart. These magnetic stresses become very great. In large power-switching devices (where solid metal bars are the conductors), it becomes an engineering problem to design the bus bars so that they will not be torn from their mountings.

In Figure 2-42, the lines of force from the magnet go from *N* to *S*, and the current in the conductor is coming toward us, so the flux around the conductor is clockwise. You could compare these lines of force to rubber bands. Thus, the lines of force of the magnetic poles tend to straighten out and push the conductor down.

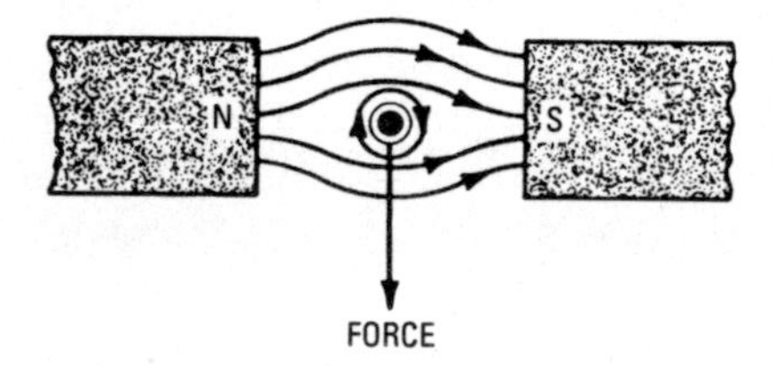

Figure 2-42 Composite of a magnetic field and a current-carrying conductor.

Figure 2-43 illustrates the *right-hand rule*. The right hand is cupped over the pole piece as shown, the thumb represents the

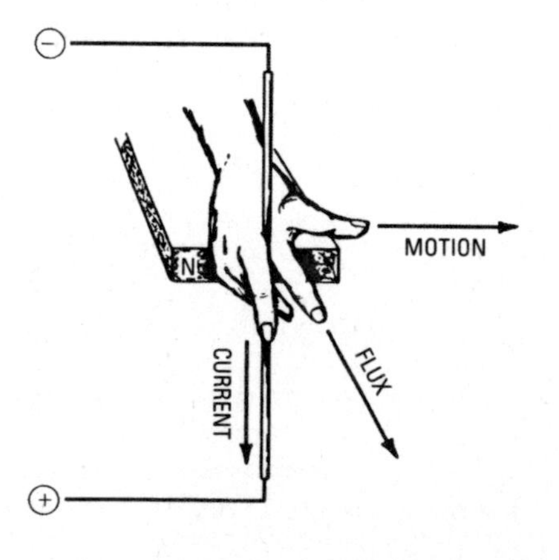

Figure 2-43 Right-hand rule.

direction of motion, the index finger the direction of the flux, and the middle finger the direction of the current.

Galvanoscope

Figure 2-32 showed a magnetic needle under a wire carrying current, as did Figure 2-38. A *galvanoscope* is such a device. Figure 2-44 will be used in the explanation. Figure 2-44A shows a simple galvanoscope, similar to the one previously illustrated. Current will deflect the magnetic needle, which is suspended by a thread. If the current is very feeble, the deflection will be hard to notice. To overcome this, more turns are added, as in Figure 2-44B. Thus, if there are 100 turns, the effect will be 100 times as much as the effect of the simple galvanoscope.

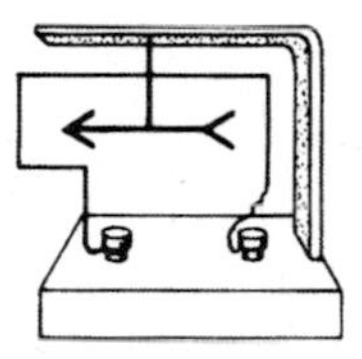

(A) Simple galvanoscope.

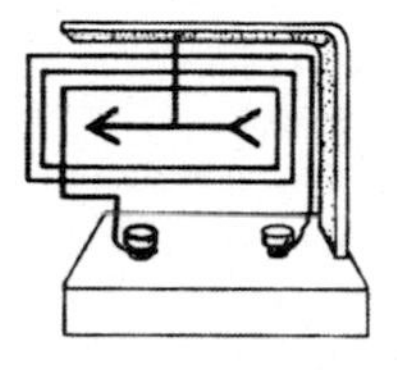

(B) Multiplying effect of many turns.

Figure 2-44 The galvanoscope.

Solenoids

In Figure 2-45 the effect of conductors in parallel is shown. Notice that conductors in parallel, with the currents in the same direction, cause the lines of force to embrace the conductors as one. In

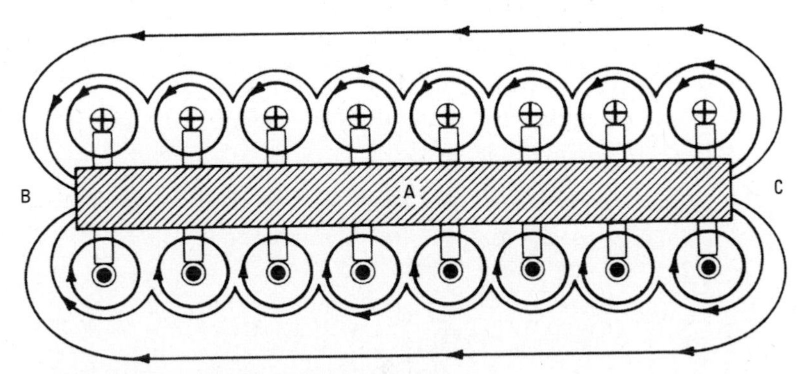

Figure 2-45 Field of an electromagnet.

solenoids and electromagnets, this phenomenon is taken advantage of to increase the strength of the solenoids and magnets.

In Figure 2-45, there is a soft iron core *A* wound with turns of wire. As you see, the turns are parallel and the current flows in the same direction in all turns. Notice that the lines of force are all in the same direction on either side, so there is an addition of the lines of force in proportion to the number of turns. The current must be in the same direction in each turn as the turns are in series. The lines of force add up and concentrate at poles *B* and *C* of the iron core.

It is worth mentioning here that a circuit doubled back on itself produces no magnetic field. This is because the magnetic field of one wire cancels out the effect of the magnetic field of the other wire, since the fields are equal and opposite.

Mechanical functions are performed by solenoids such as the one shown in Figure 2-46. The coil *M* is wound on a nonmagnetic form. Plunger *A* is drawn into coil *M* when the coil is energized, pulling lever *L* up, actuating some mechanical function. When coil *M* is de-energized, the spring *S* pulls the lever back down.

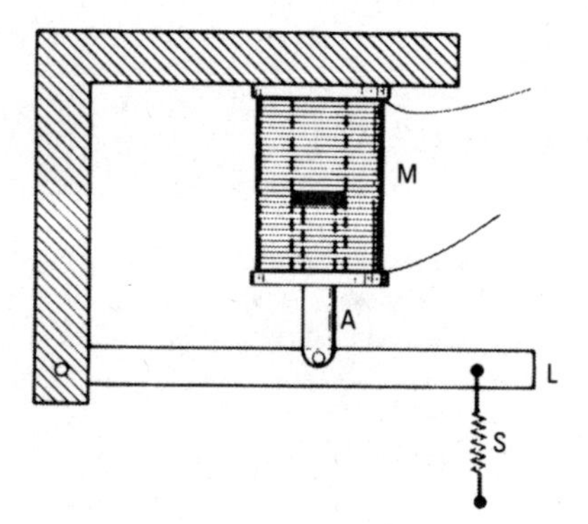

Figure 2-46 Principle of a solenoid.

Solenoids and electromagnets are used extensively in the electrical industry. Figure 2-47 illustrates the principle of a relay. The

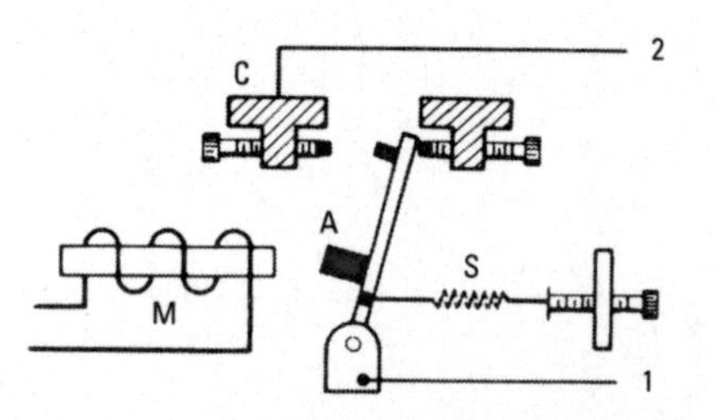

Figure 2-47 Principle of a relay.

electromagnet *M* is energized, pulling armature *A* to it, and the contact on the armature *A* makes contact with the contact *C*, closing a circuit between 1 and 2. When the magnet is de-energized, the spring *S* pulls the armature *A* away, opening the contact between *A* and *C*.

Some circuit breakers are triggered by relays, when overcurrents occur.

Summary

The three primary factors affecting any electrical circuit are voltage (electromotive force), current, and resistance. These factors control every electrical circuit everywhere. The relationship between voltage, current, and resistance is expressed by Ohm's Law, which is distilled into three formulas: $E = I \times R$, $I = E/R$, and $R = E/I$.

Electricity has its origins in atoms. When electrons are forced to move from the outer (valence) shells of atoms, they constitute an electrical current. Elements that give up electrons easily serve as conductors, and elements that do not easily give up free electrons serve as insulators.

Electrical power is measured in watts. The fundamental formula for one watt is $E \times I$. From this, 12 separate power formulas can be derived, all of which can be used to calculate power in a circuit.

Magnetism, like electricity, begins at microscopic levels. When the magnetic domains of certain materials line up together, they create a magnetic field surrounding that material. One characteristic of a magnetic field is that it has two opposite poles, which we designate as north and south. Another characteristic is that the magnetic field is arranged as curved lines of force.

Electromagnetism acts in exactly the same way as permanent magnetism, but arises from the flow of electrical current. This also makes electromagnetism exceptionally useful, since the fields can be reversed, increased, decreased, and otherwise manipulated by controlling the electrical currents that give rise to the electromagnetism.

Static electricity is created by using simple friction to remove free electrons from one substance and transfer them to another. This creates an imbalance of charges, hence a voltage.

Chapter 3 discusses circuits, including the laws that govern their usage.

Review Questions

1. What is a neutron?
2. What is a proton?
3. What is an electron?

4. What is a coulomb?
5. What is signified by the Greek letter omega?
6. One volt is the force necessary to move__________ through a resistance of one ohm.
7. Ohm's Law states that current is directly proportional to __________, and inversely proportional to __________.
8. A watt is the electrical unit of __________.
9. One horsepower equals _____ watts.
10. Like poles _____.
11. Unlike poles _____.
12. What is a magnetic substance?
13. What is magnetite?
14. What is static electricity?
15. Explain the effect that current through a conductor will have on a compass needle.
16. What happens to magnetic lines of force produced by conductors with current in both conductors in the same direction?
17. What is the effect of magnetic lines of force around two conductors, when the currents in the two conductors are flowing in opposite directions?

Exercises

1. Sketch a helium atom and label its parts.
2. Describe the molecular theory of magnetism.
3. Describe and draw an electroscope.
4. Illustrate direct current.
5. Illustrate alternating current.
6. Draw an atom with two electron shells. Label the parts and explain the numbers of electrons available in the two shells.
7. Explain why the outermost shell of an atom is especially important.
8. Sketch and explain an electromagnet.
9. Sketch and explain a solenoid.
10. Two pith balls are negatively charged and supported by a dry silk thread. Draw a sketch showing their relative positions when they are brought close to each other.

Chapter 3

Circuits

An electric circuit is a closed path for current flow. Electricity needs a complete loop to flow. With a broken path, it will not move.

This chapter discusses a variety of different types of circuits, including open and closed circuits, short circuits, series circuits, and parallel circuits. Included in the discussion are several examples, as well as an overview of Kirchhoff's voltage law and Kirchhoff's current law.

Open and Closed Circuits

Figure 3-1A shows an open circuit. Notice that there is a source of voltage (in this case, a battery) at the top of the drawing. On the bottom is a resistor. On the right-hand side is a switch in the open position. No current whatsoever will flow in this circuit. Raising the voltage will not cause current to flow, just as reducing the resistance will not cause current to flow. This is an open circuit, and electricity will not flow through it.

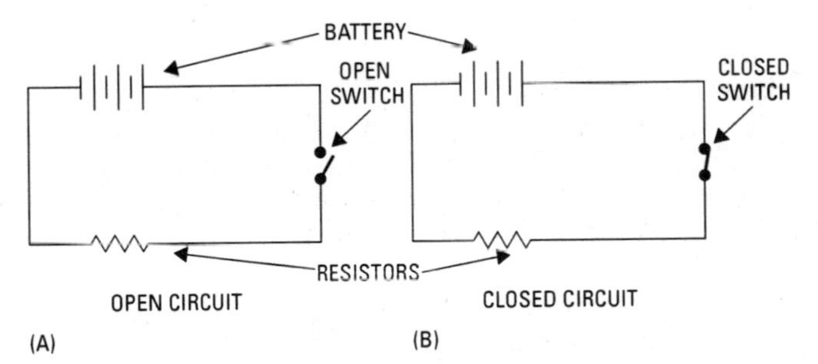

Figure 3-1 Open and closed circuits.

Figure 3-1B shows the same circuit with the switch now in the closed position. In this case, current is free to flow in the circuit. Raising the voltage will now cause more current to flow, and/or reducing the resistance will cause more current to flow.

Figure 3-2A shows the circuit of Figure 3-1B with values added. The battery voltage is given as 15 volts, and the resistance is given at 5 ohms. Figure 3-2B shows the same circuit, showing values for current and power. These were obtained by using the formulas you learned in Chapter 2. Ohm's law specifies that voltage divided by resistance (E/R) equals current (I). In this case, $15/5 = 3$ amperes.

Watt's law specifies that voltage times current equals power ($E \times I = P$). In this case, $15 \times 3 = 45$ watts.

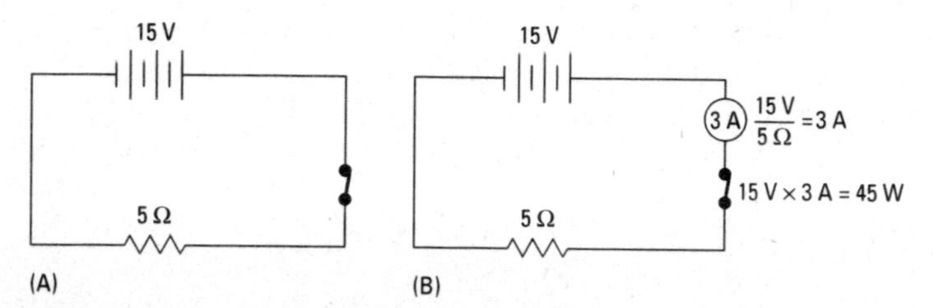

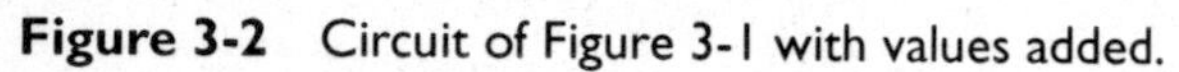

Figure 3-2 Circuit of Figure 3-1 with values added.

Series and Parallel Circuits

Series and parallel circuits are included in one classification of circuits. You will find these terms used continually. Different methods are required for calculations for series and parallel circuits. Although the concepts are not at all difficult, be sure you understand the differences between series and parallel circuits.

Series Circuits

Series circuits are circuits for which electricity has only one path on which to travel. A series circuit has one (and only one) loop for current to take.

Figure 3-3A shows a simple series circuit. It has a power source, two resistors, and a switch. There is only one possible path for current to follow. Figure 3-3B shows a much more complex circuit. There are two batteries, four resistors, and two switches. But note also that there is still only one possible path for current to take. This also is a series circuit.

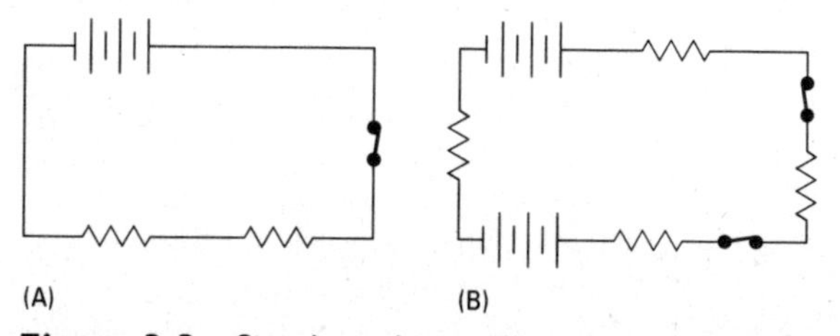

Figure 3-3 Simple and complex series circuits.

Parallel Circuits

A *parallel circuit* is one that provides more than one path for current to take.

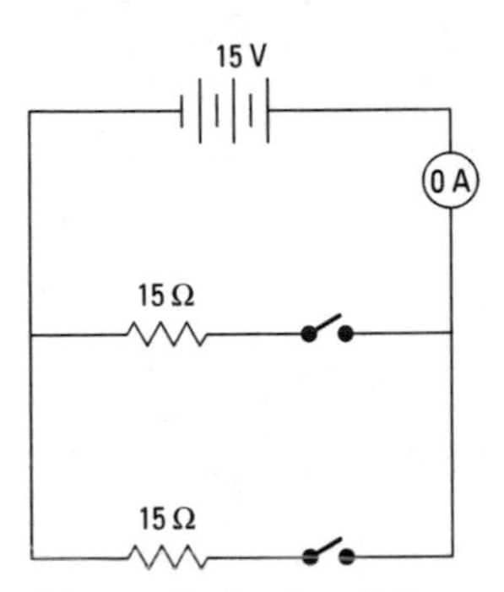

Figure 3-4 Parallel circuit.

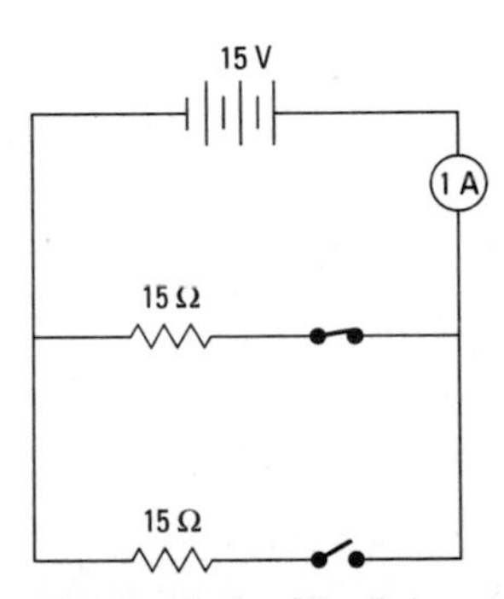

Figure 3-5 Parallel circuit with one branch open.

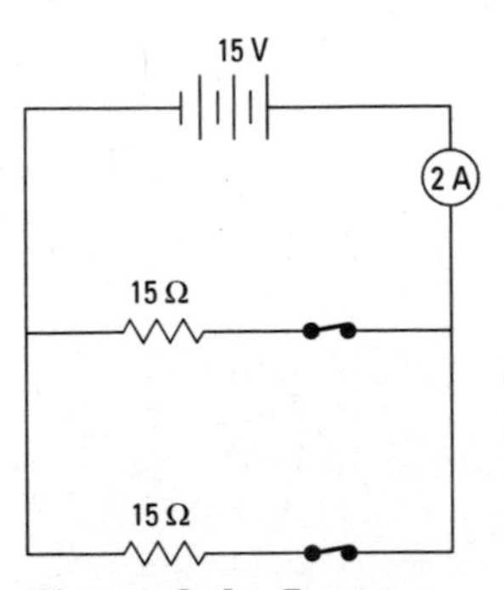

Figure 3-6 Functioning parallel circuit.

Figure 3-4 shows a basic parallel circuit. There are two *legs*, or paths, of the circuit, each with a resistor and a switch. Note also that each leg has its own independent pathway to and from the power source. In Figure 3-4, the switches in each leg are in the open position. In this case, no current will flow in the circuit. Trace the wires and see that this is so.

In Figure 3-5, the top switch has been placed in the closed position. In this state, the circuit functions as a series circuit, since the bottom switch is still open, and there is only one path for current to take. Note also that the value shown for current can be derived with Ohm's law.

In Figure 3-6, both switches are in the closed position. Now we have a functioning parallel circuit. Notice that this circuit is essentially two series circuits stacked on top of each other. Notice that the intensity of current flow is now 2 amps. Each leg of the circuit is exposed to 15 volts, and 1 ampere flows through each leg. However, since the two legs are independent of each other, the total current in the circuit rises to 2 amperes.

Parallel circuits may have any number of legs. As long as there is more than one path for current to take, it is a parallel circuit.

Series-Parallel Circuits

A *series-parallel circuit* is one that has both series and parallel elements combined into one circuit.

Figure 3-7 does not show the value in ohms of the resistors. Rather, it is identifying them only as R_1, R_2, R_3, and R_4 (that is, resistor number 1, resistor number 2, resistor number 3, and resistor number 4). The switches also are numbered in this way (that is, S_1 through S_4).

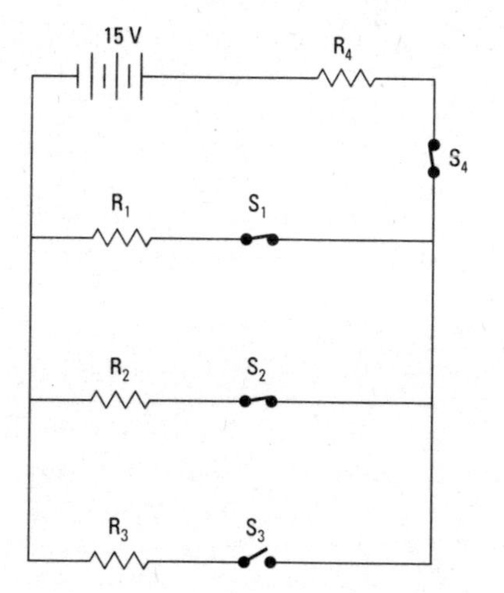

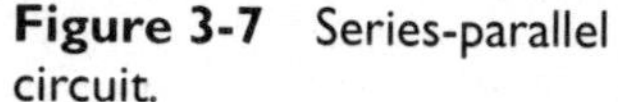

Figure 3-7 Series-parallel circuit.

Figure 3-7 has three parallel paths for current. Opening S_1 would stop current in the top path only. Opening S_2 would stop current in the middle path only. Opening S_3 would stop current in the bottom path only. However, notice that opening S_4 would shut down all current flow in the entire circuit. The top portion of this circuit is a series circuit. Every electron flowing through the circuit must go through R_4 and S_4.

The circuit of Figure 3-7, then, is a series-parallel circuit. The top of the circuit contains elements that are in series, and the bottom part of the circuit contains three legs that are in parallel.

Later in this chapter, you will learn how to deal with each of these types of circuits. But it is first critical that you understand the distinctions.

Short Circuits

While discussing types of circuits, it is important to explain short circuits. All kinds of circuits (series, parallel, series-parallel) can be exposed to *shorting*.

Figure 3-8 shows how the term *short circuit* originated, and why a short circuit can be dangerous. This drawing shows a circuit that resembles house wiring—one long circuit with many *loads* connected in parallel. (A *load* is any power-consuming device that is connected to a circuit. In your home, typical loads would be electric lights, appliances, heaters, air conditioners, and so on. In this case, the loads are represented by resistors, which is appropriate because each load

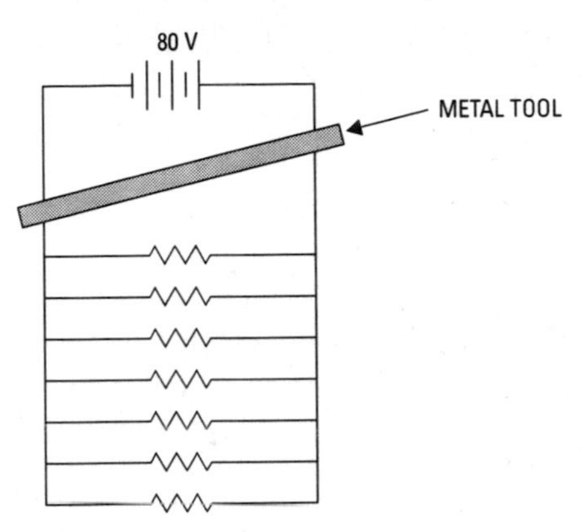

Figure 3-8 Short circuits.

has a resistance, and its resistance is the characteristic that affects the circuit.)

In Figure 3-8, notice that a metal tool is shown (accidentally) extending across the circuit. It is contacting the circuit conductors on both sides. This is a classic short circuit. Short circuits are very dangerous. In this case, there is virtually zero resistance through the short portion of the circuit. All that exists in this loop is the power source, the metal tool, and a little bit of circuit conductor (wire). Since both wires and metal tools have very low resistance, a tremendous amount of current can flow through the shorted portion of the circuit. (And you will notice that any value for the resistance of circuit conductors has been completely eliminated. This is done for simplicity, and because the values are so close to zero that they are negligible.)

If the overall resistance of these components was 0.01 ohm, the amount of current flowing in the shorted portion of this circuit would 80/0.01 = 8000 amperes ($E/R = I$). At this level of current, both the metal tool and the wire will heat so rapidly that they not only melt, but vaporize, creating a serious explosion and causing serious burns to anyone close enough to be hit by the metal vapors.

While it would be unlikely that the power source shown could actually produce 8000 amperes, the point should be clear—shorted circuits can be dangerous and cause the circuit to malfunction.

The example given previously and shown in Figure 3-8 is known as a *dead short*. (This is an old term that originated in the electrical trade, and no one is certain as to the precise derivation.) A dead short exposes the power supply to an effectively zero resistance. Immense

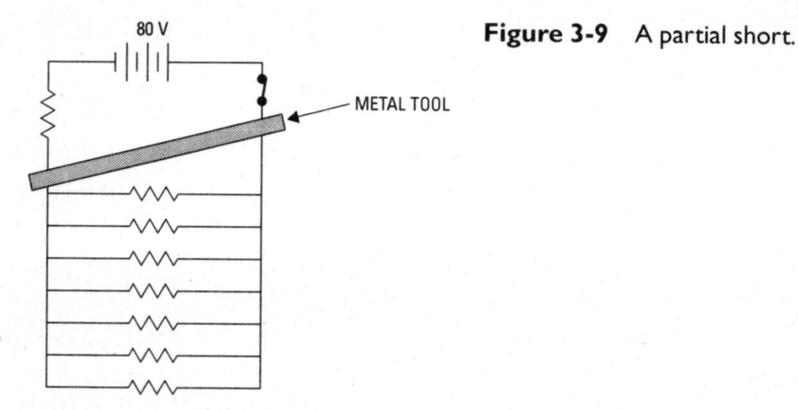

Figure 3-9 A partial short.

amounts of current will flow, the remainder of the circuit will lose voltage (since the power supply is not infinite,and is exhausted by the short), and things explode and/or burn.

Figure 3-9 shows a *partial short*. In this case, the same metal tool is shorting across the circuit conductors. However, this time there is a resistance between the power source and the short. In this situation, that power source in the resistor and the switch will be exposed to levels of current beyond those for which they were designed. But the current in the shorted circuit will not rise to the levels it would in the dead short situation of Figure 3-8. In this case, the resistor (or some type of household load, presumably) would burn out. This would create a hazard at the burning resistor, but only at that location. And, generally, this results in something that burns more slowly, then leaves the circuit open when it is physically destroyed. This type of short can lead to fires, but generally not to explosions.

Circuit Diagrams

As you have seen, specialized diagrams are used to properly depict electrical circuits. There are several types of diagrams, and it is important that you understand what you are looking at. Follow this discussion carefully.

Figure 3-10 shows a diagram of two electric lamps connected in series with a battery. We occasionally use picture diagrams, but generally work from schematic diagrams. A schematic diagram shows an electric circuit by means of graphical symbols instead of outline pictures. For example, Figure 3-11 shows two schematic diagrams. Standard electrical symbols are used to represent a battery, a lamp, and a resistor. Notice that the schematic diagram in Figure 3-2A can

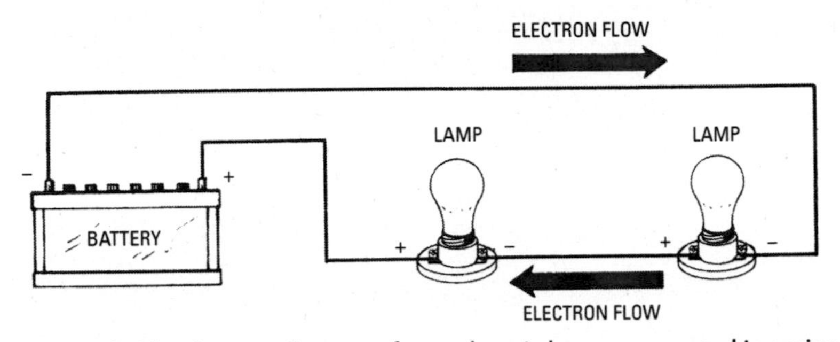

Figure 3-10 Picture diagram of two electric lamps connected in series with a battery.

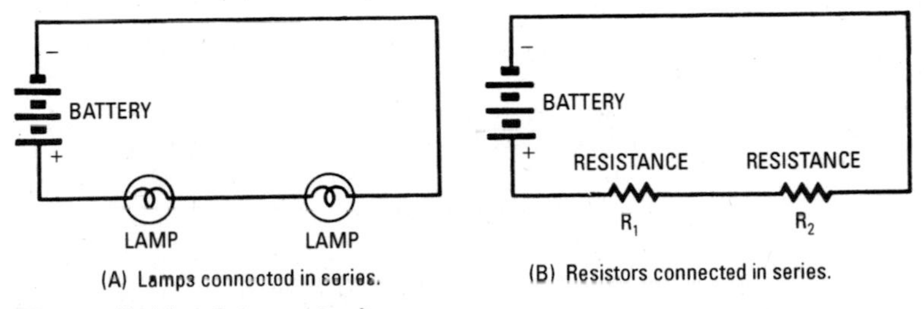

Figure 3-11 Schematic diagram.

be represented by an equivalent circuit, as seen in Figure 3-2B. An equivalent circuit has the same electrical properties as the original circuit, but an equivalent circuit does not serve the same purpose.

If R_1 and R_2 in Figure 3-11B have the same amounts of resistance as the lamps in Figure 3-11A, and if the batteries in both circuits have the same voltage, it is clear that each circuit will draw the same amount of current. However, these circuits do not serve the same purpose because the circuit in Figure 3-11A produces light and heat, while the circuit in Figure 3-11B produces heat only. Nevertheless, the equivalent circuit is useful because it is the first step in *reducing* the original series circuit into a simplified circuit.

It is evident that the load in Figure 3-11B consists of resistance R_1 plus resistance R_2. Therefore, we can combine R_1 and R_2 into a single resistor, as seen in Figure 3-12. For example, if each lamp in Figure 3-11A has a resistance of 25 ohms, then $R_1 = 25$ ohms and $R_2 = 25$ ohms in Figure 3-11B. In turn, $R_L = 50$ ohms in Figure 3-12. The circuit in Figure 3-12 represents the final reduction, or simplest possible equivalent circuit for the lamp circuit in

Figure 3-11A. Equivalent circuits are useful because they help us to better understand the operation of complicated circuits.

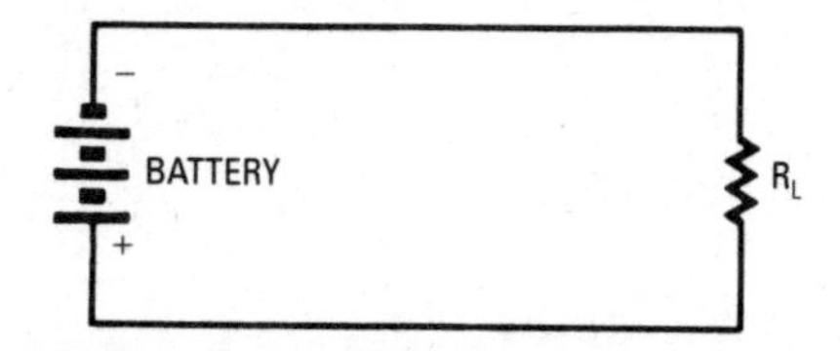

Figure 3-12 Equivalent circuit of resistor connected in series.

Kirchhoff's Voltage Law

Kirchhoff's voltage law states that *the sum of the voltage drops around a circuit is equal to the source voltage.*

Gustav Kirchoff taught at the University of Heidelberg in the mid-nineteenth century. While still a student in 1845, he discovered two fundamental electrical laws, extending Ohm's work. He also discovered several optical phenomena and even proved that electricity traveled (in a perfect conductor) at the same speed as does light.

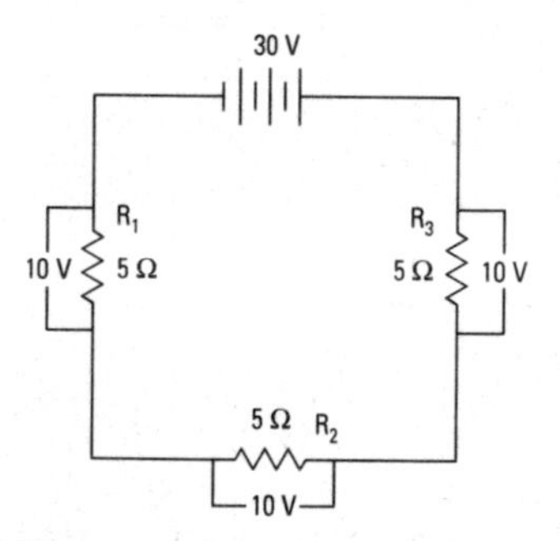

Figure 3-13 Sum of voltage drops are equal to the source voltage.

A voltage *drop* is the voltage measured across a single load in a circuits, as shown in Figure 3-13. The voltage drops measured at R_1, R_2, and R_3 add up to 30 volts, the same as the voltage of the source. This was Kirchhoff's discovery.

Kirchhoff's voltage law is actually a summary of Ohm's law as applied to all the resistance in a circuit. Although we do not need to use Kirchhoff's voltage law in describing the action of simple circuits, this law will be found very useful in solving complicated circuits.

Figure 3-14 shows one way this law can be useful. In this figure, the values for R_1 and R_2 are given. We are also given the information that R_1 and R_3 are equal. Given this information, we can calculate values for the entire circuit. Since the voltage drop across R_1 must be the same (equal resistances in a series circuit), the voltage drop across R_3 must be 10 volts. With this information, we know (using Kirchhoff's voltage law) that the source voltage is 50 volts. We can then use Ohm's Law to determine the total current in the circuit.

Sensibly enough, the voltage drops measured at the resistors (or whatever load) would be opposite in polarity to the source voltage.

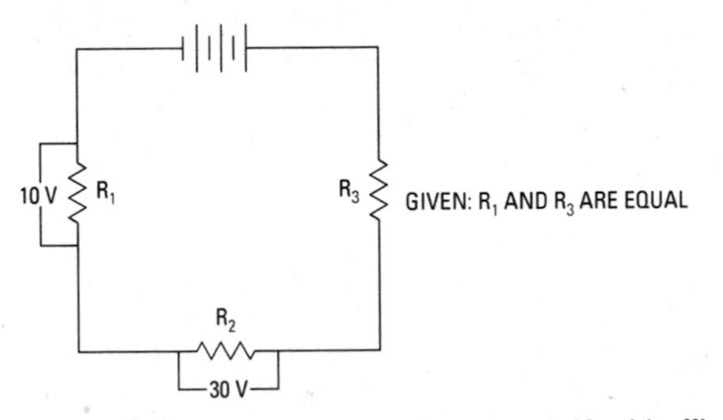

Figure 3-14 Circuit values found with Kirchhoff's voltage law.

Observe the polarities shown in Figure 3-10. Electrons flow from the negative terminal of the battery to the right-hand lamp in the diagram. Of course, there is a voltage drop across this lamp. Since the right-hand terminal of the lamp is connected to the negative terminal of the battery, and the left-hand terminal of the lamp eventually returns to the positive terminal of the battery, it is obvious that the right-hand terminal of the lamp has negative polarity with respect to its left-hand terminal.

To show why we need to observe the polarity of a voltage, consider the connection of a voltmeter across a battery, as shown with an old-style voltmeter in Figure 3-15. To measure the battery voltage, we must connect the positive terminal of the voltmeter to the positive terminal of the battery, and we must connect the negative terminal of the voltmeter to the negative terminal of the battery. The pointer will move up-scale on the voltmeter. If we make a mistake and reverse the polarity of the meter connections, the pointer will not move up-scale on the voltmeter; instead, the pointer will move off-scale to the left. Therefore, voltmeters have their positive and negative terminals indicated.

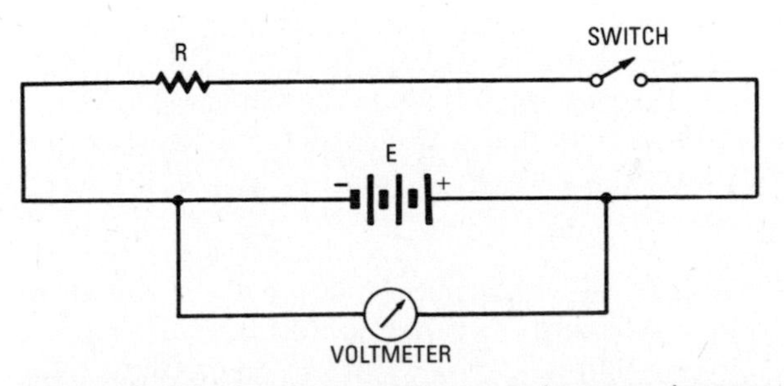

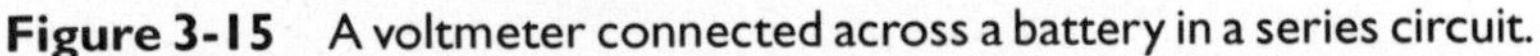
Figure 3-15 A voltmeter connected across a battery in a series circuit.

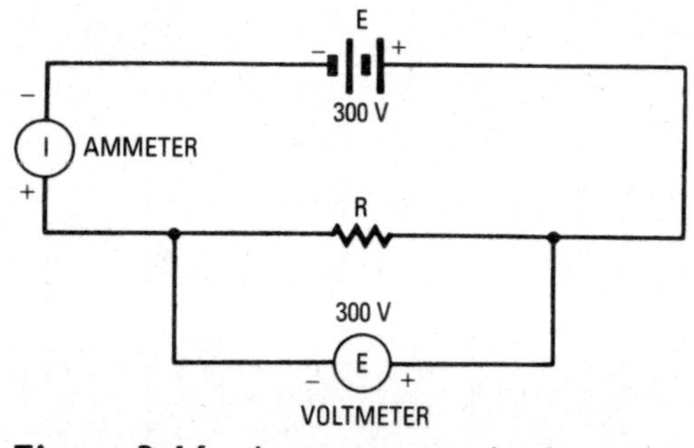

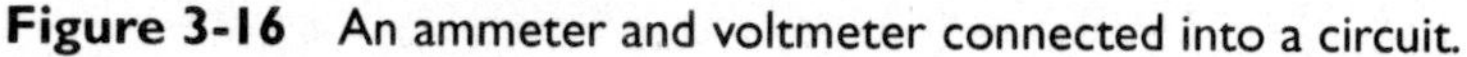

Figure 3-16 An ammeter and voltmeter connected into a circuit.

An ammeter also has its positive and negative terminals indicated. An ammeter must be connected into a circuit in proper polarity, as shown in Figure 3-16. This is true for instrument-type ammeters, but not for clamp-on types. The voltmeter is connected to read the voltage drop across R. Note carefully that the ammeter must *not* be connected across R; in effect, the battery would thereby be short-circuited through the ammeter. The ammeter would probably be damaged. Even if the ammeter did not burn out, the short circuit would soon ruin the battery. Therefore, you should keep the following rules in mind:

- An ammeter is always connected in *series* with a circuit.
- A voltmeter is always connected *across* the battery, resistor, or other device to measure a voltage drop.

Voltage Measurements with Respect to Ground

Most circuits are (in one way or another) connected to ground. In the case of power wiring, one conductor in the system is literally connected to a rod that is driven into the ground. In the case of electronic circuits, the chasis serves as the ground and is usually connected to the electrical system's grounding electrode conductor.

Figure 3-17 shows a series circuit with a ground. The circuit consists of battery E and resistors R_1, R_2, and R_3. Voltmeter V_1 measures the voltage drop across R_1; V_2 is the voltage across the battery and R_1. Note carefully that to find voltage E, we *subtract* voltage V_1 from voltage V_2, as shown here:

$$E = V_2 - V_1$$

Next, we observe that the voltage drop across R_2 is found by subtracting V_3 from V_2. Finally, V_3 is the voltage drop across R_3. For example, if $R_1 = 1$ ohm, $R_2 = 2$ ohms, $R_3 = 3$ ohms, and

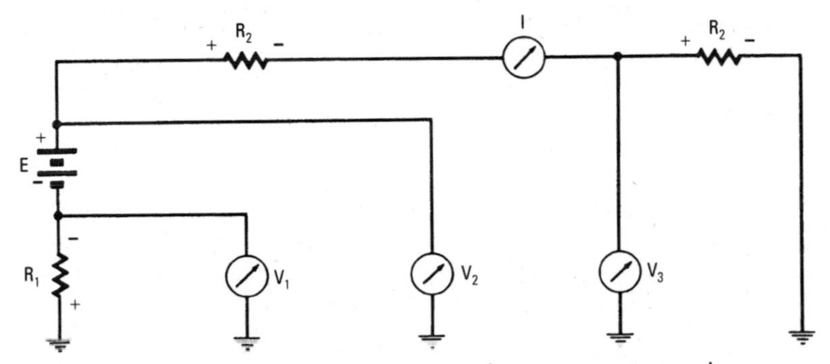

Figure 3-17 Voltage measurements with respect to ground.

$E = 6$ volts, the current flow $I = 1$ ampere. In turn, $V_1 = 1$ volt, $V_2 = 5$ volts, and $V_3 = 3$ volts. Note that we can disregard the voltage drop across ammeter I in Figure 3-17 because the resistance of an ammeter is very small.

Note also in Figure 3-17 that the letters E and V are used to indicate voltages. We commonly use E to indicate a *source voltage*, and use V to indicate a *voltage drop*. This is helpful, because it prevents confusion when we are working out a circuit problem. However, many electricians simply use E to indicate any voltage in a circuit, and other electricians use V to indicate any voltage in a circuit. It does not make any difference whether we write V or E or both, as long as we remember which voltage the letter stands for.

Batteries

Since we are using batteries repeatedly in diagrams, it make sense to give them a bit of explanation. What we call a battery is actually a battery of electrochemical cells. The word battery originally meant "an array of."

Batteries of electrical cells (or individual cells) produce electricity by chemical means. Groups of chemicals (usually metals and acids) are put together in such a way as to cause electrons to form at one end of the cell. This creates a difference of electrical *potential* between one side of the cell and the other. One side of the battery is negatively charged and the other becomes positively charged.

Batteries have a limited life because producing electricity changes the chemicals. In effect, the energy is drained from them. Frequently, forcing electricity through the battery in a reversed direction (by applying a voltage higher than the voltage the battery produces and of opposite polarity) will cause the chemical reaction to reverse, thus recharging the battery. Not all batteries will function this way, so don't try it unless you are sure.

The voltage and current potentials of any battery are dependent upon their internal chemical reactions. So, to produce more current, you connect batteries in parallel, and to produce more voltage, you connect them in series.

Any battery has a certain amount of resistance, called its *internal resistance.* In many circuits, we can disregard the internal resistance of a battery without getting into practical difficulties. On the other hand, we will find various circuits for which it is necessary to take the internal resistance of a battery into account. Figure 3-18A

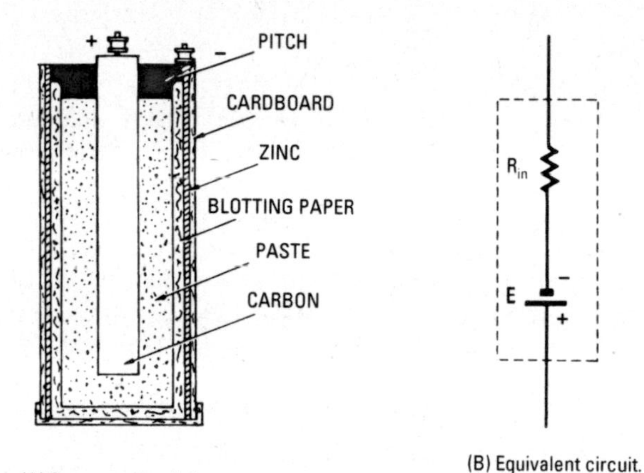

(A) Cross-sectional view.

(B) Equivalent circuit.

(C) Load resistor connected to equivalent circuit.

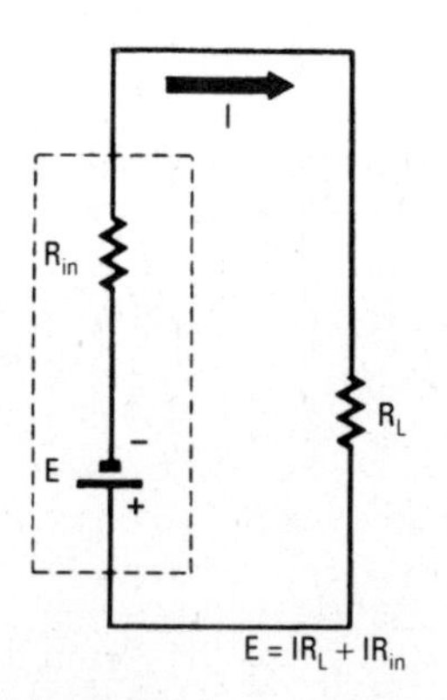

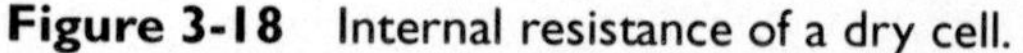
Figure 3-18 Internal resistance of a dry cell.

illustrates the internal construction of a dry cell. The paste contains a chemical solution called an *electrolyte*. This electrolyte has a small amount of resistance in a brand new cell. In a nearly dead cell, the resistance will be very great. Therefore, any cell can be represented by the equivalent circuit shown in Figure 3-18B. The cell operates electrically as if a perfect cell E were connected in series with an internal resistance R_{in}.

When a cell is connected to a load resistor, as shown in Figure 3-18C, the voltage across R_L will be less than E. There is a voltage drop inside the cell equal to IR_{in}. Therefore, the voltage across R_L is equal to $E - IR_{in}$. We call E the *electromotive force* (*emf*) of the cell. On the other hand, the voltage drop across R_L in Figure 3-18C is equal to the *terminal* voltage of the cell. We will find that E remains about the same (1.5 volts) even in a nearly dead cell. However, as noted previously, the resistance of R_{in} becomes greater as the cell becomes weaker. It follows that the terminal voltage of a cell is the same as its emf when no current is being drawn, even if the cell is nearly dead.

Therefore, we cannot test a cell properly with a voltmeter alone, because a voltmeter draws a very small amount of current. Instead, we must connect a load resistance across the cell as shown in Figure 3-18C and measure the voltage drop across the load resistor. A large dry cell has more electrode area than a small flashlight cell. A large cell can normally supply more current than a small cell. This means that a large cell should be tested with a load resistor that has a comparatively small value.

Efficiency and Load Power

Efficiency means output power divided by input power. For example, if we get half as much power out of a machine as we put into it, we say that the efficiency of the machine is 50 percent. The same principle applies to electric circuits. The power that we get out of the circuit shown in Figure 3-19 is the *power* in the lamp filament

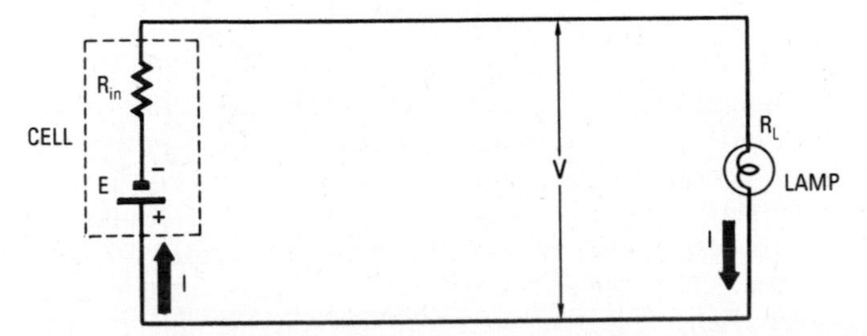

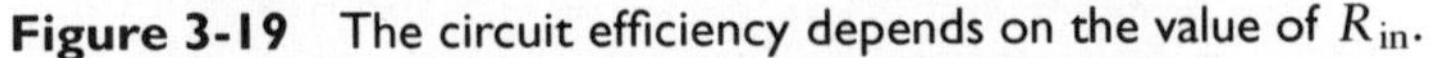
Figure 3-19 The circuit efficiency depends on the value of R_{in}.

R_L; the power that goes into the circuit is supplied by the cell. We know that the power in the lamp is equal to VI watts, and the power supplied by the cell is equal to EI watts. Therefore, the efficiency of the circuit is equal to VI/EI, or V/E. In terms of percentage, the efficiency of the circuit is equal to $100V/E$ percent.

The efficiency of the circuit in Figure 3-19 could be 100 percent only if R_{in} were zero. Since R_{in} can never be zero, the circuit efficiency must always be less than 100 percent. We will find that the efficiency of a circuit is greatest when the load is very light (that is, when a very small amount of current is drawn by the load). Let us see why this is so. If R_L has a very high value, then IR_L is much greater than IR_{in}. This is just another way of saying that V is large when R_L is large. Since E remains the same, it follows that the circuit efficiency V/E is high when R_L has a high value.

Let us consider how we can get the greatest power out of the circuit shown in Figure 3-19. We will find the value of R_L that makes VI as large as possible. It can be shown that the lamp will have the largest number of watts when $R_L = R_{in}$. This is a surprising answer at first glance. Let us note the following facts:

- If R_L were zero, maximum current would flow in the circuit. However, V would then be zero, and the power in the load would be zero.
- If R_L were infinite, maximum voltage would be dropped across the load. However, I would then be zero, and the power in the load would be zero.
- The load power has its greatest value when the load resistance is equal to the internal resistance of the cell.

Figure 3-20 shows a practical example. Notice in Figure 3-20A that the source voltage is 100 volts, and the source resistance is 5 ohms. As the resistance of the load is changed, the load voltage, circuit current, load power, and circuit efficiency change as shown in Figure 3-20B. When the load resistance is equal to the source resistance (5 ohms), the load power has its greatest value. If the load resistance is less or greater than 5 ohms, the load power decreases. We say that *maximum power transfer* occurs when the load resistance is equal to the source resistance. Note that the circuit efficiency is only 50 percent when maximum power transfer is obtained. Figure 3-20C shows how the efficiency, load voltage, circuit current, and load power change as the load resistance is changed.

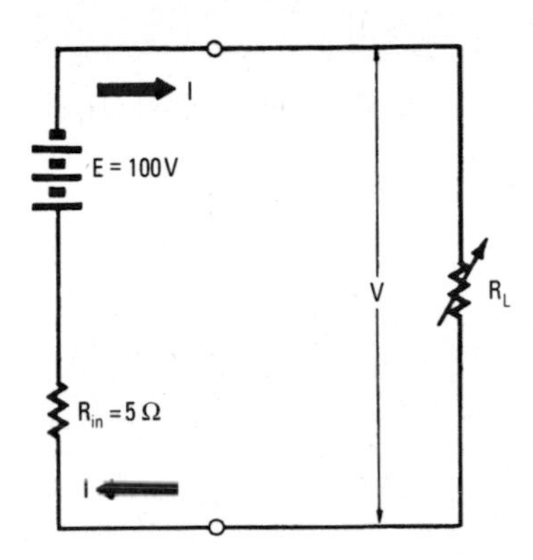

E = OPEN-CIRCUIT VOLTAGE OF SOURCE
R_{in} = INTERNAL RESISTANCE OF SOURCE
V = TERMINAL VOLTAGE
R_L = RESISTANCE OF LOAD
P_L = POWER USED IN LOAD
I = CURRENT FROM SOURCE
% EFF = PERCENTAGE OF EFFICIENCY

(A) Circuit.

R_L	V	I	P_L	% EFF
0	0	20	0	0
1	16.6	16.6	267.6	16.6
2	28.6	14.3	409	28.6
3	37.5	12.5	468.8	37.5
4	44.4	11.1	492.8	44.4
5	50	10	500	50
6	54.5	9.1	495.4	54.5
7	58.1	8.3	482.2	58.1
8	61.6	7.7	474.3	61.6
9	63.9	7.1	453.7	63.9
10	66	6.6	435.6	66
20	80	4	320	80
30	87	2.9	252	87
40	88	2.2	193.6	88
50	91	1.82	165	91

(B) Chart.

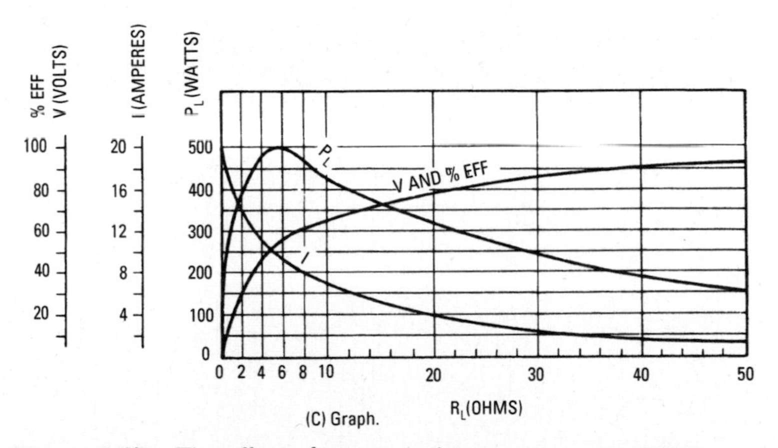

(C) Graph.

Figure 3-20 The effect of source resistance on power output.

Circuit Voltages in Opposition

Up to now we have considered only cells connected in series-aiding. However, in practical electrical work, cell voltages may be connected in series-opposing. Figure 3-21 shows a 1.5 volt source connected in series-opposing with a 3-volt source. The 1.5 volts subtracts from the 3 volts, and the voltmeter reads 1.5 volts. Note the voltmeter polarity. Automobile electricians are concerned with series-opposing voltages in storage-battery charging circuits. Therefore, let us briefly consider the properties of storage batteries and battery-charging circuits.

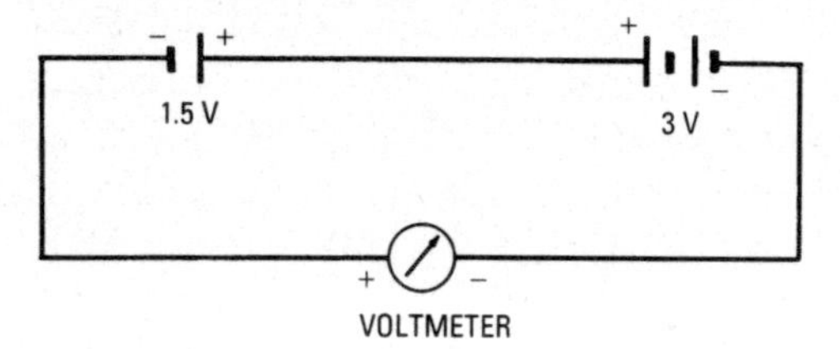

Figure 3-21 Circuit voltages in opposition.

A storage battery is also called a *secondary battery*, whereas a dry cell is called a *primary battery*. Secondary cells are different from primary cells in that a secondary cell can be recharged, whereas a primary cell cannot. The basic storage cell consists of a pair of lead plates immersed in a solution of sulfuric acid. Figure 3-22 depicts a simple storage cell and charging circuit. An electric generator is used to supply the charging voltage and current. Before the storage cell is charged, it has very little resistance, and the circuit current is practically 3 amperes.

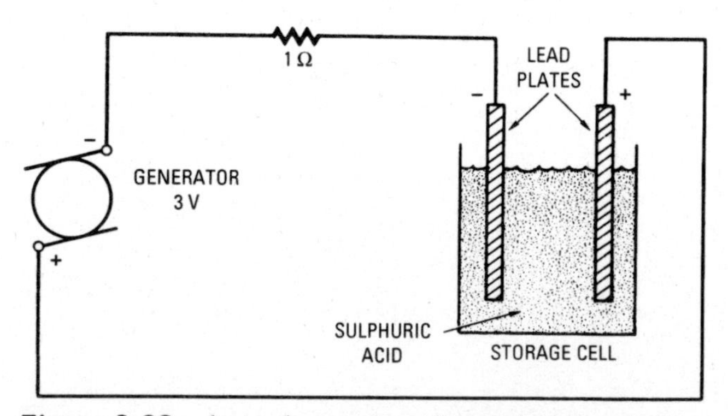

Figure 3-22 A simple storage cell and charging circuit.

As the charging process continues in Figure 3-22, chemical changes take place at the surfaces of the lead plates, and the storage cell builds up a voltage between its terminals. This voltage has the polarity shown in the diagram, and it opposes the generator voltage. Therefore, the charging current becomes less as the storage cell charges. Figure 3-23 shows how the terminal voltage of a storage cell increases during charge. The terminal voltage rises to 2.2 volts, then remains comparatively constant for a time, and finally rises to a maximum of 2.6 volts. At this point, the effective voltage in the circuit of Figure 3-22 is $3-2.6 = 0.4$ volt.

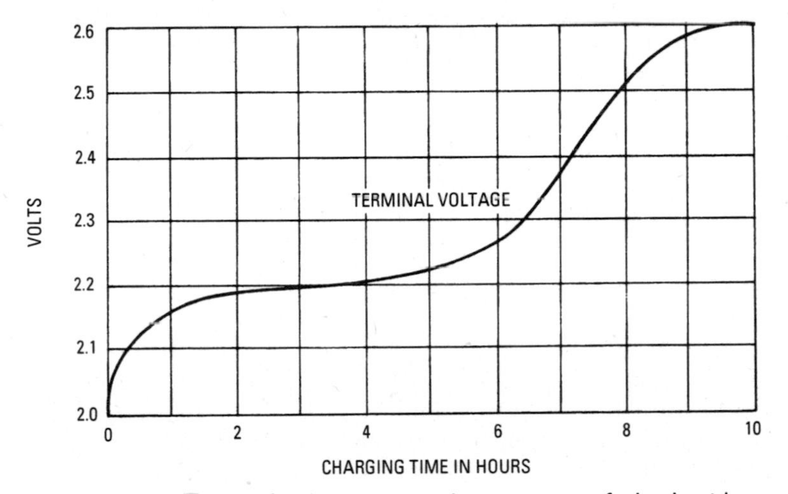

Figure 3-23 Terminal voltage versus charging time of a lead-acid storage cell.

When a storage cell is removed from the charging circuit, its terminal voltage falls to approximately 2 volts. The cell can then be used as a voltage source until the chemical film on its plates has been used up. The voltage of the cell remains at practically 2 volts until it is almost discharged and then falls rapidly. After the cell is discharged, it can be recharged as has been explained. We measure the *capacity* of a storage cell in ampere-hours. Its ampere-hour capacity is equal to the number of amperes supplied by the cell on discharge multiplied by the number of hours that the cell can supply current. Commercial storage cells and batteries have the construction shown in Figure 3-24.

Another type of storage cell, called the *nickel-cadmium* cell, is in wide use. This type of cell has a terminal voltage of approximately

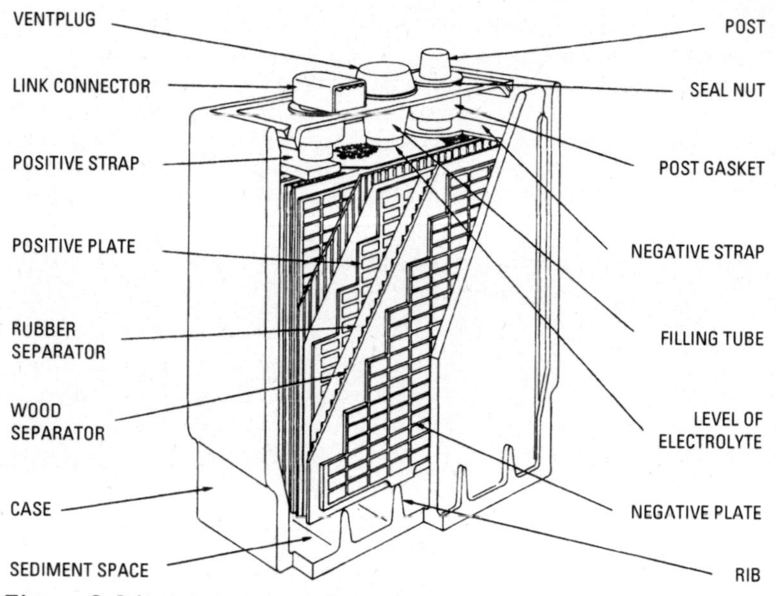

Figure 3-24 Lead-acid cell and battery.

1.2 volts. It is lighter and has a longer life than a lead storage cell. A nickel-cadmium cell also requires less attention and care. It can be completely discharged and left uncharged for an indefinite time. This abusive treatment would ruin a lead cell.

Principles of Parallel Circuits

In a parallel circuit, each load is connected in a *branch* across the voltage source, as shown in Figure 3-25. Therefore, there are as many paths for current flow as there are branches. The loads in a parallel circuit are sometimes said to be connected in *multiple*. We know that when more loads are connected into a series circuit, the total circuit resistance increases. On the other hand, when more loads are connected into a parallel circuit, the total resistance decreases.

We observe in Figure 3-25 that the voltage across all branches of a parallel circuit (from *a* to *b*) is the same because all branches are connected directly to the voltage source *E*. Each load resistor draws current independently of the other load resistors. Each branch current depends only on the load resistance in that branch. Therefore, we find the current in each branch by dividing its load resistance into

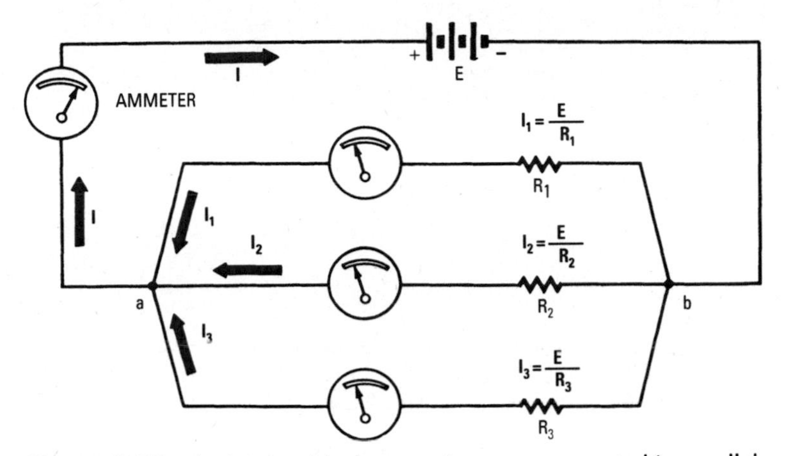

Figure 3-25 A circuit with three resistances connected in parallel.

the source voltage. The application of Ohm's law in each branch is shown in Figure 3-25. Since the source voltage E is applied across each branch, it follows that we may write the following:

$$E = I_1 R_1 = I_2 R_2 = I_3 R_3$$

Also, the three branch currents must be supplied by the total current I in Figure 3-25. Therefore, we write the following:

$$I = I_1 + I_2 + I_3$$

or,

$$I = \frac{E}{R_1} + \frac{E}{R_2} + \frac{E}{R_3}$$

Next, let us find an equivalent circuit for the parallel circuit in Figure 3-25. We will replace R_1, R_2, and R_3 with an equivalent resistor R_{eq}. Since $I = E/R_{eq}$, we will write the equation as follows:

$$\frac{E}{R_{eq}} = \frac{E}{R_1} + \frac{E}{R_2} + \frac{E}{R_3}$$

We observe that E cancels out in the foregoing formula, leaving:

$$\frac{1}{R_{eq}} = \frac{1}{R_1} + \frac{1}{R_2} + \frac{1}{R_3}$$

To show how the foregoing formula is used, let us take a practical example. Suppose that $R_1 = 5$ ohms, $R_2 = 10$ ohms, and $R_3 = 30$ ohms in Figure 3-25. Then, we find the value of R_{eq} as follows:

$$\frac{1}{R_{eq}} = \frac{1}{5} + \frac{1}{10} + \frac{1}{30} + \frac{10}{30} = \frac{1}{3}$$

or,

$$\frac{1}{R_{eq}} = 0.2 + 0.1 + 0.033 = 0.333$$

Therefore,

$$R_{eq} = 0.2 + 0.1 + 0.033 = 0.333$$

Our equivalent circuit is drawn as shown in Figure 3-26, and the value of R_{eq} is approximately 3 ohms. This is the method used by most electricians to find the equivalent resistance of several resistors connected in parallel. It makes no difference how many branches there may be in a parallel circuit. You simply add up the reciprocals of all the load resistors to find the reciprocal of the equivalent resistance. Then, the equivalent resistance is the denominator of this fraction.

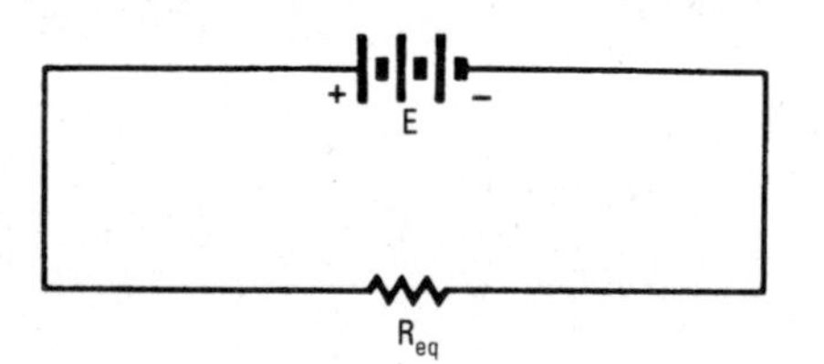

Figure 3-26 Equivalent circuit for Figure 3-25.

Shortcuts for Parallel Circuits

Shortcuts are always nice, but care must be taken to apply them correctly. When learning shortcut methods of doing things, make sure that you understand where and where not to use the shortcut.

The first shortcut applies to any number of parallel resistors, provided each resistor has the same resistance value. In this case, the equivalent resistance is found by dividing the resistance of one resistor by the number of resistors that are connected in parallel. For example, suppose that we have five 10-ohm resistors connected in parallel, as shown in Figure 3-27. In turn, we write the following:

$$R_{eq} = \frac{10}{5} = 2 \text{ ohms}$$

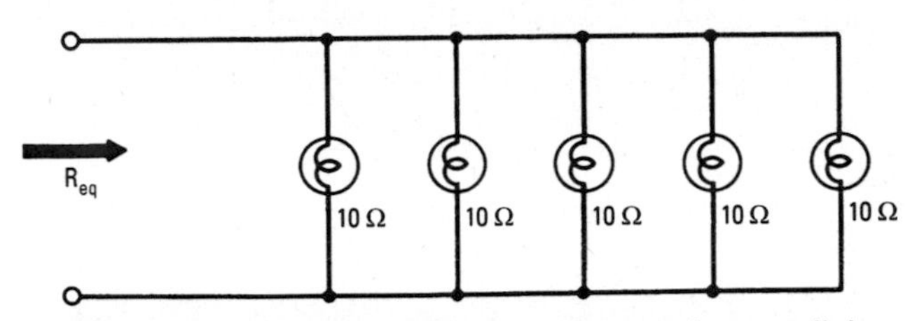

Figure 3-27 Five 10-ohm loads connected in parallel.

The second shortcut applies when we have two resistors with different values connected in parallel. In this case, the equivalent resistance is equal to the product of the two resistances divided by their sum. For example, if we have two resistors with values of 3 ohms and 6 ohms connected in parallel as shown in Figure 3-28, their equivalent resistance is found as follows:

$$R_{eq} = \frac{R_1 R_2}{R_1 + R_2} = \frac{3 \times 6}{3 + 6} = \frac{18}{9} = 2 \text{ ohms}$$

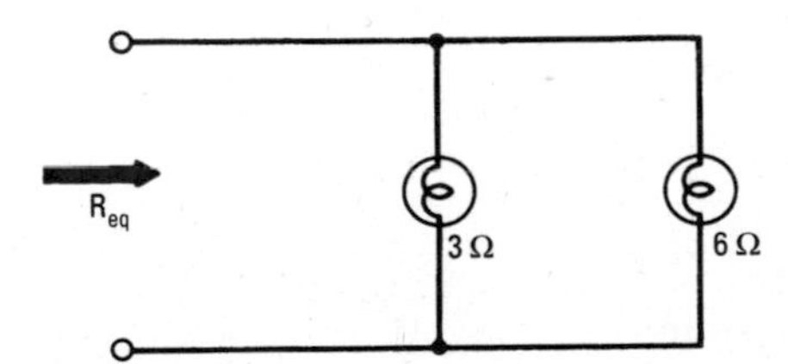

Figure 3-28 Equivalent circuit for Figure 3-25.

Kirchhoff's Current Law for Parallel Circuits

Kirchhoff's current law for parallel circuits states that *at any junction of conductors, the algebraic sum of the currents is zero*. This is just another way of saying that just as much electricity must leave a junction as there is electricity entering the junction. (Remember, we are discussing current now, not voltage as in Kirchhoff's other law.) For example, let us consider junction *a* in Figure 3-29. We assume that the current flowing toward junction *a* is positive, and that the currents flowing away from junction *a* are negative. We consider that I_t is positive, and that I_1, I_2, and I_3 are negative. In turn, we write Kirchhoff's current law as follows:

$$+I_t - I_1 - I_2 - I_3 = 0$$

Or, for the example shown in Figure 3-29, we write the following:

$$+10 - 6 - 3 - 1 = 0$$

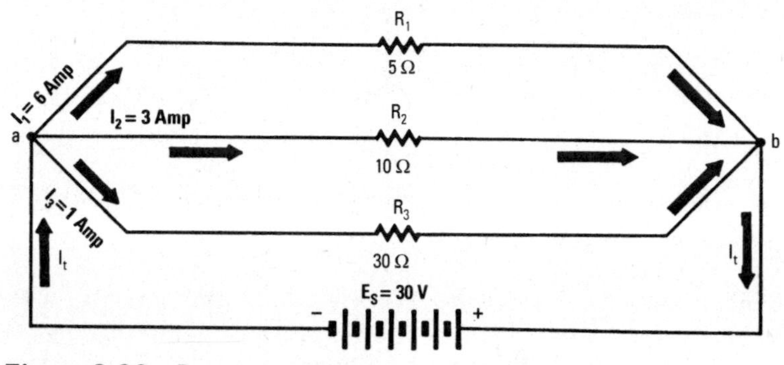

Figure 3-29 Resistors in parallel.

The importance of Kirchhoff's current law will be shown later by means of practical examples. We will find that, just as in a series circuit, the total power consumed in a parallel circuit is equal to the sum of the power values in each resistor. For example, the power P_1 in R_1 of the circuit in Figure 3-26 is written as follows:

$$P_1 = EI_1 = 30 \times 6 = 180 \text{ watts}$$

Next, power P_2 in R_2 is found:

$$P_2 = EI_2 = 30 \times 3 = 90 \text{ watts}$$

Similarly, power P_3 in R_3 is found:

$$P_3 = EI_3 = 30 \times 1 = 30 \text{ watts}$$

The total power consumed by the parallel circuit in Fig 3-29 is simply the sum of the power values in each branch:

$$T = P_1 + P_2 + P_3 = 180 + 90 + 30 = 300 \text{ watts}$$

We can check this answer by using the total current value in the power formula:

$$T = EI_1 = 30 \times 10 = 300 \text{ watts}$$

Practical Problems in Parallel Circuits

Let us consider the parallel circuit shown in Figure 3-30. This circuit has two branches, *a* and *b*. Branch *a* has three lamps in parallel. L_1 takes a power of 50 watts, L_2 takes 25 watts, and L_3 takes 75 watts. Branch *b* also has three lamps in parallel. L_4 takes 150 watts, L_5 takes 200 watts, and L_6 takes 250 watts. The source voltage is

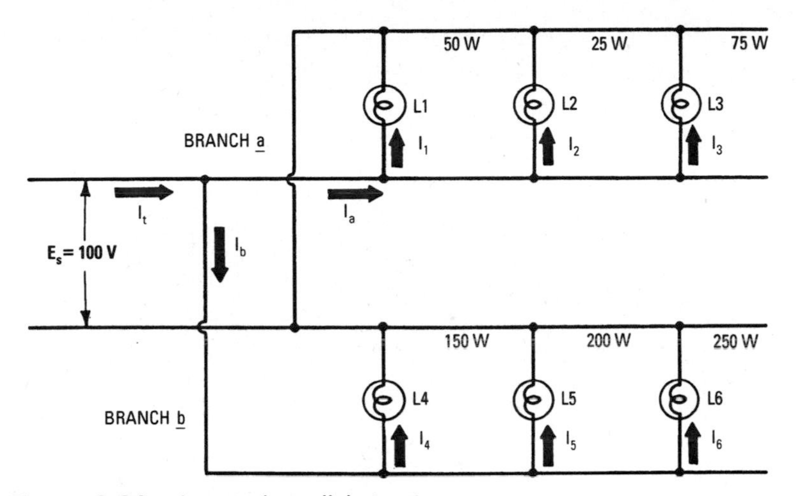

Figure 3-30 A typical parallel circuit.

100 volts. Our problem is to:

1. Find the current in each lamp.
2. Find the resistance of each lamp.
3. Find the current in branch *a*.
4. Find the current in branch *b*.
5. Find the total circuit current.
6. Find the total circuit resistance.
7. Find the total power in the circuit.

This problem is solved by means of Ohm's law, the power equations, and Kirchhoff's current law. We proceed as follows:

1. The current in L_1 is $I_1 = P_1/E_s = 50/100 = 0.5$ ampere. Similarly, $I_2 = 0.25$ ampere, $I_3 = 0.75$ ampere, $I_4 = 1.5$ amperes, $I_5 = 2$ amperes, and $I_6 = 2.5$ amperes.
2. The resistance of L_1 is $R_1 = E_s/I_1 = 100/0.5 = 200$ ohms. Similarly, $R_2 = 400$ ohms, $R_3 = 133$ ohms, $R_4 = 66.7$ ohms, $R_5 = 50$ ohms, and $R_6 = 40$ ohms.
3. The current in branch *a* is $I_1 + I_2 + I_3 = 0.5 + 0.25 + 0.75 = 1.5$ amperes.
4. The current in branch *b* is $I_4 + I_5 + I_6 = 1.5 + 2.0 + 2.5 = 6$ amperes.

5. The total circuit current is $I_a + I_b = 1.5 + 6.0 = 7.5$ amperes.
6. The total circuit resistance is $R_t = E_t/I_t = 100/7.5 = 13.3$ ohms.
7. The total power supplied to the circuit is $50 + 25 + 75 + 150 + 200 + 250 = 750$ watts. To check the total power in the circuit, note that we can write:

$$P_t = EI_t = 100 \times 7.5 = 750 \text{ watts}$$

Parallel Connection of Cells

Cells may be connected in parallel, as shown in Figure 3-31, when the load draws a greater amount of current than can be supplied by a single cell. As a practical example, let us assume that dry cells are to be used to supply 1.5 volts to a load that draws a steady current of $^1/_2$ ampere. A No. 6 dry cell will supply $^1/_8$ ampere over an extended time, such as several hours. To meet this requirement, we will connect cells in parallel, as shown in Figure 3-32A. In a parallel connection, all positive cell terminals are connected to one side of the line, and all negative cell terminals are connected to the other side of the line. The voltage across the line is the same as the voltage of one cell, or 1.5 volts. However, each cell can contribute its maximum allowable current of $^1/_8$ ampere to the line.

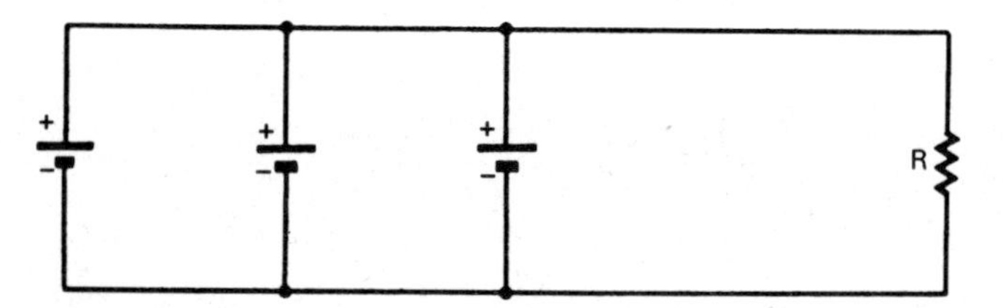

Figure 3-31 Circuit arrangement with cells connected in parallel.

Since the load in Figure 3-32B draws $^1/_2$ ampere, we use four cells in this example. Each cell contributes $^1/_8$ ampere to the load. Storage cells may also be connected in parallel when a comparatively heavy current is demanded by a load. If we use lead-acid storage cells, the voltage across the load will be approximately 2 volts. It is interesting to note that a commercial lead-acid storage cell contains several negative plates and an equal number of positive plates connected in parallel, as seen in Figure 3-33. When cells are connected in parallel, the result is the same as if we used a single cell with a much larger electrode area.

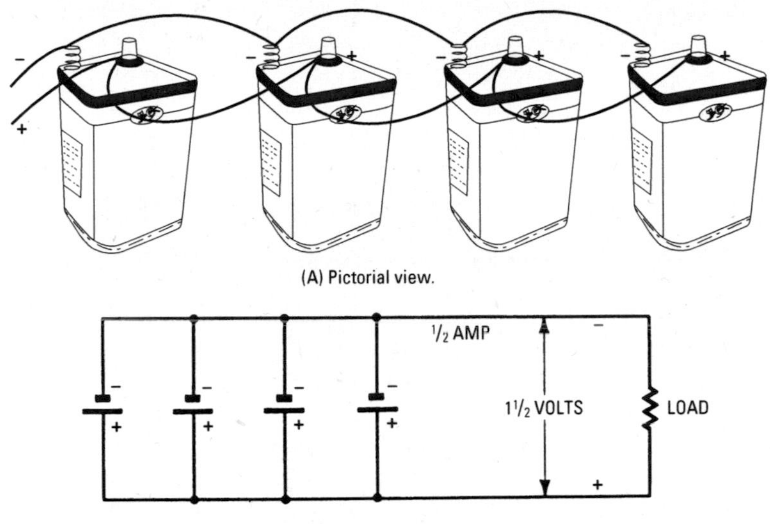

Figure 3-32 Dry cells connected in parallel.

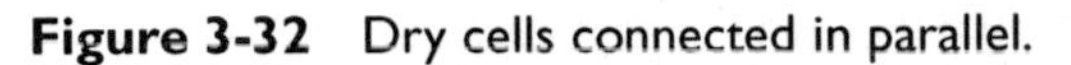

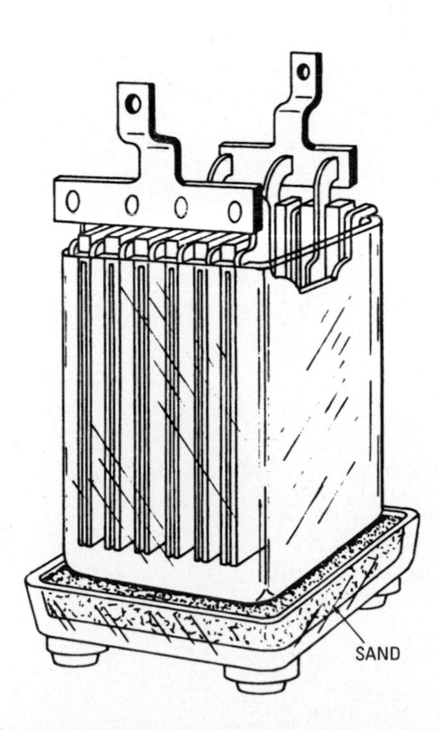

Figure 3-33 Glass jar of a storage cell supported on sand.

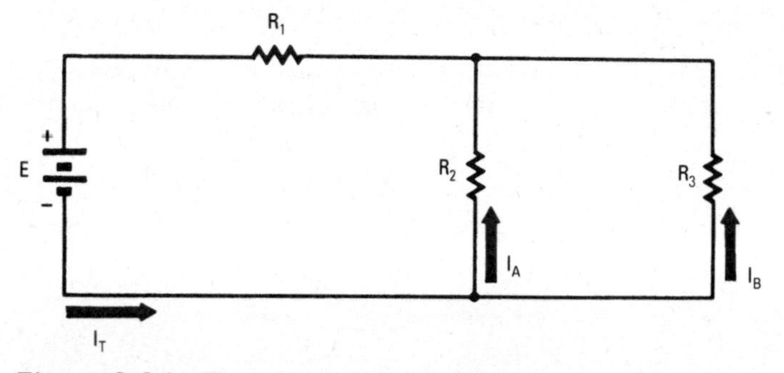

Figure 3-34 The total current and the branch current.

Principles of Series-Parallel Circuits

A series-parallel circuit has branch currents and a total current in the same way that a parallel circuit does. For example, Figure 3-34 shows the total current and the branch currents in a simple series-parallel circuit. Let us see how we can find the amounts of these currents in a practical arrangement. If the battery in Figure 3-34 supplies 22 volts, and the resistances of R_1, R_2, and R_3 are 1 ohms, 2 ohms, and 3 ohms, respectively, we reason as follows:

1. R_2 and R_3 are connected in parallel; therefore, their equivalent resistance is 6/5, or 1.2 ohms.
2. R_1 is connected in series with the equivalent resistance of R_2 and R_3; therefore, the total circuit resistance is $1 + 1.2$ ohms, or 2.2 ohms.
3. Ohm's law states that $I = E/R$; therefore, I_T is equal to 22/2.2, or 10 amperes.
4. Ohm's law also states that $E = IR$; therefore, the voltage drop across R_1 is equal to 10×1, or 10 volts.
5. The voltage applied to R_2 and R_3 is equal to E minus the voltage drop across R_1; therefore, the voltage applied to R_2 and R_3 is 22/2.2, or 10 amperes.
6. Since 12 volts are applied across R_2, I_A is equal to 12/2, or 6 amperes; similarly, I_B is equal to 12/3, or 4 amperes.

It is important to observe that the total current in Figure 3-34 is equal to the sum of the branch currents I_A and I_B. That is, we write:

$$I_T = I_A + I_B$$

or,

$$10 \text{ amperes} = 6 + 4 \text{ amperes}$$

Kirchhoff's Current Law for Series-Parallel Circuits

The foregoing example illustrates an important law of electricity called Kirchhoff's current law. This law states that *the sum of the currents leaving a junction is equal to the current entering the junction.* A *junction* is a circuit point at which a current splits up, or branches, as shown at P in Figure 3-35. In this diagram, current I_T is entering junction P, and currents I_A and I_B are leaving junction P. We have seen that I_T is equal to the sum of I_A and I_B.

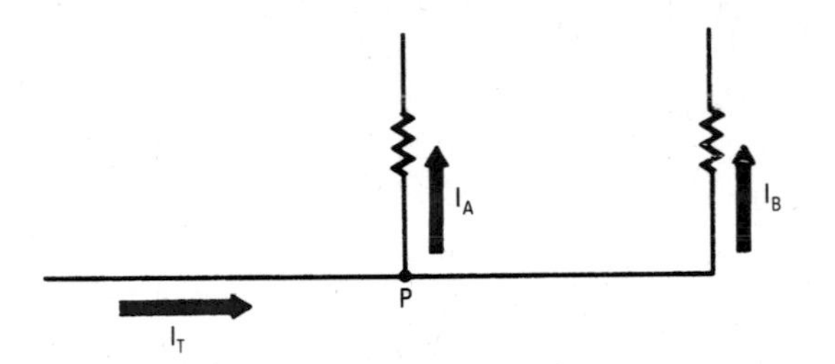

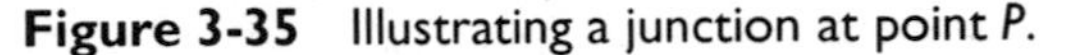

Figure 3-35 Illustrating a junction at point *P*.

Kirchhoff's current law is important because it helps us to find the amount of branch current that flows in a circuit such as the one shown in Figure 3-36. In this example, we are given the series resistance and the voltage drop across it, and three of the branch currents. Our problem is to find the fourth branch current (I_B). We must start by finding the total current. It follows from Ohm's law that we can write the following:

$$I_T = \frac{10}{5} = 2 \text{ amperes}$$

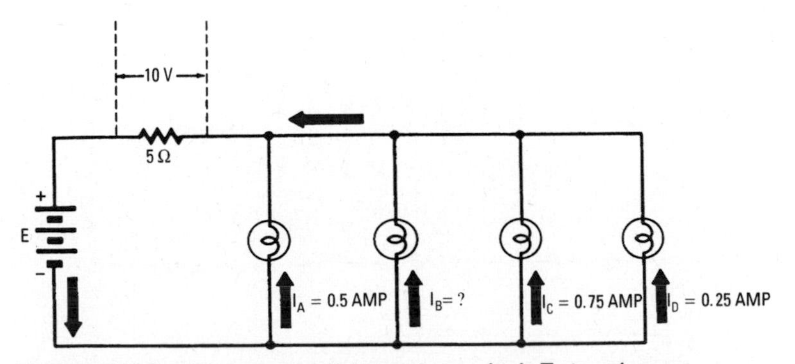

Figure 3-36 A series-parallel circuit in which T_B is unknown.

Since we know the total current, we can use Kirchhoff's current law to find I_B:

$$I_A + I_B + I_C + I_D = I_T$$

or,

$$0.5 + I_B + 0.75 + 0.25 = 2$$

Therefore,

$$I_B = 2 - 0.5 - 0.75 - 0.25 = 0.5 \text{ amperes}$$

Series-Parallel Connection of Cells

Cells can be connected in series to obtain a greater source voltage. We also know that cells can be connected in parallel to obtain a greater current capacity. In turn, we will find that cells can be connected in series-parallel to obtain a greater source voltage with a greater current capacity.

There are two ways that this can be done, as shown in Figure 3-37. We can connect series groups of cells in parallel, or we can connect

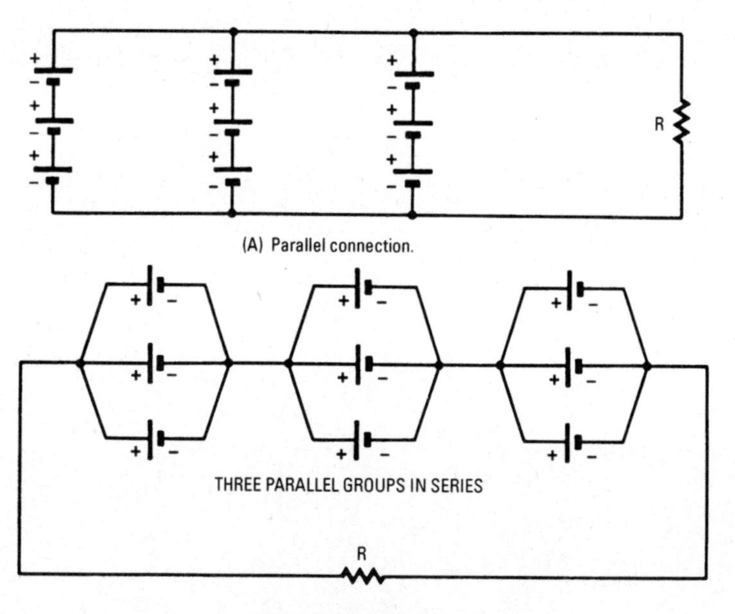

Figure 3-37 Connecting dry cells in series-parallel.

groups of cells in series. One circuit operates exactly the same as the other. Therefore, it makes no difference in practice which method of connection we may use.

Circuit Reduction

The reduction of a series-parallel circuit to a simple equivalent circuit is based on the methods used for series circuits and for parallel circuits. For example, reduction of the circuit shown at A-1 in Figure 3-38 is very simple if we keep the following facts in mind:

- Any number of resistances connected in series can be replaced by a single resistance with a value equal to the sum of the individual resistances.

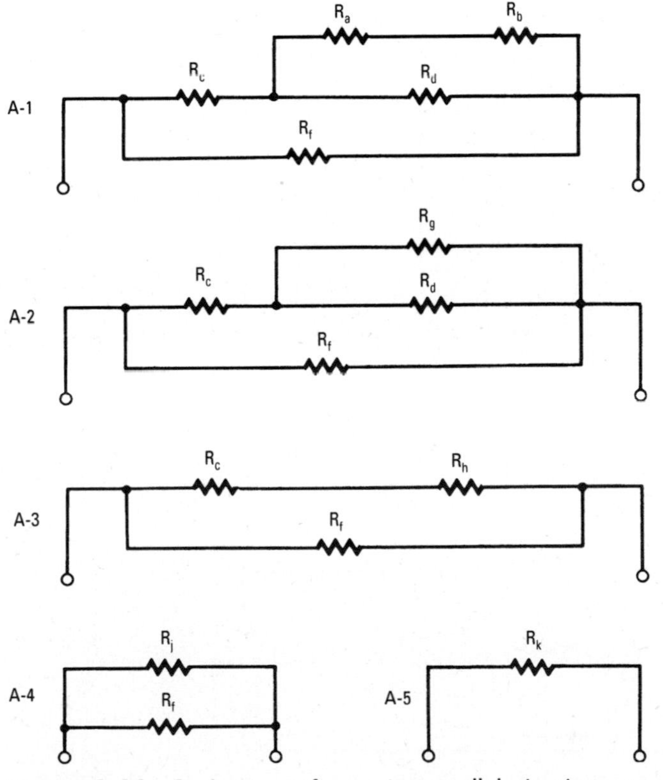

Figure 3-38 Reduction of a series-parallel circuit to an equivalent resistance.

- Any number of resistances connected in parallel can be replaced by a single resistance with a value equal to the reciprocal of the sum of the reciprocals of the individual resistances.

For example, if we have three resistances connected in series, we write the following:

$$R_{\text{eq}} = R_1 + R_2 + R_3$$

On the other hand, if we have three resistances connected in parallel, we write:

$$R_{\text{eq}} = \frac{1}{\frac{1}{R_1} + \frac{1}{R_2} + \frac{1}{R_3}}$$

We observe that circuit A-1 in Figure 3-38 consists of resistors R_a and R_b in series, and that this series combination is in parallel with R_d. In turn, this series-parallel combination is connected in series with R_c, and the resulting combination is connected in parallel with R_f. To find the equivalent circuit, we first replace R_a and R_b with their equivalent resistance R_g, as shown in A-2. The next step is to combine R_g and R_d, replacing them by their equivalent resistance R_h, as seen in A-3. We then replace R_c and R_h by their equivalent resistance R_j, as depicted in A-4. Finally, we replace R_j and R_f by their equivalent resistance, and obtain the equivalent resistance R_k, as shown in A-5.

Electricians use different ways to reduce resistances in parallel to a single equivalent resistance. The method that is used is simply a matter of personal preference. For example, let us consider the

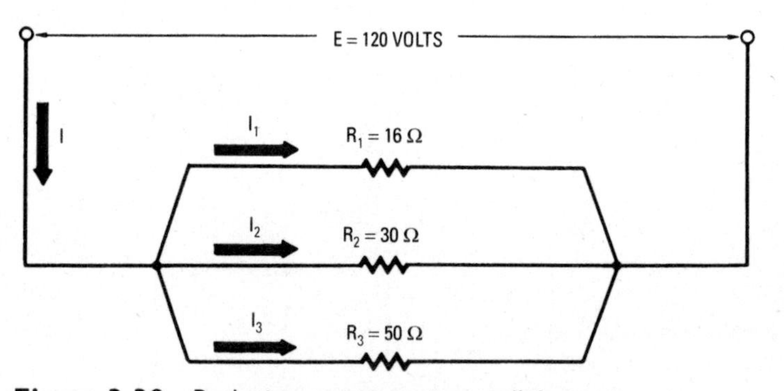

Figure 3-39 Reducing resistance in parallel circuits.

circuit shown in Figure 3-39. We can find the equivalent resistance by any of the following methods:

- *Reduction by pairs*—In reduction by pairs, we take the resistors two at a time. Thus, we may select R_1 and R_2, and find their equivalent resistance:

  $$\mathrm{R_{eq}} = \frac{R_1 R_2}{R_1 + R_2} = \frac{16 \times 30}{16 + 30} = \frac{480}{46} = 10.434 \text{ ohms}$$

  Next, we will take R_{eq} and R_3 and find their equivalent resistance:

  $$R_{eq} = \frac{R_{eq} R_3}{R_{eq} + R_3} = \frac{1200}{139} = \frac{521.7}{60.434} = 8.63 \text{ ohms, approx.}$$

- *Reduction by product-and-sum formula*—If we prefer, we can take all three resistors at the same time in the product-and-sum formula. Accordingly, we write the following:

  $$R_{eq} = \frac{R_1 R_2 R_3}{R_1 R_2 + R_1 R_3 + R_2 R_3} = \frac{16 \times 30 \times 50}{480 + 800 + 1500}$$

  $$= \frac{24,000}{2780} = \frac{1200}{139} = 8.63 \text{ ohms, approx.}$$

- *Reduction by reciprocals*—To use the method of reduction by reciprocals, we write the reciprocal of the equivalent resistance as the sum of the reciprocals of the individual resistances:

  $$\frac{1}{R_{eq}} = \frac{1}{R_1} + \frac{1}{R_2} + \frac{1}{R_3} = \frac{1}{16} + \frac{1}{30} + \frac{1}{50}$$

  or,

  $$\frac{1}{R_{eq}} = \frac{75}{1200} + \frac{40}{1200} + \frac{24}{1200} = \frac{139}{1200}$$

  $$\frac{1}{R_{eq}} = \frac{1200}{139} = 8.63 \text{ ohms, approx.}$$

- *Reduction by conductances*—In reduction by conductances, we write the equivalent conductance as the sum of the individual conductances:

  $$G_{eq} = G_1 + G_2 + G_3 = 0.0625 + 0.0333 + 0.02$$
  $$= 0.1158 \text{ ohms, approximately}$$

or,

$$R_{\text{eq}} = \frac{1}{G_{\text{eq}}} = \frac{1}{0.1158} = 8.63 \text{ ohms, approx.}$$

Power in a Series-Parallel Circuit

The power consumed by the loads in a series-parallel circuit is equal to the sum of the power values consumed by each load. For example, Figure 3-40 shows a 3000-watt (3 kW) load consisting of a heater, hot plate, and flatiron, each of which consumes 1000 watts. However, the total power in this circuit also includes the power loss to the line caused by line drop. The line drop in this example is 117 – 107, or 10 volts. To find the power loss in the line, we must first find the current drawn by the load. Accordingly, we write the following:

$$1071 = 3000 \text{ watts}$$

or,

$$I = \frac{3000}{107} = 28 \text{ amperes, approx.}$$

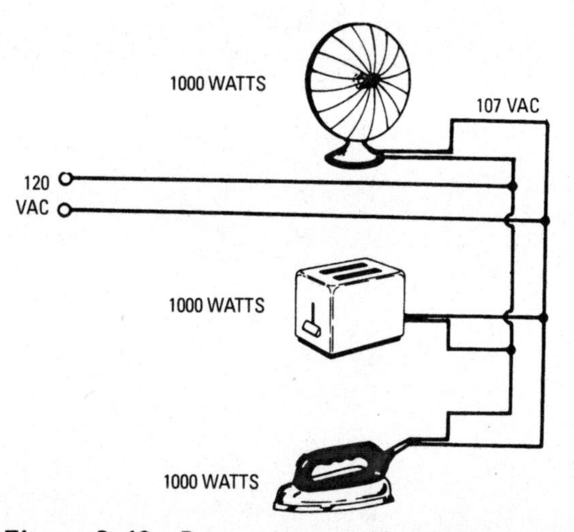

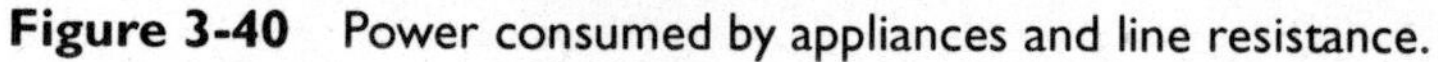

Figure 3-40 Power consumed by appliances and line resistance.

The power loss in the line is equal to the line drop multiplied by the line current:

$$P_{\text{line}} = 10 \times 28 = 280 \text{ watts, approximately}$$

Therefore, the power consumption of the circuit shown in Figure 3-40 is equal to 3280 watts. The power input to the line is the product of the input voltage and the line current:

$$P_{in} = 117 \times 28 = 3276 \text{ watts, approximately}$$

Note that we find a value of 3276 watts for the power input, but that we found a value of 3280 watts for the circuit power. The reason is that both of these answers are approximate, because we "rounded off" the decimals in our arithmetic.

Next, the *efficiency* of the circuit shown in Figure 3-40 is equal to the load power divided by the input power:

$$\text{Efficiency} = \frac{3000}{3276} = 0.91 = 91\% \text{ approx.}$$

Three-Wire Distribution Circuit

A series-parallel circuit of particular importance to electricians is called a *three-wire distribution circuit*. This type of circuit often operates at a total voltage of 240 volts, while the loads operate at 120 volts. As shown in Figure 3-41, a *positive feeder*, a *negative feeder*, and a neutral wire are used in the basic arrangement. The loads are connected between the negative feeder and the neutral, and between the positive feeder and the neutral. When the loads are *unbalanced* (unequal), the neutral wire carries a current equal to the difference of the currents in the positive and negative feeders.

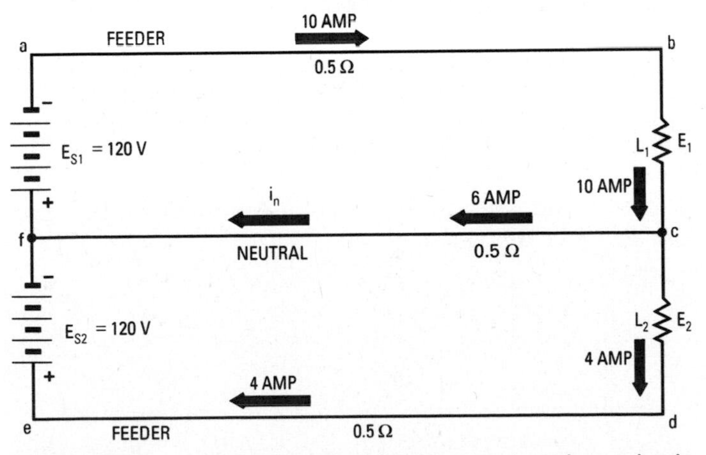

Figure 3-41 A three-wire distribution circuit with two loads.

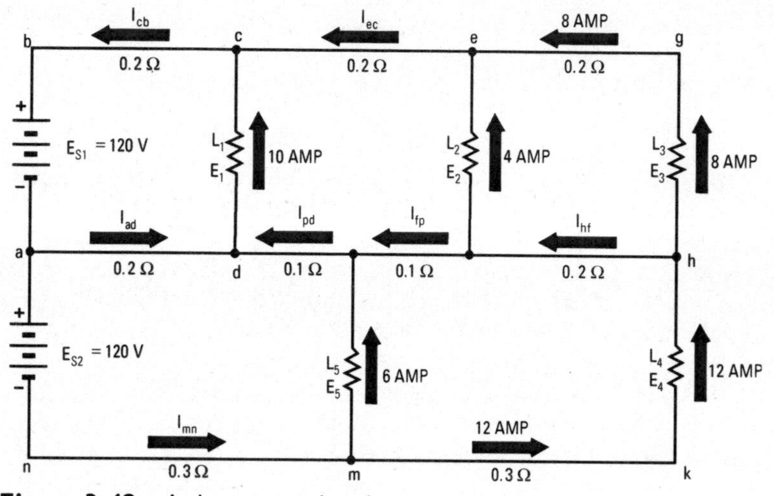

Figure 3-42 A three-wire distribution circuit with five loads.

In the example of Figure 3-42, load L_1 draws 10 amperes, load L_2 draws 4 amperes, and the neutral wire carries a current of $10 - 4 = 6$ amperes. The direction of current flow in the neutral wire is always the same as that of the smaller of the currents in the positive and negative feeders. Note that the current flow is to the left in the positive feeder, and that this current is smaller than the current in the negative feeder. In turn, the current in the neutral wire is in the same direction as in the positive feeder.

With respect to junction *c* in Figure 3-41, we observe that the algebraic sum of the currents entering and leaving the junction is zero:

$$+10 - 4 - 6 = 0$$

Next, let us see how to find the load voltage E_1 in Figure 3-41. Since the algebraic sum of the voltages around the circuit *fabcf* must be zero, we write the following:

$$+120 - 10 \times 0.5 - E_1 - 6 \times 0.5 = 0$$

or,

$$E_1 = 120 - 5 - 3 = 112 \text{ volts}$$

Thus, the voltage across load L_1 is 112 volts. The source voltage is 120 volts, and the line drop is 8 volts. In other words, the line drop in the negative feeder is equal to the line resistance (0.5 ohm) times the line current (10 amperes), or 5 volts. The line drop in the neutral wire is equal to the line resistance (0.5 ohm) times the line current (6 amperes), or 3 volts. Thus, the total line drop between E_{S1} and L_1 is 8 volts.

Load voltage E_2 in Figure 3-41 is found in a similar manner. We write the following:

$$+120 + 6 \times 0.5 = E_2 - 4 \times 0.5 = 0$$

or,

$$E_2 = 120 + 3 - 2 = 121 \text{ volts}$$

In tracing the circuit from f to c, note that we proceed against the direction of the current arrow, and, therefore, the IR drop of 6×0.5 volts has a plus sign. The load voltage (121 volts) is 1 volt greater that the source voltage (120 volts). The total source voltage is 240 volts, and the total load voltage is $112 + 121 = 233$ volts. Observe that this total load voltage is also equal to the difference between the total source voltage and the sum of the voltage drops in the positive and negative feeders, or, $240 - (2 + 5) = 233$ volts.

When we have a balanced load on the positive and negative sides of a three-wire system, the current in the neutral is zero, and the currents in the feeders are equal. However, when the loads are unbalanced, the unbalance current flows in the neutral. Therefore, the voltage decreases on the heavily loaded side, and the voltage increases on the lightly loaded side. We see that if the neutral wire had zero resistance, there would be no unbalance in the load voltages. For this reason, it is desirable to use a low-resistance neutral wire when unbalanced loads are present in a system.

A more complicated three-wire circuit is shown in Figure 3-42. The source voltage is 120 volts between each outside wire and the neutral wire. Load currents in the upper side of the system are 10, 4, and 8 amperes, respectively, for L_1, L_2, and L_3 in this example. In the lower side of the system, the load currents for loads L_1 and L_5 are 12 and 6 amperes, respectively. To find the load voltages, we must find the currents in each outside wire and in the neutral wire. Since the resistances of these wires are given, the voltage drops and the load voltages can be found after the currents are found.

To find the currents, it is best to start at the load farthest from the source in Figure 3-42. Electrons flow out of the negative terminal at n and return to the positive terminal at b. It is standard practice to assume that currents flowing toward a junction are positive, and that currents flowing away from a junction are negative. Now, let us apply Kirchhoff's current law at junction h; the neutral current

I_n (flowing from h to f) is evidently:

$$12 - 8 - I_{bf} = 0$$

or,

$$I_{bf} = 12 - 9 = 4 \text{ amperes}$$

In the same way, at junctions f, e, p, m, d, and c, we write the following at f:

$$4 - 4 - I_{fp} = 0$$

or,

$$I_{fp} = 4 - 4 = 0 \text{ amperes}$$

At e, we write the following:

$$4 + 8 - I_{ec} = 0$$

or,

$$I_{ec} = 4 + 8 = 12 \text{ amperes}$$

At p, we write the following:

$$6 + 0 - I_{pd} = 0$$

or,

$$I_{pd} = 6 + 0 = 6 \text{ amperes}$$

At m, we write the following:

$$+\mathrm{I}_{mn} - 6 - 12 = 0$$

or,

$$\mathrm{I}_{mn} = 6 + 12 = 18 \text{ amperes}$$

At d, we write the following:

$$\mathrm{I}_{ad} + 6 - 10 = 0$$

or,

$$\mathrm{I}_{ad} = -6 + 10 = 4 \text{ amperes}$$

At c, we write the following:

$$\mathrm{I}_{cb} + 10 + 12 = 0$$

or,

$$\mathrm{I}_{c} = -10 - 12 = 22 \text{ amperes}$$

Therefore E_{S1} in Figure 3-42 supplies 22 amperes and E_{S2} supplies 18 amperes. The electron flow in all parts of the lower wire is outward from the source, and the electron flow in all parts of the upper wire is back toward the source. The current in the neutral wire is always equal to the difference in the currents in the two outside wires, and the electron flow is in the direction of the smaller of these two currents.

In Figure 3-42, the neutral current in section *ad* is 3 amperes, which is the difference between 19 amperes and 22 amperes. Also, it is in the direction of the smaller current in section *mn*. The neutral current in section *pd* is 6 amperes, which is the difference between 18 amperes and 12 amperes. It is in the same direction as the 12 amperes in section *ec*. The neutral current in section *fp* is zero because the current in each outside wire in that section is 12 amperes. The neutral current in section *hf* is 4 amperes, which is the difference between 12 amperes and 8 amperes. It is in the direction of the smaller outside current in section *ge*.

To find the load voltages in Figure 3-42, we apply Kirchhoff's voltage law to the various individual circuits. Thus, to find the voltage E_1 across L_1, the algebraic sum of the voltages around the circuit *abcda* is equated to zero. Starting at *a*:

$$-120 + 22 \times 0.2 + E_1 \times 0.2 = 0$$

or,

$$E_1 = 120 - 4.4 - 0.8 = 114.8 \text{ volts}$$

To find load voltage E_2, we trace circuit *dcefpd*. Starting at *d*:

$$-114.8 + 12 \times 0.2 + E_2 + 0 \times 0.1 - 6 \times 0.1 = 0$$

or,

$$E_2 = 114.8 - 2.4 - 0 + 0.6 = 113 \text{ volts}$$

To find voltage E_3, we trace circuit *feghf*. Starting at *f*:

$$-113 + 8 \times 0.2 + E_3 - 4 \times 0.2 = 0$$

or,

$$E_3 = 113 - 1.6 + 0.8 = 112.2 \text{ volts}$$

To find load voltage E_4, we trace circuit *nadpfhkmn*. Starting at *n*:

$$-120 - 4 \times 0.2 + 6 \times 0.1 + 4 \times 0.2 + E_4 + 12 \times 0.3 + 18 \times 0.3 = 0$$

or,

$$E_4 = 120 + 0.8 - 0.6 - 0 - 0.8 - 3.6 - 5.4 = 110.4 \text{ volts}$$

To find load voltage E_5, we trace circuit *nadpmn*. Starting at *n*:

$$-120 - 4 \times 0.2 + 6 \times 0.1 + E_5 + 18 \times 0.3 = 0$$

or,

$$E_5 = 120 + 0.8 - 0.6 - 5.4 = 114.8 \text{ volts}$$

The foregoing is an example of a comparatively involved series-parallel circuit problem. However, it is a very practical type of problem for the electrician. We recognize that this problem would have been very difficult to solve if we did not understand Kirchhoff's current law and Kirchhoff's voltage law.

Thevenin's Theorum

One of the more difficult parts of electronics is dealing with very complex circuits, such as the one shown in Figure 3-43 The easiest way to deal with such circuits is by using *Thevenin's theorum.* Leon Thevenin discovered this method in 1863, although it seems to have been originally discovered by Hermann von Helholtz 10 years earlier.

Thevenin's theorum is actually a fairly simple trick. Thevenin merely says that if you want to find information on only two terminals (points on the circuit), you can simply treat the rest of the circuit as a black box, which can be represented by a voltage and a resistance. Figure 3-43A shows the circuit, Figure 3-43B shows the two terminals we are concerned with, and Figure 3-43C shows the black box represented as a simple voltage and resistance. The subscript "th" identifies the voltage and resistance as the "Thevenin" equivalent voltages.

We will run through an example of this.

Figure 3-44 shows the circuit we want to analyze. We are particularly interested in the way the circuit will affect the 8 kilo-ohms resistor. So, we begin by separating the 8 kilo-ohms resistor from the rest of the circuit to expose the two terminals we want to analyze. Bear in mind that we're looking into the black box in this example so that you can better understand how this works.

Of course, the easiest way to determine the Thevenin voltage and resistance is simply to pull out a meter and to measure it. This is shown in Figures 3-45 and 3-46. However, the Thevenin voltage and resistance can also be calculated.

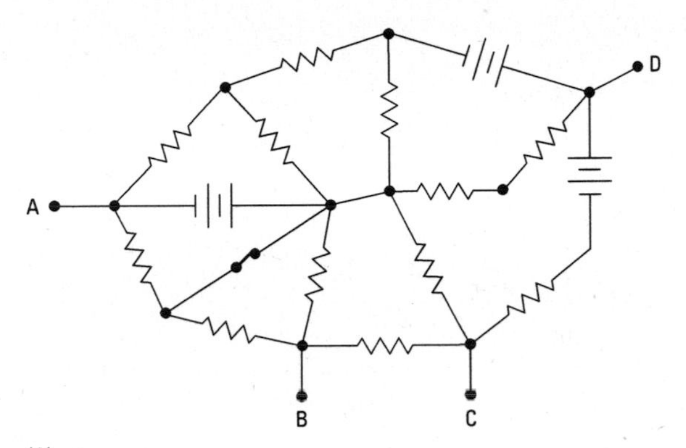

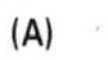
(A)

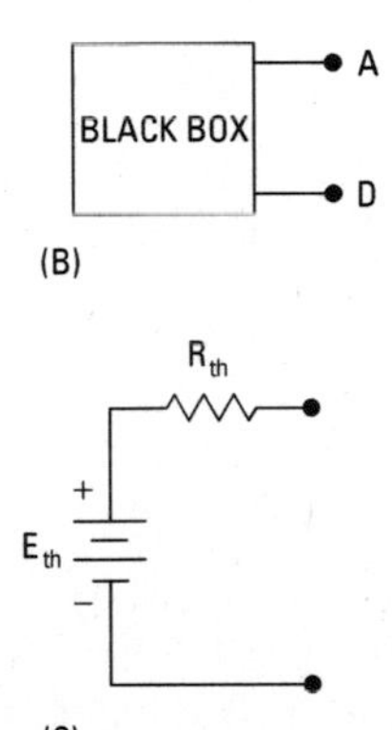

(C)

Figure 3-43 Thevenin's theorum solution for a very complex circuit.

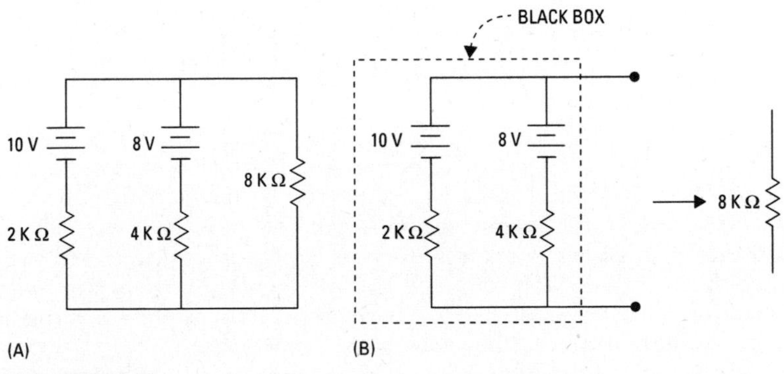

Figure 3-44

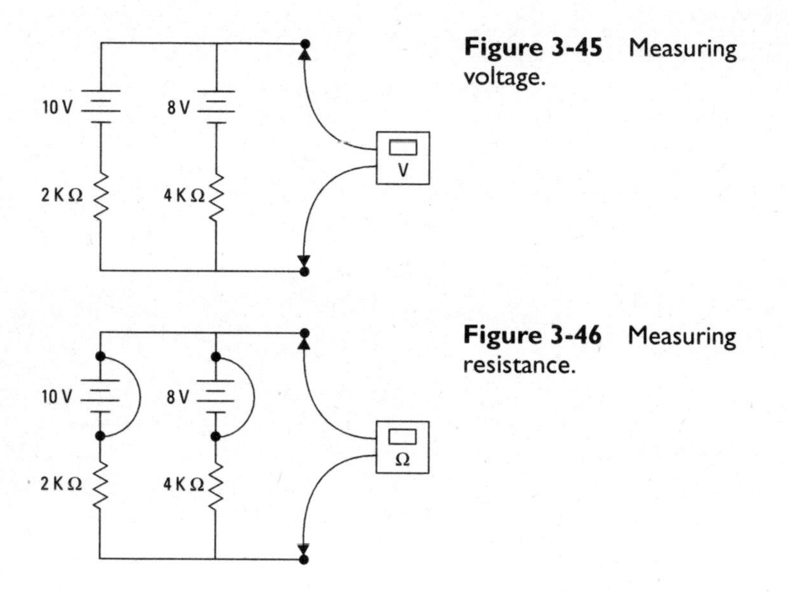

Figure 3-45 Measuring voltage.

Figure 3-46 Measuring resistance.

First we calculate the current through the loop. We assume the courses to be opposing, and the calculations look like this:

$$I = \frac{10V - 8V}{2K\Omega + 4K\Omega} = .333 \text{ mA}$$

We can then calcuate the current through each branch by using Ohm's law:

10-volt branch: $10 - .333 \text{ mA} \times 2 \text{ k}\Omega = 9.333$ volts

8-volt branch: $8 + .333 \text{ mA} \times 4 \text{ k}\Omega = 9.333$ volts

So, our Thevenin voltage, V_{th}, is 9.33 volts. (The previous figures are very slightly rounded.)

To get our Thevenin resistance, we could, of course simply measure it, as is shown in Figure 3-46. Note that in this drawing we are showing a jumper around the batteries. This is necessary to remove the power supply from the resistance test. A resistance test is always made with power off, otherwise it will distort the test and likely damage the meter. (Ohmmeters are very sensitive.) However, although schematic drawings show a jumper wire connected across the battery, be aware that in real life this would short out the battery, possibly leading to an explosion, and certainly wearing out the battery very quickly. In real life, you must disconnect the battery first, then connect the jumper in its place.

As with voltage, we can calculate our Thevenin resistance if we have enough information. This is calculated in the standard method:

$$\frac{1}{R_{th}} = \frac{1}{2K\Omega} + \frac{1}{4K\Omega} \qquad R_{th} = 1333\ \Omega$$

Or, it can be calculated with the product-over-sum method:

$$R_{th} = \frac{2K \times 4K}{2K + 4K}$$

$$R_{th} = 1333\ \Omega$$

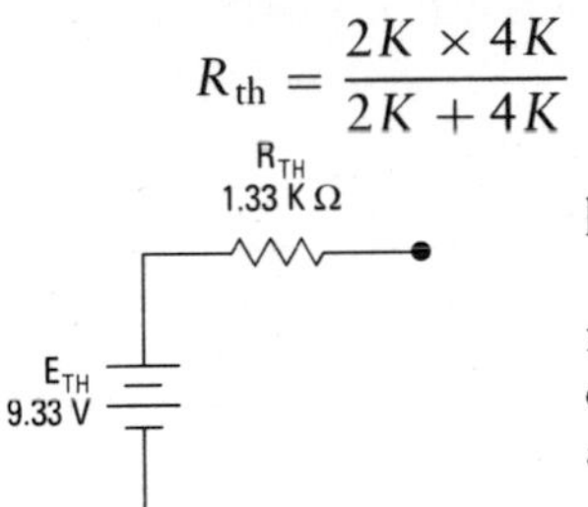

Figure 3-47 Thevenin equivalent.

You can see that out Thevenin equivalent resistance, R_{th}, is 1333 Ω.

With these equivalent values we can now represent this Thevenin equivalent circuit as a simple voltage and resistance, as is shown in Figure 3-47.

To complete our original problem, we now simply reattach the 8 kΩ resistor we removed at the beginning, which gives us the circuit shown in Figure 3-48. We can calculate accurate values for this circuit using our standard formulas.

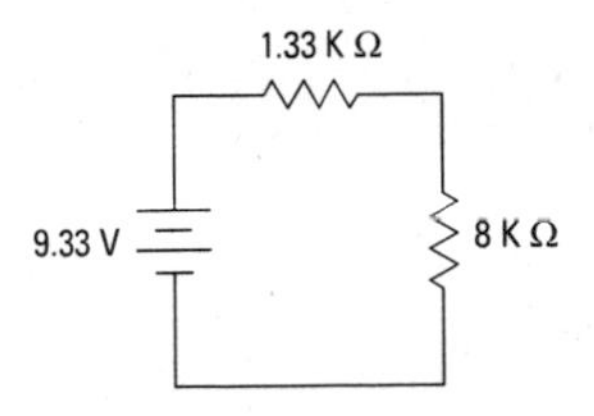

Figure 3-48 Completed circuit.

Summary

Circuits must be closed in order for current to flow. That is, they must form a closed loop. If not, they are called open circuits, and no current flows in them.

Functioning circuits may be of either series or parallel design. A series circuit has a single loop—only one possible path for current to take. Parallel circuits have more than one path that can be taken by current. Series-parallel circuits have some elements that are in series, and others that are in parallel.

A short circuit is the result of some wiring error or accident. Shorts prevent the correct operation of a circuit by creating a new path for current to follow. We classify short circuits as "partial" or "dead" shorts.

Kirchhoff's voltage law states that the sum of voltages in a closed loop always equals zero. In other words, the voltage drops around a series circuit will always equal the voltage of the source, although they will be of opposite polarity.

Kirchhoff's current law says that the amount of current that arrives at a junction must be equal to the amount of current that leaves that junction.

Analyzing circuits is a critical skill. When doing this analysis, it is helpful to simplify (or reduce) circuits to their simplest representation. Since electronic circuits can become so very complex, special methods of analysis are used, such as Thevenin's theorum.

Chapter 4 discusses the properties and charcteristics of alternating current.

Review Questions

1. What is the definition of a series circuit?
2. Do resistances add or subtract in a series circuit?
3. What does Kirchhoff's voltage law state?
4. If a voltage source is supplying maximum power to a load, what is the circuit efficiency?
5. What are circuit voltages operating in series-opposing?
6. What is the definition of a parallel circuit?
7. Why would we connect cells in parallel?
8. How are the positive terminals and the negative terminals of cells connected in a parallel arrangement?
9. What is the definition of a series-parallel circuit?
10. What is Kirchhoff's current law?
11. What is a junction?
12. Why must the branch voltages in a series-parallel arrangement of cells be equal?
13. How do we calculate the load power in a series-parallel circuit? The total power?
14. What is Thevenin's theorum?
15. When would you use Thevenin's theorum?

Exercises

1. Show a group of batteries connected together to supply maximum current. Show the values.
2. Show a group of batteries connected together to produce 150 volts. Show the values.
3. Draw a series circuit, show voltage and resistance, then calculate the current. Show your calculations.
4. Draw a series circuit, show voltage and current, then calculate the resistance. Show all of your calculations.
5. Draw a parallel circuit, show voltage and resistance, then calculate the current. Show all of your calculations.
6. Draw a series-parallel circuit, show all values and your calculations.

Chapter 4

Alternating Current

An *alternating current* (AC) is usually defined as a current that charges in strength according to a sine curve. An alternating voltage reverses its polarity on each alternation, and an alternating current reverses its direction of flow on each alternation. Figure 4-1 shows the development of an AC voltage. The point of maximum positive voltage occurs at 90 degrees, and is also called the *crest* voltage or *peak* voltage of the sine curve. The maximum negative voltage occurs at 270 degrees.

In the representation in Figure 4-1, current would be moving in one direction from 0 degrees through 180 degrees, then running in the opposite direction from 180 degrees through 360 (0) degrees.

Alternating Current Characteristics

You should be familiar with several AC characteristics, including frequency, instantaneous and effective voltages, Ohm's law in AC circuits, and power laws in resistive AC circuits. You should also be familiar with the ways in which you combine AC circuits.

Frequency

The *frequency* of an AC voltage of current is its number of cycles per second. For example, electricity supplied by public utility companies in the United States has a frequency of 60 cycles per second. Cycles per second (cps) are also called *hertz* (abbreviated as Hz). Thus, 60 cps is the same as 60 Hz. Figure 4-2 shows the relation between time and voltage at a frequency of 1 Hz. Each alternation is completed in $^1/_2$ second. In this example, the time for one complete cycle (called the *period* of the AC voltage) is 1 second. If an AC voltage has a frequency 60 Hz, its period is $^1/_{60}$ second.

Instantaneous and Effective Voltages

An *instantaneous* voltage is the value of an AC voltage at a particular instant. For example, the maximum voltage shown in Figure 4-1 is an instantaneous voltage; the positive maximum voltage occurs at 90 degrees, and the negative maximum voltage occurs at 270 degrees. The electrical industry generally considers that the most important instantaneous AC voltages are the *maximum* (peak) voltage and the *effective* voltage. We will first explain what is meant by the effective voltage, and then observe why it is so important. The effective voltage is 70.7 percent of the peak voltage and occurs at 45 degrees, as shown in Figure 4-3.

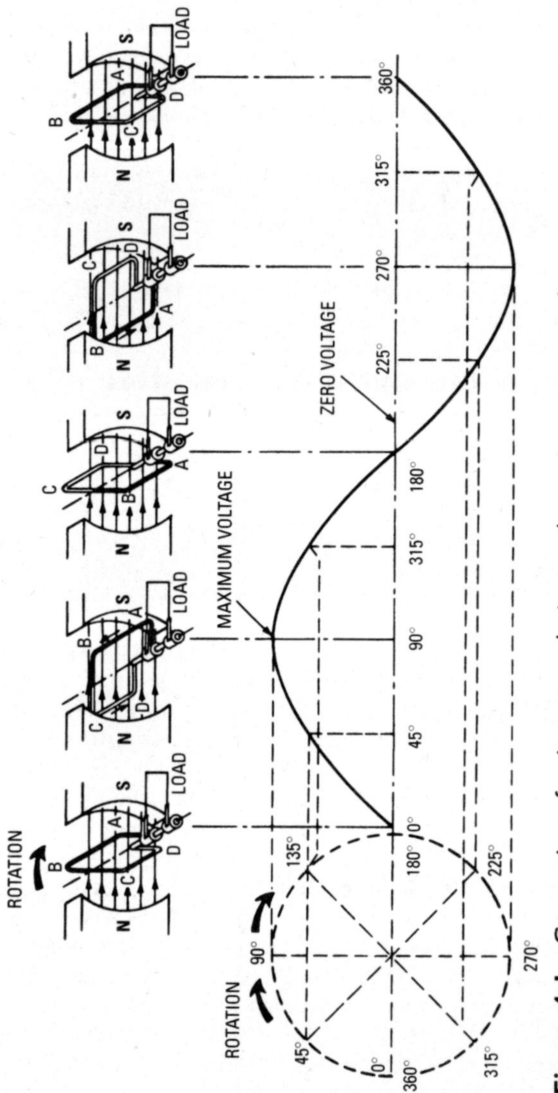

Figure 4-1 Generation of a sine curve during an alternating current cycle.

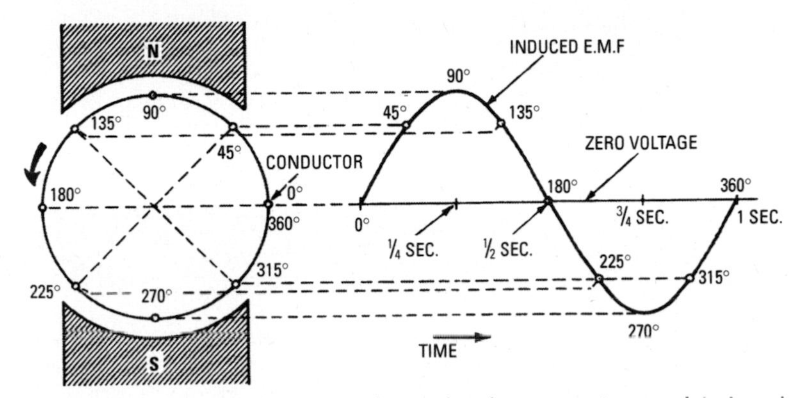

Figure 4-2 Curve showing relationship between time and induced voltage in an alternating-current circuit.

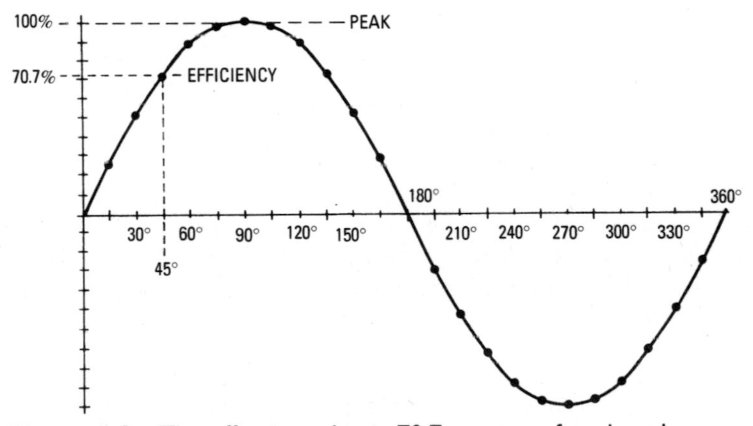

Figure 4-3 The effective value is 70.7 percent of peak and occurs at 45 degrees.

Practically all AC voltmeters read the effective value of an AC voltage. An effective AC voltage may also be called a *root-mean-square* (rms) voltage. Since nearly all AC voltmeters indicate effective values, this particular value is always understood when someone speaks of 120 volts, 240 volts, and so on. The word effective is understood, although it is not commonly stated in electrical diagrams and instruction sheets. When speaking of a peak voltage, we will state "169 volts peak." An effective voltage of 120 volts has a peak voltage of approximately 169 volts.

To illustrate why effective AC volts are so important in practical work, Figure 4-4 shows the arrangement of a resistance element

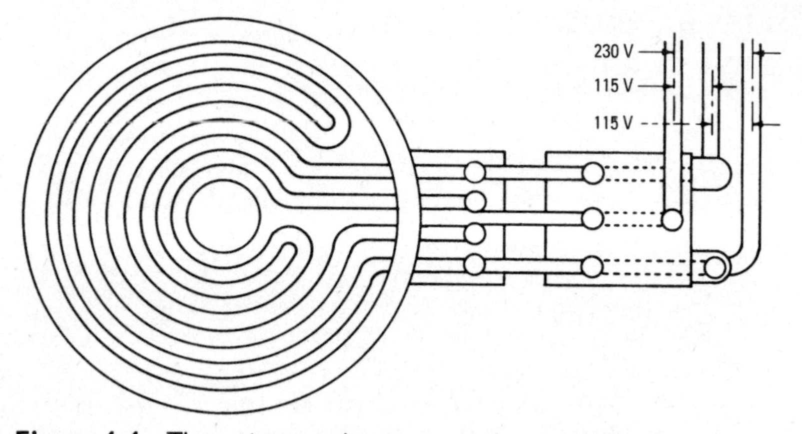

Figure 4-4 The resistance element in an electric range.

in an electric range. A three-wire distribution system is used, with 230 volts between the feeders and 115 volts between each feeder and the neutral wire. We will find that the resistance element will have the same temperature, whether we use 230 DC volts between the feeders or use 230 effective AC volts between the feeders. In other words, *a given effective AC voltage provides the same amount of power as the same value of DC voltage.*

As another practical example, consider an electric light bulb that is rated for operation on 120 volts AC. This means that the bulb is rated for operation on 120 effective volts of AC. We will find that the bulb provides the same amount of light if it is operated on 120 volts DC. Similarly, a soldering iron, toaster, percolator, or space heater draws the same amount of power from either a 120-volt DC line or a 120-effective-volt AC line. The power in watts consumed by a resistance is equal to E^2/R, whether E is taken in DC volts or in effective AC volts.

Ohm's Law in AC Circuits

In most practical electrical work with AC circuits, we use effective values of voltage and current in Ohm's law. Just as an effective voltage is 70.7 percent of the peak voltage, so is an effective current 70.7 percent of the peak current. In terms of effective (rms) values of current and voltage, we write Ohm's law as follows:

$$I_{\text{eff}} = \frac{E_{\text{eff}}}{R}$$

or,

$$I_{\text{rms}} = \frac{E_{\text{rms}}}{R}$$

As we proceed with our discussion of AC circuits, we will simply write I and E with the understanding that the letters stand for I_{eff} and E_{eff}, which is the same thing as I_{rms} and E_{rms}.

You may sometimes work with peak voltages (crest voltages). As stated previously, effective voltage is 70.7 percent (0.707) of the peak voltage. Therefore, a peak voltage is 1.414 times the effective voltage. Similarly, a peak current is 1.414 times the effective current. Therefore, we may write Ohm's law for peak values as follows:

$$1.414 I_{\text{eff}} = \frac{1.414 E_{\text{eff}}}{R}$$

or,

$$I_{\text{peak}} = \frac{E_{\text{peak}}}{R}$$

Once in a while we work with instantaneous voltages and currents. We write Ohm's law for instantaneous voltages and currents as follows:

$$i = \frac{e}{R}$$

where small letters i and e represent instantaneous values.

In other words, Ohm's law is true for effective, peak, or instantaneous values of voltage and current. The only thing that we must watch for is to use the *same kind* of values in Ohm's law. We would get incorrect answers if we used a peak voltage value and an effective current value in Ohm's law. With reference to Figure 4-3, various instantaneous values are indicated by dots along the sine curve. We see that since the instantaneous value at 45 degrees is the same as the effective value, Ohm's law for effective values is merely a special case of Ohm's law for instantaneous values. Similarly, since the instantaneous value at 90 degrees is the same as the peak value, Ohm's law for peak values is merely a special case of Ohm's law for instantaneous values.

The armature of a basic AC generator (Figure 4-1) is represented by means of an arrow, as shown in Figure 4-5. This arrow is called a *vector.* As the arrow rotates, it generates a sine curve of voltage, just as a loop of wire rotating in a magnetic field generates a sine curve of voltage. A sine curve is usually called a *sine wave* because its shape suggests a water wave. Note that the length of the vector E in Figure 4-5 represents the peak voltage of the sine wave, and

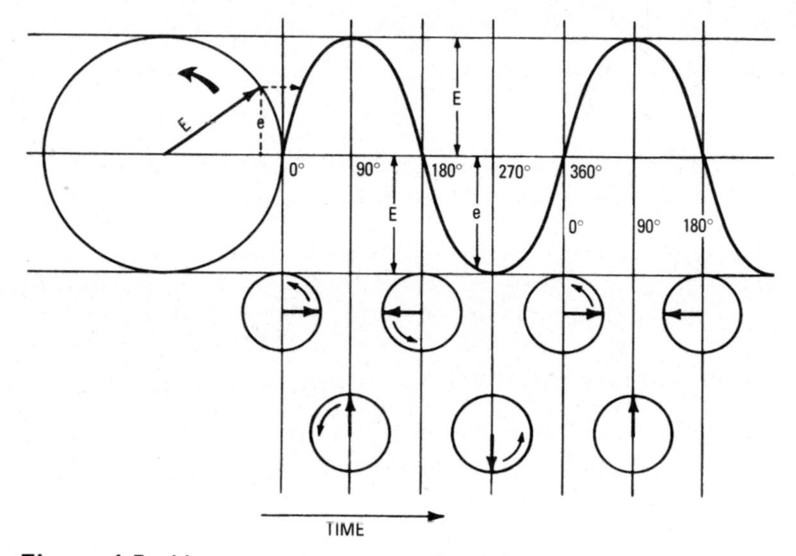

Figure 4-5 Vector representation of an AC voltage.

the dotted line (which has a length e) represents the instantaneous voltage of the sine wave.

When an AC voltage is applied across a resistance, a sine-wave current flows through the resistance, as shown in Figure 4-6B. Note that the voltage and current vectors rotate together, and that the current rises and falls in step with the voltage. For example, at time t_1, the voltage in Figure 4-6A has an instantaneous value indicated by point P_e. At this instant, the current in Figure 4-6B has an instantaneous value indicated by point P_1. Figure 4-6C shows how the voltage and current vectors rotate together.

In the foregoing example, the peak voltage is 10 volts, which is applied across a resistance of 7.5 ohms, as shown in Figure 4-6D. Since $I = E/R$, the peak current is 1.33 amperes. Since the frequency is assumed to be 60 cps (60 Hz), the current and voltage go through their peak positive values each $^1/_{60}$ second, and go through their peak negative values each $^1/_{60}$ second. Thus far, we see that Ohm's law is applied to resistive AC circuits in the same way that it is applied to resistive DC circuits.

Power Laws in Resistive AC Circuits

Consider the power that is consumed in the 7.5-ohm resistor in Figure 4-6. Because AC voltage and current have instantaneous, effective, and peak values, it follows that we can find three power values in the 7.5-ohm resistor. The basic power laws are written as follows:

$$\text{Power} = EI = I^2R = \frac{E^2}{R}$$

Now, consider the use of effective voltage and current values in the power laws. This is the most important case, because effective power corresponds to DC power, as we have learned. In other words, if we find the power consumed by a resistor when an effective voltage is applied, the same power will be consumed if we apply a DC voltage equal to the effective voltage. In terms of effective voltages and currents, the power laws are written as follows:

$$\text{Effective power} = E_{\text{eff}}\, I_{\text{eff}} = I_{\text{eff}}^2\, R = \frac{E_{\text{eff}}^2}{R}$$

Therefore, the effective power in the 7.5-ohm resistor (Figure 4-6) can be found by multiplying the effective voltage by the effective current. The effective (rms) voltage is equal to 10 × 0.7071, or 7.071 volts. The effective current is equal to 1.33 × 0.7071 or 0.94 ampere. This is written as follows:

$$\text{Watts}_{\text{eff}} = 7.071 \times 0.94 = 6.65\text{, approximately}$$

Of course, the power in the resistor rises and falls. The *peak power* in the resistor is given by the following equation:

$$\text{Peak power} = E_{\text{peak}}\, I_{\text{peak}}$$

Therefore, the peak power in the example of Figure 4-6 is found as follows:

$$\text{Watts}_{\text{peak}} = 10 \times 1.33 = 13.3\text{, approximately}$$

Note that *effective power is equal to 1/2 of peak power.* Since the power rises and falls in the resistor, we have at any instant an *instantaneous power* value in the resistor:

$$\text{Instantaneous power} = ei$$

or,

$$\text{Watts}_{\text{instantaneous}} = ei$$

or,

$$w = ei$$

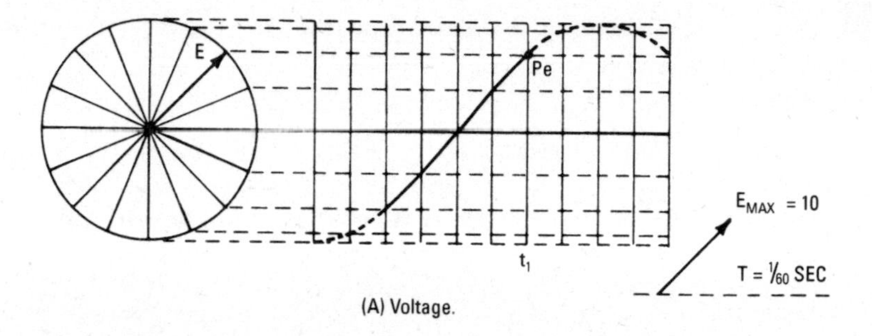

(A) Voltage.

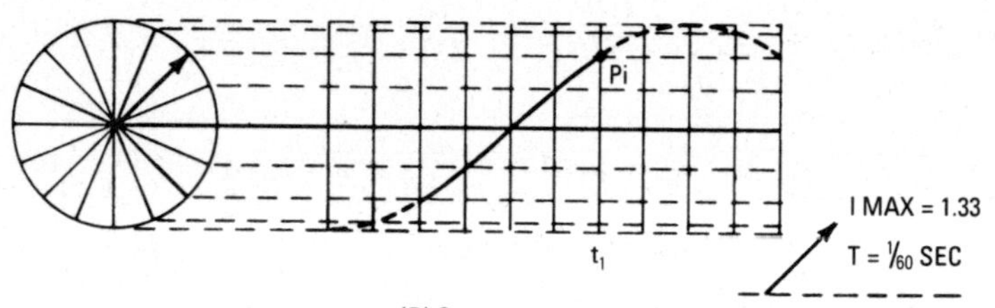

(B) Current.

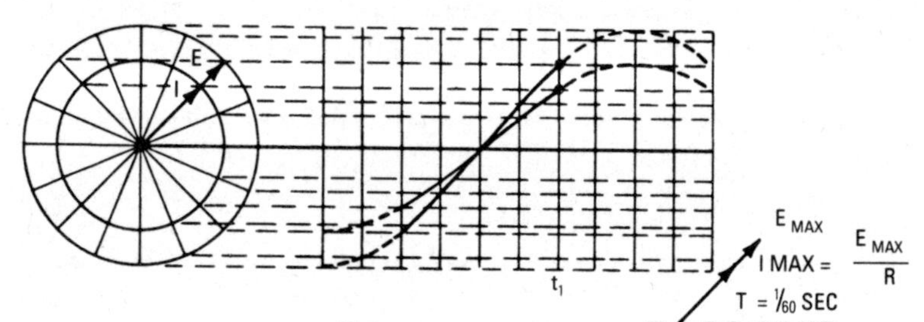

(C) Current and voltage.

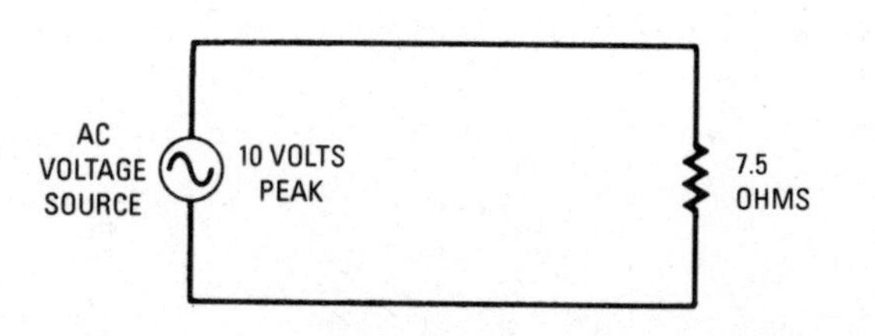

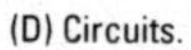

(D) Circuits.

Figure 4-6 Vector representation of AC voltage and current.

Instantaneous power is represented by a small letter w, just as e represents instantaneous voltage, and i represents instantaneous current. It is obvious that the instantaneous power value at 45 degrees in Figure 4-6 is the same as the effective power value and that the instantaneous power value at 90 degrees is the same as the peak power value. Note that power does not depend on frequency. The power value is the same, regardless of frequency.

Combining AC Voltages

We know how to combine DC voltages by connecting cells in series to form a battery. We also know how to combine voltage drops when we trace around a DC circuit according to Kirchhoff's voltage law. There are also circuits that combine AC voltages. To understand how this is done, consider the simple example shown in Figure 4-7. Two lamps are connected in series with an AC source. The source supplies 234 volts. The filament resistances of the lamps are the same, and there is a voltage drop of 117 volts across each lamp.

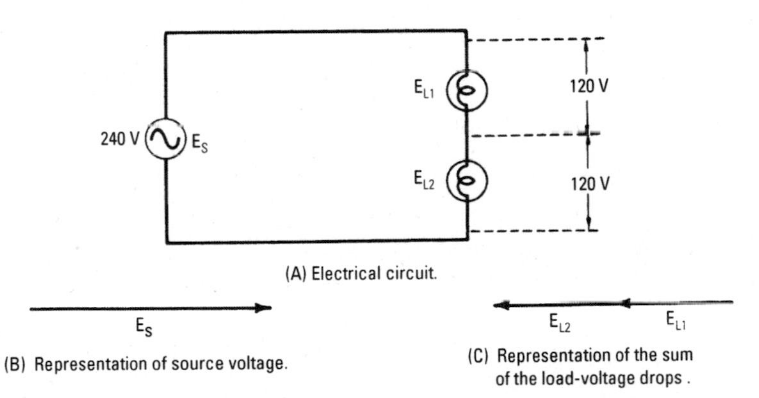

(A) Electrical circuit.

(B) Representation of source voltage.

(C) Representation of the sum of the load-voltage drops.

Figure 4-7 Two lamps connected in series with an AC source.

We see that Kirchhoff's voltage is applied to the circuit in Figure 4-7 just as if it were a DC circuit. It will be helpful for us to observe how the vector voltages combine. The source voltage E_S is represented by a vector as shown in Figure 4-7B. We know that the *polarity of the load* voltage opposes the *polarity of the source* voltage at any given instant. Therefore, the load-voltage vectors are drawn as shown in Figure 4-7C. Note that each vector is half the length of the source vector, and each load-voltage vector points in the *opposite* direction to the source vector. We add the load-voltage vectors by placing them end-to-end. Observe that since $E_{L1} + E_{L2}$

has the same length as E_S, but points in the opposite direction, the vectors cancel. This shows that the algebraic sum of the voltage drops around the circuit is equal to zero.

Note that E_S, E_{L1}, and E_{L2} have the same frequency in Figure 4-7. Vectors can be used to represent AC voltages, provided the voltages are sine waves, and provided that they have the same frequency. Vectors can also be used to combine sine-wave voltages that have the same frequency but that are more or less out of step.

For example, let us consider the two AC voltage sources shown in Figure 4-8. A pair of sine-wave voltage sources E_1 and E_2 are shown

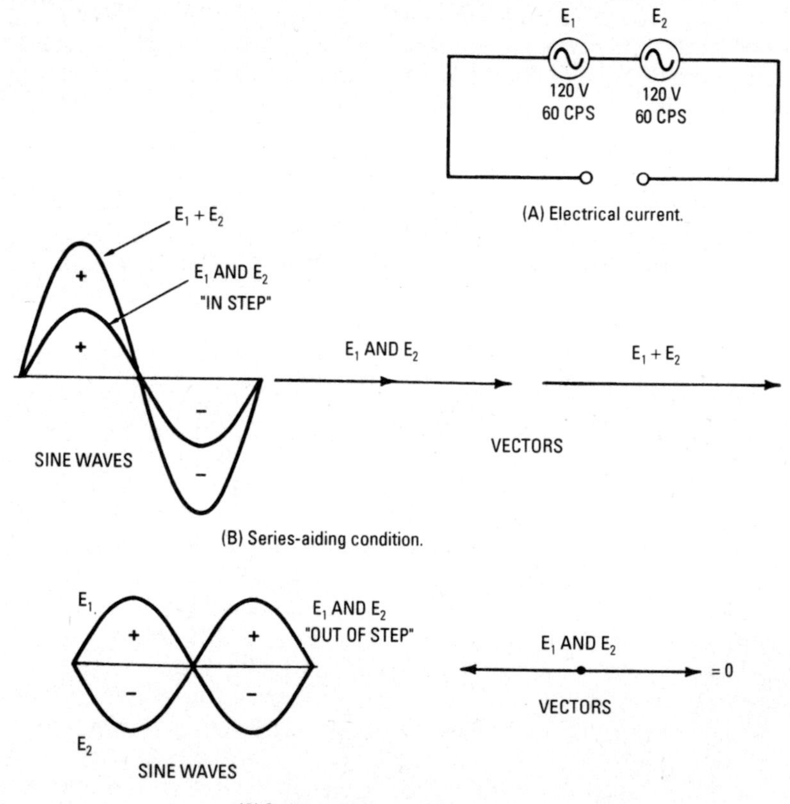

Figure 4-8 Two AC voltage sources with the same frequency, connected in series.

in Figure 4-8A. We know that cells can be connected in series-aiding or in series-opposing. If the two AC voltage sources are connected so that the sine waves are in step, as shown in Figure 4-8B, the voltages are in series-aiding and the voltages add. On the other hand, if the sine waves are completely out of step, as shown in Figure 4-8C, the voltages are in series-opposing and the voltages subtract.

The foregoing examples are very simple. However, electricians may have to work with equipment in which a pair of sine-wave voltages that have the same frequency are halfway out of step, as shown in Figure 4-9. The peaks of the two sine-wave voltages are 90 degrees apart. In this example, the AC voltage sources are represented as a pair of generators. However, we will find that other devices (such as capacitors or coils) produce sine waves that have peaks 90 degrees apart. It is easier to understand such situations if we start with generators such as those shown in Figure 4-9.

The sine-wave voltages generated in loops *a* and *b* (Figure 4-9A) are 90 degrees apart because the loops are located 90 degrees apart on the two-pole armatures. These armatures are assumed to be mounted on a common shaft in this example. Note that when loop *a* is cutting squarely across the magnetic field, loop *b* is moving parallel to the field and is not cutting magnetic lines of force. Therefore, the voltage in loop *a* is maximum when the voltage in loop *b* is zero. When two voltages are 90 degrees apart, electricians say that the two voltages are 90 degrees *out of phase*. Or, they will say that the *phase angle* between E_a and E_b is 90 degrees, as shown in Figure 4-9B.

If loops *a* and *b* in Figure 4-9A are connected in series, and the maximum voltage generated in each loop is 10 volts, these voltages do *not* combine to give a maximum voltage of 20 volts simply because the maximum voltages of the loops do not occur at the same time. The peak voltages are separated by 90 degrees, or by 1/4 cycle. We cannot merely add the two voltages because they are out of phase. However, we can add E_a and E_b vectorially, as shown in Figure 4-9B, and their vectorial sum is given by E_c.

In Figure 4-9B, E_c is the vector sum of E_a and E_b; E_c is the diagonal of the parallelogram, the sides of which are E_a and E_b. In this example, the parallelogram is a square, because $E_a = E_b$. Therefore, E_c is equal to 2 times 10 volts, or approximately 14.14 volts. Let us observe Figure 4-9C. This diagram shows how the sine waves E_a and E_b add to give E_c. In other words, if we go along point-by-point and add the instantaneous values of E_a and E_b, we will get the sine wave E_c. We observe that it is much easier and quicker to find the vector sum of E_a and E_b.

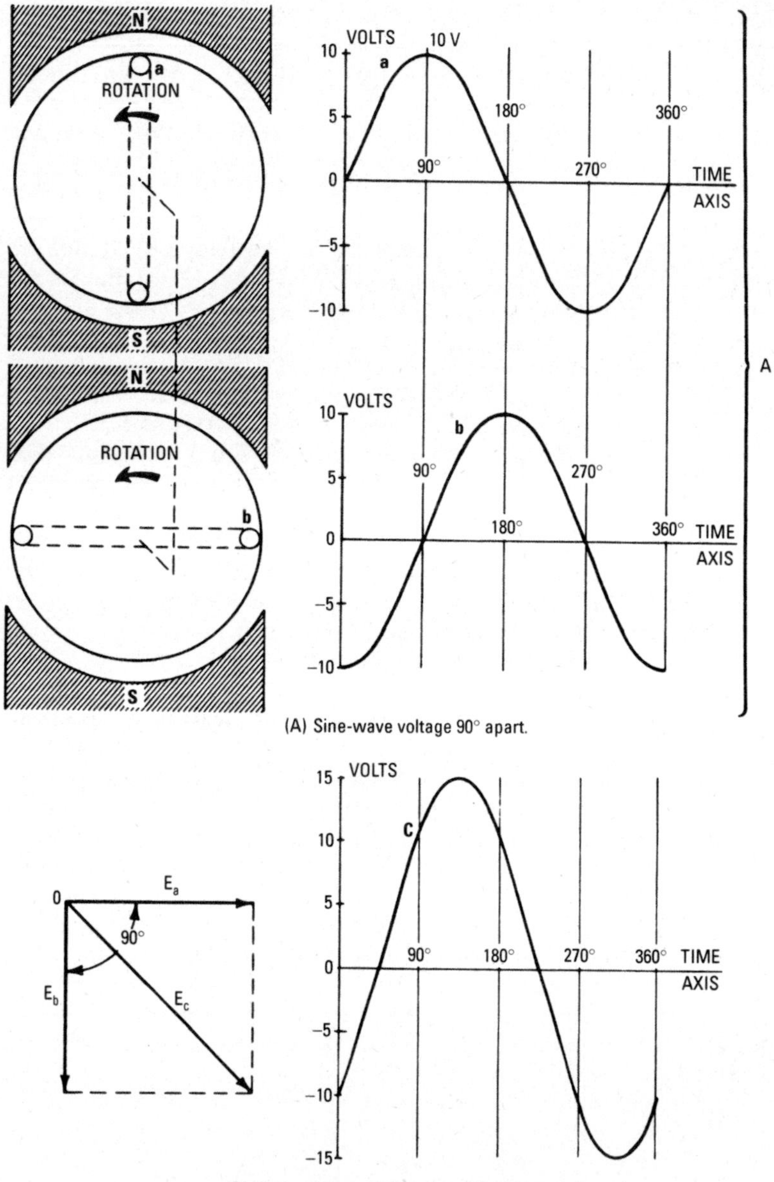

(A) Sine-wave voltage 90° apart.

(B) Vector representation of an AC voltage.

Figure 4-9 Two AC voltages of the same frequency that are halfway out of step.

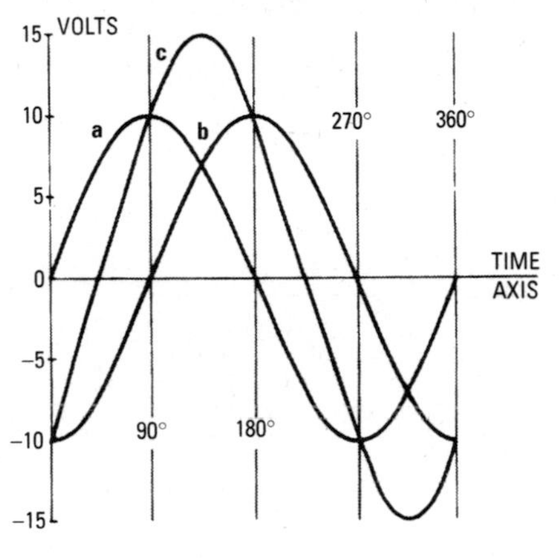

(C) The sum of a and b.

Figure 4-9 *(Continued)*

Figure 4-10 shows how the square on the hypotenuse of a right-angled triangle is equal to the sum of the squares on the other two sides. This is a very important fact for us to remember, because it shows us how two AC voltages are combined when the voltages are not the same. For example, let us suppose that two AC voltages of the same frequency have peak values of 8 volts and 6 volts and are 90 degrees different in phase. As shown in Figure 4-10, the vector sum of E_a and E_b is 10 volts. If we prefer, we can find E_c by writing the following:

$$E^2_c = E^2_a + E^2_b$$

or,

$$E_c = \sqrt{E^2_a + E^2_b}$$

In the foregoing example, $E_a = 8$ volts, and $E_b = 6$ volts. Therefore:

$$E_c = \sqrt{8^2 + 6^2} = \sqrt{64 + 36} = \sqrt{100} = 10 \text{ volts}$$

Whether we use vectors or arithmetic, we find that E_c has a peak voltage of 10 volts. To find the effective (or rms) value of E_c, we multiply 10 by 0.707, to obtain 7.07 volt rms. The principle of vector sums is the most important principle of AC to keep in mind.

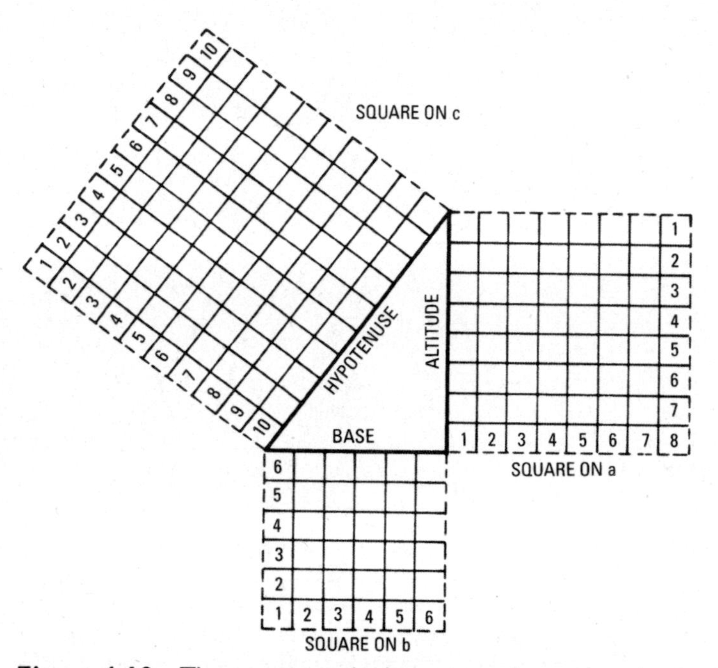

Figure 4-10 The square on the hypotenuse is equal to the sum of the squares on the sides of the right-angled triangle.

Inductive and Capacitive Circuits

Induction is a magnetic effect. As we mentioned earlier, a magnetic field forms around a conductor any time a current flows through it. We also mentioned that a magnetic field moving through a conductor causes a current. With alternating current, the magnetic fields around wires never stop moving, which gives rise to many circuit effects that we call *induction*.

Inductive AC circuits are used in many, many types of equipment. For example, induction coils are used in theaters and motion-picture studios for lighting control equipment. The basic principle of lighting control is shown in Figure 4-11. To control the brightness of the lamp, it is connected in series with a coil having adjustable inductance. The coil has least inductance when the iron core is withdrawn. The coil has greatest inductance when the core is fully inserted. In turn, the circuit current is greatest when the inductance value is least, and the current is small when the inductance is large.

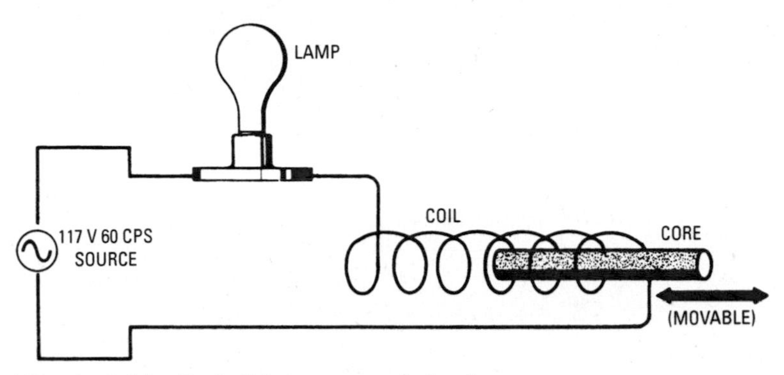

Figure 4-11 Basic lighting-control circuit.

Note that the coil has very little resistance, because it is simply a coil of copper wire. But when inserted into an AC circuit, this device will restrict current flow, just as if it were a resistor. The same coil inserted into a DC circuit would make no difference in the operation of the lamp.

A coil with an adjustable core is called a *variable reactor* or a *variable inductor.* Inductance opposes the flow of AC current because self-induction causes the inductance to have *reactance.* Inductance is measured in *henrys.* (Joseph Henry was the man who discovered that magnetic fields can be used to generate alternating currents.) Although reactance is different from resistance, reactance is nevertheless measured in ohms.

Inductive Circuits

Inductance opposes any change in current flow. If we increase the voltage across an inductor, an increase in current flow builds up gradually because of the self-induced voltage that opposes current change. On the other hand, if we decrease the voltage across an inductor, the current flow falls gradually because of the self-induced voltage. In other words, the most noticeable difference between a resistive circuit and an inductive circuit is in their response speeds. The current flow in a resistive circuit changes immediately if the applied voltage is changed. On the other hand, the current flow in an inductive circuit is delayed with respect to a change in applied voltage.

We know that inductance is the result of magnetic force lines cutting the coil turns when the magnetic field strength changes. Therefore, the inductance of an air core is greatly increased if we insert an

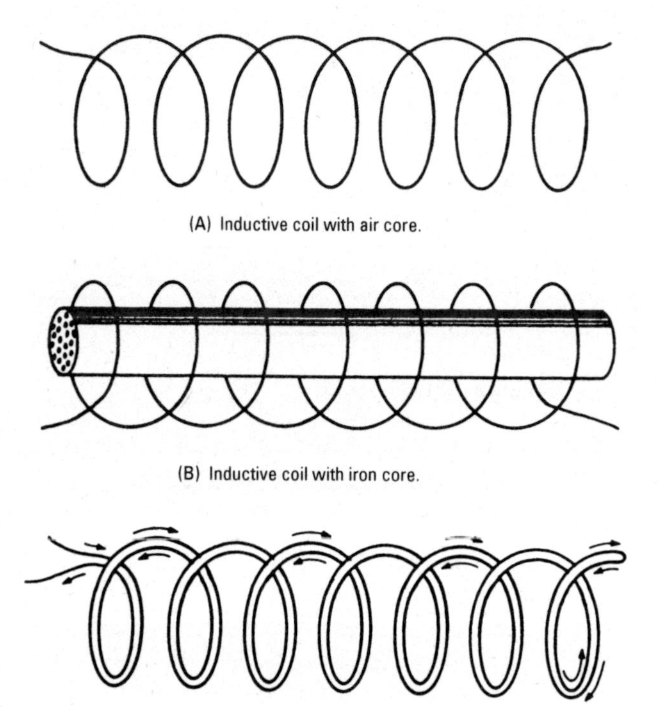

Figure 4-12 The inductance of an air-core coil is increased by an iron core. If the wire is doubled back on itself, the inductance is zero.

iron core (Figure 4-11). Note that if the coil wire is doubled back on itself, as shown in Figure 4-12, the magnetic fields will cancel, and the inductance will be zero. This is called a *noninductive coil*, and it is used in electrical equipment in which we wish to employ wire-wound resistors that have no inductance.

An inductor has an inductance of 1 henry if the current flow increases at the rate of 1 ampere per second when 1 volt is applied across the inductor. The inductance of a coil increases as the square of the number of turns. For example, with reference to Figure 4-13, a coil with two layers of wire has four times as much inductance as a coil with one layer. A coil with three layers of wire has nine times as much inductance as a coil with one layer. Let us see why this is so.

If we wind a single-layer coil, we have a certain number of turns that are cut by a magnetic field of a certain strength. If we wind two layers on the coil, we have twice as many turns that are cut by

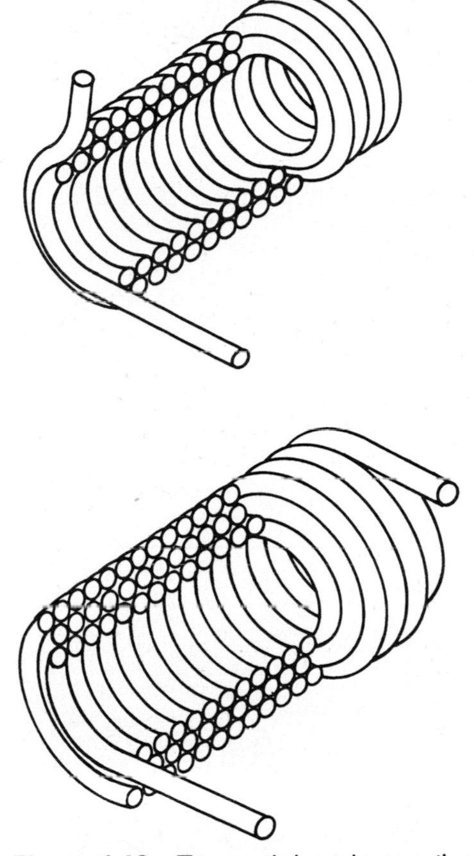

Figure 4-13 Two- and three-layer coils.

a magnetic field that is twice as great. Therefore, the self-induced voltage is four times as great as in a single-layer coil. This is just another way of saying that the inductance of a coil increases very rapidly as we add more layers to its winding. We will now assume that we have constructed a 1-henry coil, and we will ask how much opposition it has to AC current flow. To simplify our question, we will assume that the coil has been wound with large wire so that its DC resistance can be neglected.

Measurements with an AC voltmeter and ammeter will show that Ohm's law for an inductor is slightly more complicated than Ohm's law for DC with a resistor. We will find that Ohm's law for an inductor is written as follows:

$$I = \frac{E}{2\pi fL} \text{amperes}$$

where the following is true:

I represents AC amperes

E represents AC volts

π is 3.1416

f is the frequency in Hz

L is henrys of inductance

Therefore, if we apply an AC voltage of 117 volts across an inductance of 1 henry, and the frequency is 60 Hz, as shown in Figure 4-14, the AC current flow is approximately 0.31 ampere. It is helpful to write Ohm's law for an inductive circuit as follows:

$$\text{AC Current} = \frac{\text{AC Voltage}}{\text{Reactance}}$$

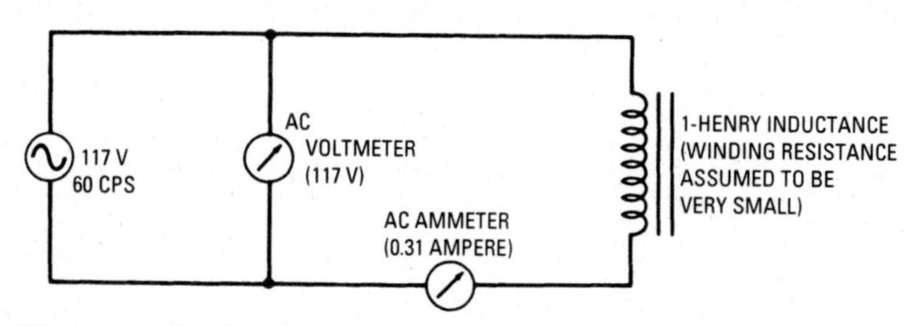

Figure 4-14 Illustration of Ohm's law for a purely inductive circuit.

We observe that reactance can be compared with resistance in Ohm's law. In other words, $2\pi fL$ is measured in AC ohms. Electricians use X_L to represent AC ohms. Thus, $X_L = 2\pi fL$. Therefore, Ohm's law for inductive circuits is written as follows:

$$I = \frac{E}{X_L}$$

It is obvious that if we used a 117-volt DC source in Figure 4-14 instead of a 117-volt AC source, an extremely large current would flow because the wire resistance of the 1-henry coil was assumed to be very small.

Note how the current varies in the circuit of Figure 4-14 when the frequency is changed. Ohm's law for AC in an inductive circuit tells us if we increase the frequency from 60 cps to 120 Hz, the current will be reduced one-half. On the other hand, if we decrease the frequency from 60 Hz to 30 Hz, the current will be doubled.

If we keep the frequency at 60 Hz, but double the inductance to 2 henrys, the current will be reduced one-half. On the other hand, if we reduce the inductance to $^1/_2$ henry, the current will be doubled. Of course, if we keep the frequency and the inductance the same, the current will double if we double the applied AC voltage.

Power in an Inductive Circuit

The power in a DC resistive circuit is equal to volts multiplied by amperes. This is also true in an AC resistive circuit, as shown in Figure 4-15. On the other hand, in a circuit that has inductance only as shown in Figure 4-15, the *true power* is zero. This fact might seem puzzling until we observe the circuit action. We recognize that the current flow through an inductor is delayed when voltage is applied due to the self-induced voltage of the inductor, which opposes the applied voltage. It can be shown that the AC current in an inductor is delayed 90 degrees with respect to the applied voltage, as seen in Figure 4-15.

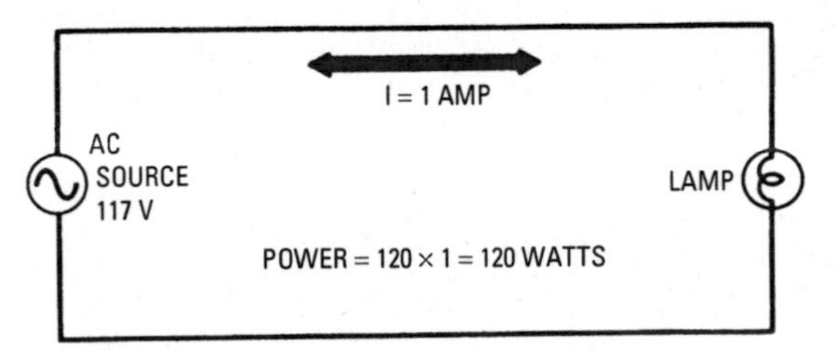

Figure 4-15 AC power in a resistive circuit.

If we multiply instantaneous voltages by instantaneous currents, we will find the instantaneous power. Over a complete cycle we see that half of the power is below the axis (negative power) and the other half of the power is above the axis (positive power). Therefore, the average power in an inductor is half positive and half negative; no power is consumed by the inductor. So, we can simply say that an inductor takes power from the source during the positive alteration and returns this power to the source during the negative alternation.

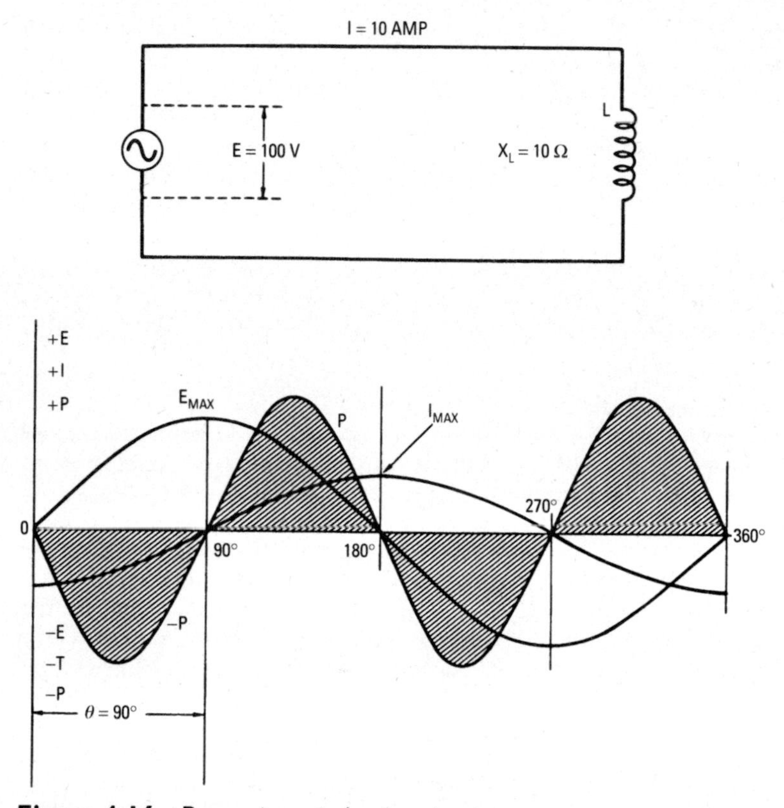

Figure 4-16 Power in an inductive circuit.

However, although no power is taken on the average by the inductor from the source in Figure 4-16, we see that power is surging back and forth in the circuit. The inductor first stores power in its magnetic field and then returns this stored power to the circuit. Thus, the circuit draws current, but *consumes no power.* This is just another way of saying that the source does no work in this circuit, or that the inductor does not actually load the source. The inductor is an apparent load. Current that flows into the inductor on the positive alternation is stored, and this current is then returned to the circuit on the negative alternation.

The product of voltage and current in a purely inductive circuit is called *apparent power* because it is floating power that does no work. Apparent power is also called *reactive power*, and is measured in *vars*. We write vars to represent volt-amperes reactive. In other words, for Figure 4-16 we write the following:

$$\text{Vars} = E_{\text{rms}} I_{\text{rms}}$$

Electricians measure vars with a varmeter, as shown in Figure 4-17. A varmeter looks like a wattmeter, but it is constructed so that it indicates apparent power instead of true power. Note that if the load were a resistance, such as a lamp, the varmeter would read zero. On the other hand, if the load were a pure inductance, the varmeter would read the product of the rms voltage and the rms current.

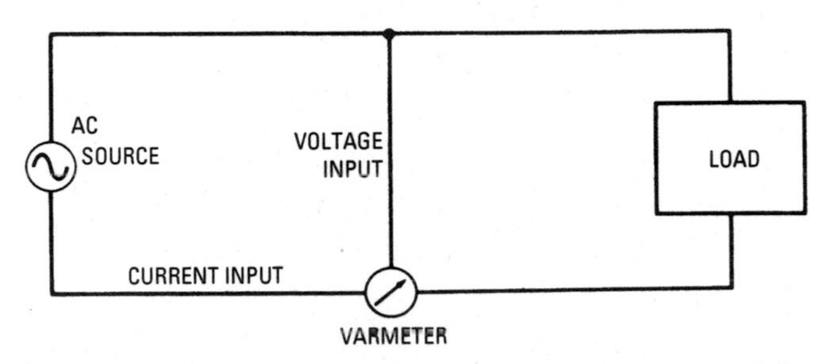

Figure 4-17 Measurement of apparent power in an AC circuit.

Resistance in an AC Circuit

We know that in a circuit containing only resistance (Figure 4-18), the current and voltage are in phase. Therefore, there is true power in the resistor, which appears as heat. True power is also called *real power*. We also know that this true power in watts is equal to $E_{\text{rms}} I_{\text{rms}}$. The shaded area in Figure 4-18 represents the product of instantaneous voltages and instantaneous currents. The entire shaded area is positive and accordingly represents true power. Note that when the voltage is negative, the current is also negative, and, therefore, their product must be positive.

Next, we will ask what the AC circuit action will be when inductance and resistance are connected in series, as shown in Figure 4-19. The AC current I_o flows through both the inductor and the resistor. Accordingly, a voltage drop is produced across the inductor and across the resistor. We observe that the resistive voltage drop E_R is in phase with the current but that the inductive voltage drop E_L is 90 degrees out of phase with the current. E_R is 50 volts and E_L is 86.6 volts in this example.

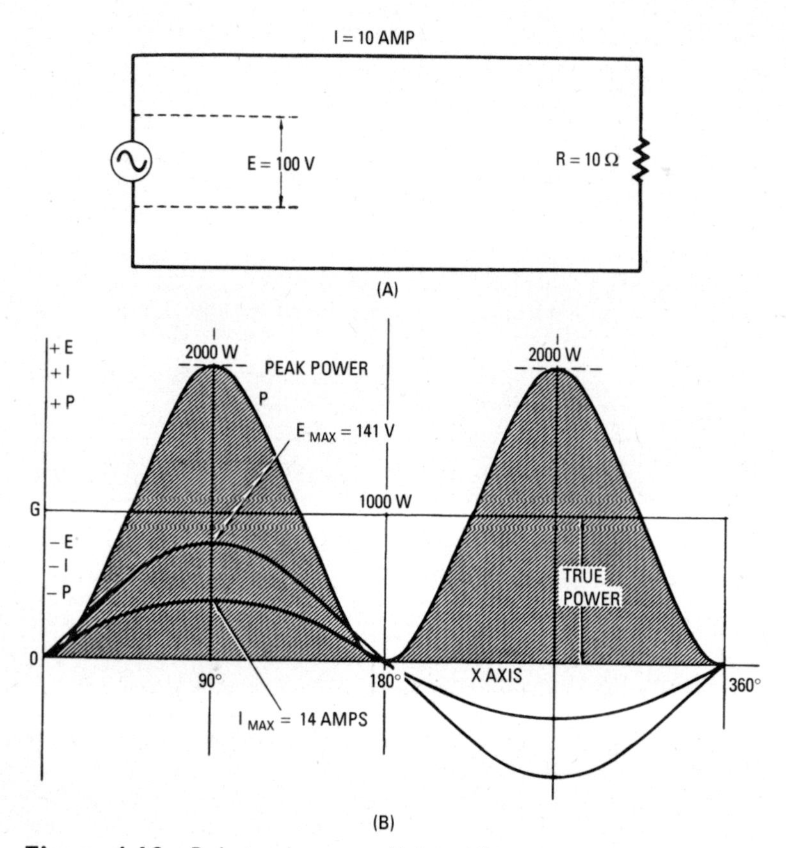

Figure 4-18 Relation between *E*, *I*, and *P* in a resistive circuit.

Note that the applied voltage E_o in Figure 4-19 is the hypotenuse of a right triangle, the sides of which are 50 and 86.6. In turn, we write the following:

$$E_v = \sqrt{50^2 + 86.6^2} = 100 \text{ volts}$$

If we were to use a protractor in Figure 4-19B, we would find that E_o makes an angle of 60 degrees with I_o. This is called the *power-factor* angle of the circuit. It is evident that if we keep L constant, and increase R, the power-factor angle will become less.

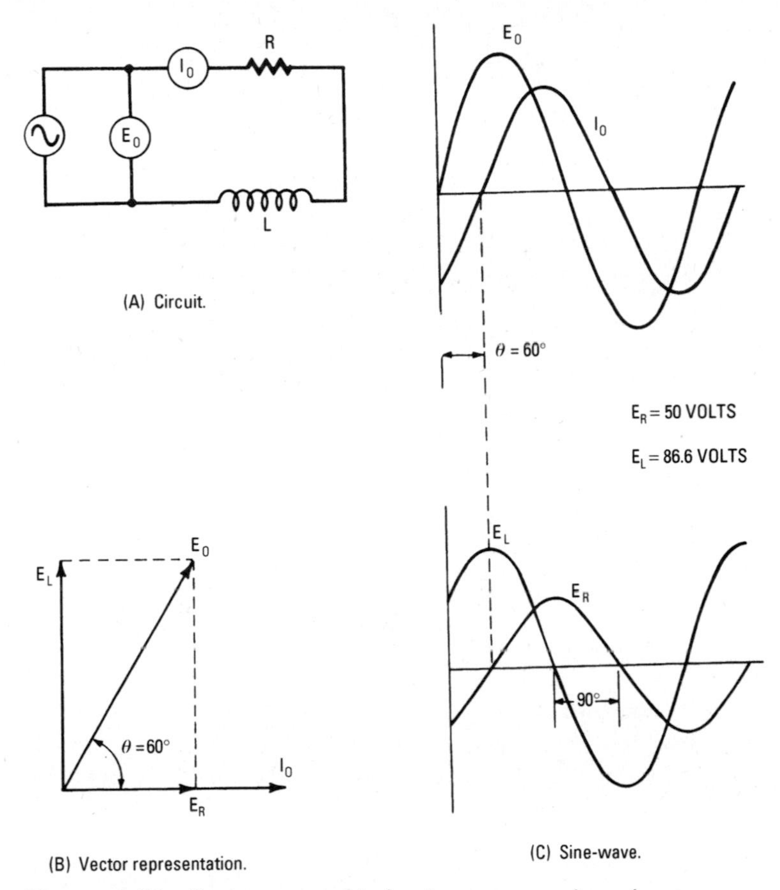

Figure 4-19 Resistance and inductive reactance in series.

Impedance

The circuit in Figure 4-19 has both resistance and inductive reactance. The total opposition to AC current flow is called *impedance.* Impedance is measured in ohms, just as resistance and reactance are measured in ohms. If an AC circuit contains reactance only, the circuit impedance is the same as the circuit reactance. Or, if an AC circuit contains resistance only, the circuit impedance is the same as the circuit resistance. When a circuit has both resistance and reactance, we find the circuit impedance as follows:

$$Z = \sqrt{R^2 + X^2}$$

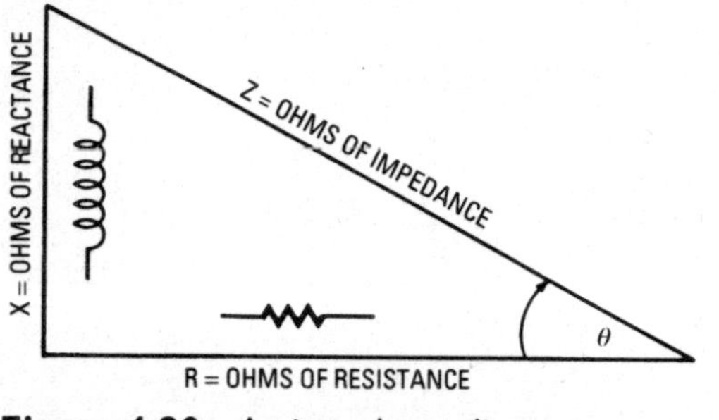

Figure 4-20 An impedance diagram.

where the following is true:

Z represents the ohms of impedance
R represents the ohms of resistance
X represents the ohms of reactance

We can draw an impedance diagram as a right triangle, such as shown in Figure 4-20. The power-factor angle is the angle theta (θ) between R and Z in the diagram.

Power and Impedance

When inductance and resistance are connected in series with an AC source (as shown in Figure 4-21), there are three power values to be considered. We will find a true power in *watts*, a reactive power in *vars*, and an apparent power in *volt-amperes*. These power values are found as follows:

1. The impedance of the circuit is 14.14 ohms.
2. The current is 100/14.14, or 7.07 amperes.
3. The true power is I^2R, or $50 \times 10 = 500$ watts.
4. The reactive power is I^2X, or $50 \times 10 = 500$ vars.
5. The apparent power is EI, or $100 \times 7.07 = 707$ volt-amperes.

Only the true power does useful work in the circuit. For example, if the resistor represents a lamp filament (as in Figure 4-11), the true power produces light. The reactive power merely surges back and forth in the circuit. The apparent power provides reactive power and true power, as shown in the power diagram of Figure 4-22.

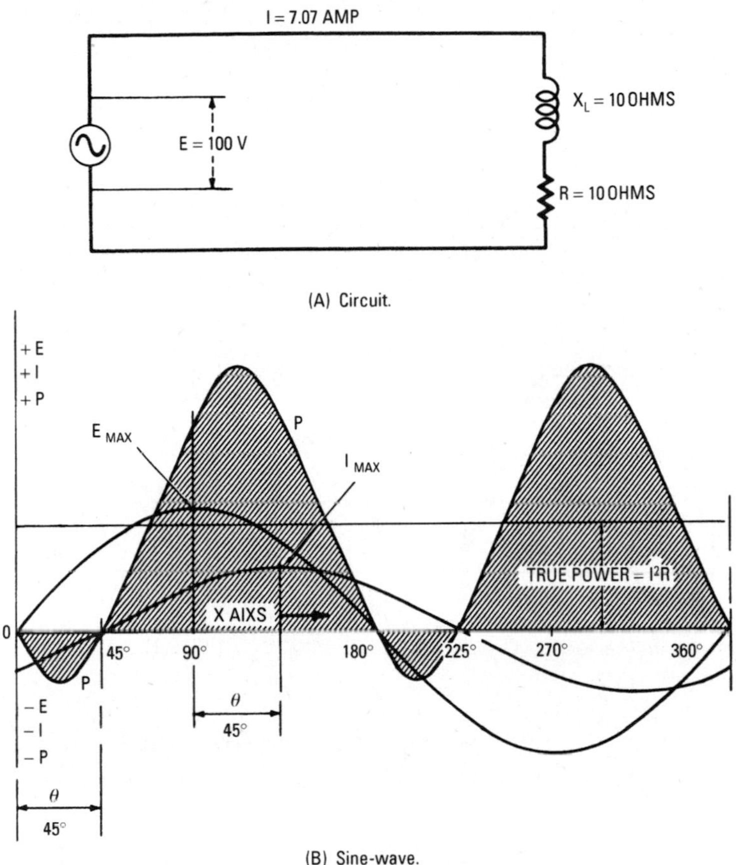

(B) Sine-wave.

Figure 4-21 Power in a circuit containing *L* and *R* in series.

Capacitive Reactance

Electricians often work with capacitive reactance as well as with inductive reactance. Capacitance is measured in *farads* with a capacitor tester, such as illustrated in Figure 4-23. A 1-farad capacitor will store 1 coulomb of charge when 1 volt is applied across the capacitor. A *coulomb* is the amount of charge produced by 1 ampere flowing for 1 second. Since the farad is a very large unit of capacitance, we generally work with *microfarads*; a microfarad is one millionth of a farad (10^{-6} farad). The symbol μ or the letter *m* is

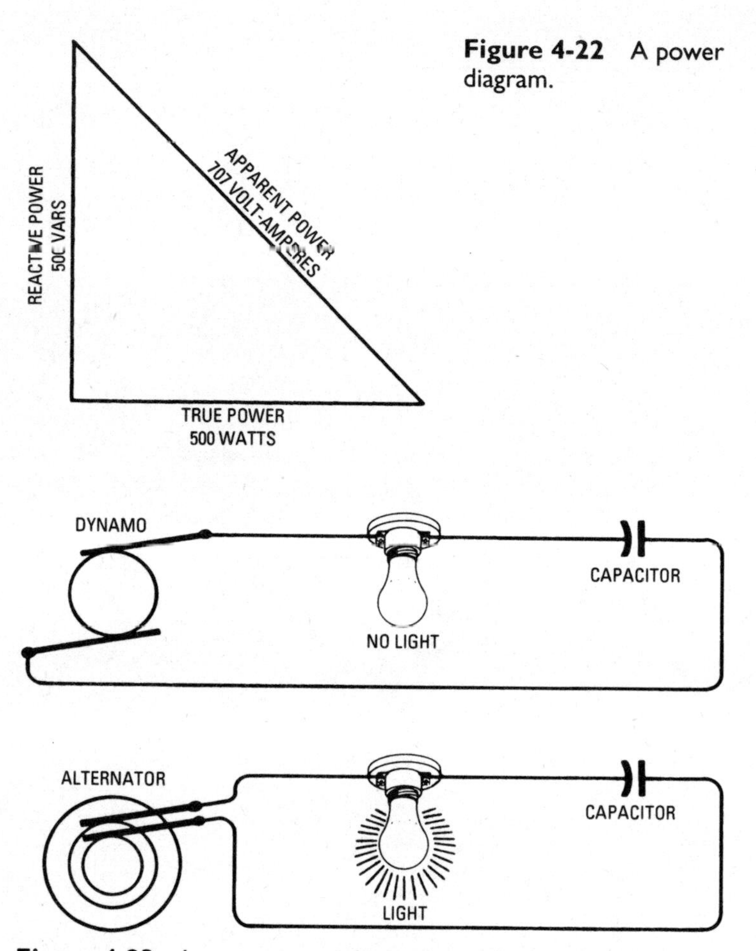

Figure 4-22 A power diagram.

Figure 4-23 Lamp connected in series with a capacitor.

used to represent micro. For example, 50 microfarads is commonly written as 50 μf, 50 μfd, 50 mf, or 50 mfd.

If we connect an inductor across a DC voltage source, the inductor short-circuits the source. On the other hand, if we connect a capacitor across a DC voltage source, the capacitor is an open circuit. Although there is a momentary rush of DC current into the capacitor to charge its plates, the current flow thereafter is zero. Capacitance is essentially the opposite of inductance in a DC circuit. We will find that they are opposites in other respects also.

Figure 4-23 shows a simple experiment of basic importance. The lamp connected in series with a capacitor and a DC voltage source does not glow. On the other hand, when an AC source is used, the lamp then glows. If we increase the capacitance, the lamp will glow brighter. By comparing Figure 4-23 with Figure 4-11, we will recognize that the circuit actions are opposite. When we increase the capacitance in Figure 4-23, the lamp glows brighter.

It is evident that AC current does not really flow through a capacitor. What actually happens is the current flows into the capacitor on the first alternation and charges the capacitor plates. Electricity is then stored in the capacitor, and this electricity returns to the circuit on the second alternation as the capacitor discharges.

We can compare a capacitor with an inductor in the sense that a capacitor does not consume power. In other words, the power in a capacitor is reactive power and is measured in vars. Therefore, the AC circuit in Figure 4-23 has a reactive power value in the capacitor that is measured in vars, a real (true) power value in the lamp that is measured in watts, and an apparent power value supplied by the alternator that is measured in volt-amperes.

Because a large capacitance stores more electricity than a small capacitance, it also has less opposition to AC current flow. The amount of AC current flow will depend, moreover, on how many times a second the capacitor is charged and discharged.

Therefore, the AC current flow increases when the frequency is increased. In its simplest form, Ohm's law for a capacitor is written as follows:

$$I = \frac{E}{X_C}$$

where the following is true:

I represents the number of AC amperes
E represents the number of AC volts
X_C represents the number of capacitive ohms

We know that the number of capacitive ohms (reactive ohms) will depend on the number of farads of the capacitor and on the AC frequency. In turn, we write the following:

$$X_C = \frac{1}{2\pi fC}$$

where the following is true:

X_C represents the number of capacitive ohms
f is the frequency in Hz
C is the capacitance in farads

Thus, Ohm's law for a capacitor may be written in the the following form.

$$I = 2\pi fCE$$

For example, if a capacitor has a capacitance of 133 microfarads (133 μf, or 1.33×10^{-4} farad), its reactance at 60 Hz will be:

$$X_C = \frac{E}{2\pi fC} = \frac{1}{6.28 \times 60 \times 1.33 \times 10^{-4}} = 20 \text{ ohms}$$

Let us observe Figure 4-24. When a sine-wave AC voltage is applied to a capacitor, the current is greatest when the voltage starts to rise from zero. This is so because there is no charge in the capacitor at this instant, and, therefore, there is no back voltage to oppose the flow of current. On the other hand, the current is zero when the voltage reaches its peak value. This is because the capacitor is then fully charged and its back voltage is equal and opposite to the applied peak voltage. This is just another way of saying that the current in a capacitive circuit goes through its peak value before the applied voltage goes through its peak value.

Electricians say that the current in a capacitive circuit *leads* the applied voltage in time. This is just the opposite of an inductive circuit in which the current is delayed with respect to the applied voltage, or the current *lags* the voltage in an inductive circuit. There is a 90-degree phase difference between voltage and current in a capacitor. However, we have a *leading phase* in a capacitor, whereas we have a *lagging phase* in an inductor.

Figure 4-25 shows the power in a capacitor, which is the product of instantaneous voltages and currents. We observe that the shaded areas represent positive power and negative power on successive alternations. In other words, the average power in the capacitor is zero. All of the capacitor power is reactive power and is measured in vars:

$$\text{Vars} = R_{\text{rms}} I_{\text{rms}}$$

Thus, a capacitor is like an inductor in that it is a *reactor* and consumes no real power. However, capacitance is the opposite of inductance in that capacitive current leads the applied voltage by

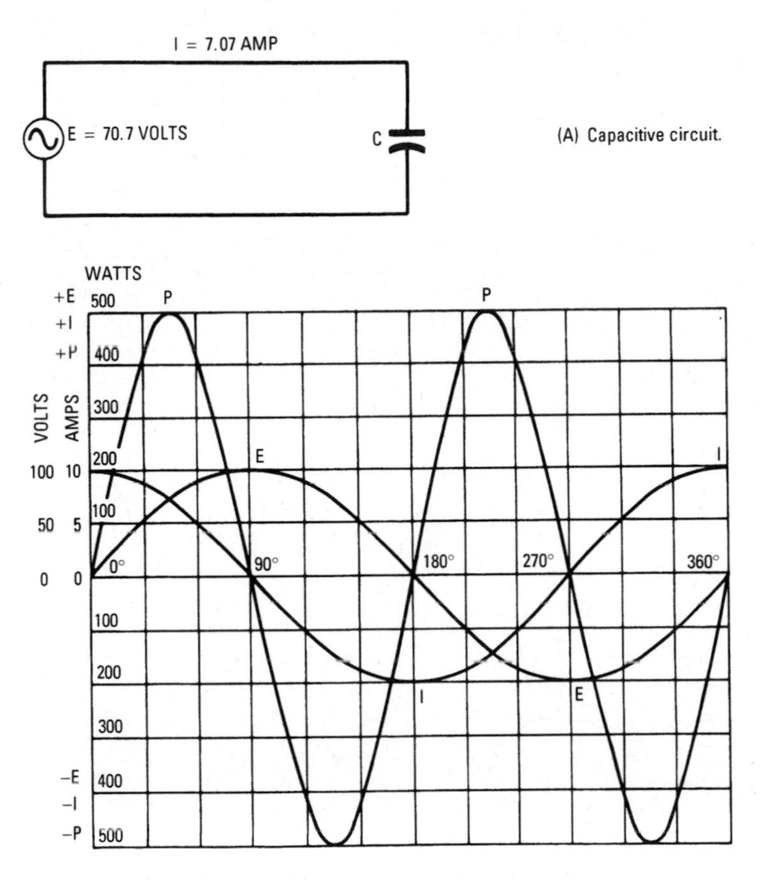

(A) Capacitive circuit.

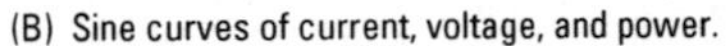

(B) Sine curves of current, voltage, and power.

Figure 4-24 Voltage, current, and power in a capacitor.

90 degrees, whereas inductive current lags the applied voltage by 90 degrees.

Capacitive Reactance and Resistance in Series

Figure 4-26 shows capacitance and resistance connected in series with an AC source. To find the current in this circuit, we proceed as follows:

1. The impedance of the circuit is

$$Z = \sqrt{R^2 + X^2{}_C} = 44.66$$

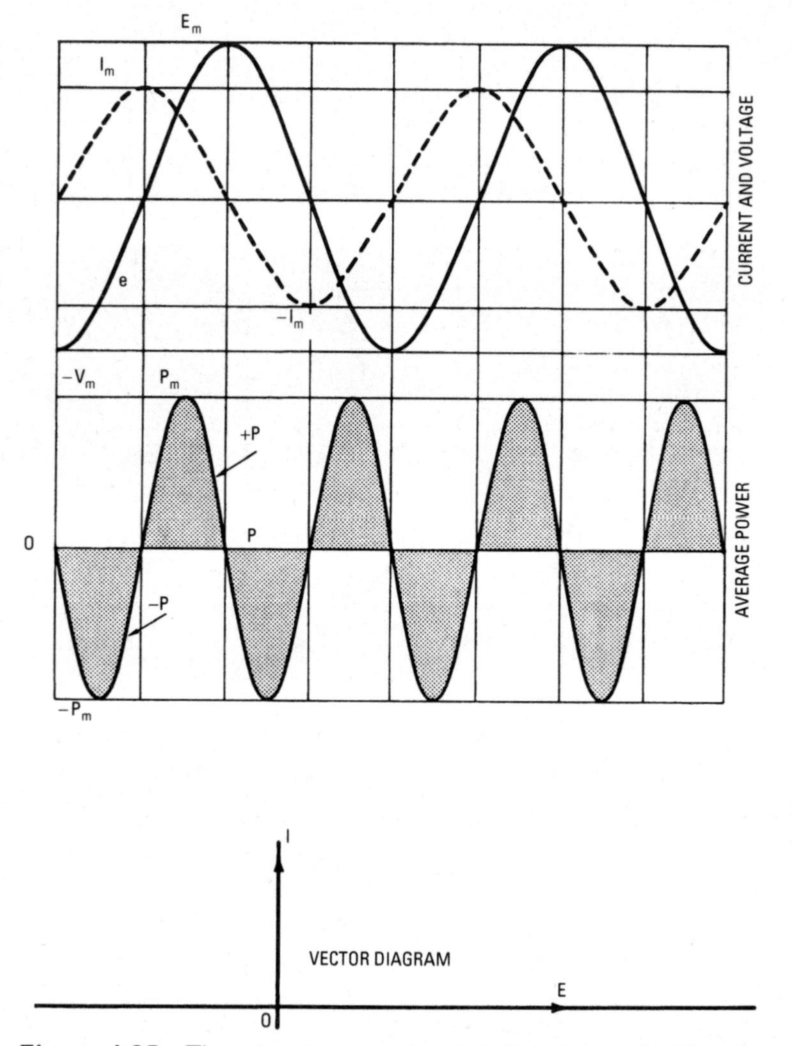

Figure 4-25 The power in a capacitor is half positive and half negative.

2. The current is $I = E/Z = 3$ amperes.
3. The resistive voltage drop is $E_R = IR = 60$ volts.
4. The capacitive voltage drop is $E_C = IX_C = 120$ volts.

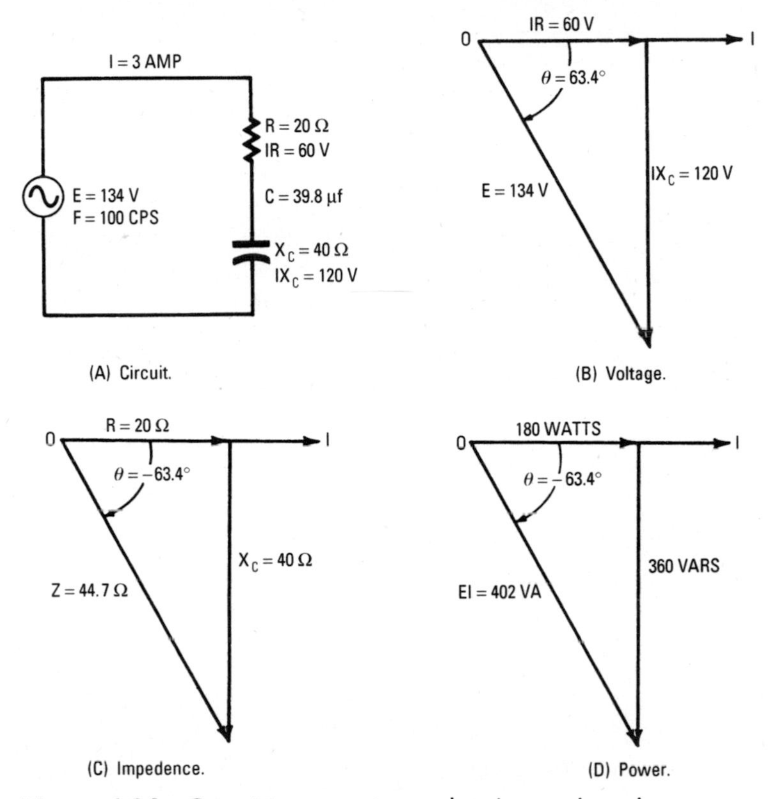

Figure 4-26 Capacitive reactance and resistance in series.

Next, let us find the three power values in the circuit of Figure 4-26:

1. The true power is $P = I^2R = 180$ watts.
2. The reactive power is $P = I^2X_C = 360$ vars.
3. The apparent power is $P = EI = 402$ volt-amperes.

Figure 4-21 showed the development of power in a circuit with inductance and resistance. The development of power in a circuit with capacitance and resistance is much the same, as seen in Figure 4-27. Basically, the only difference between an RC circuit and an RL circuit is in the lead or lag of the circuit current with respect to the applied voltage in the circuit. RC circuits are those

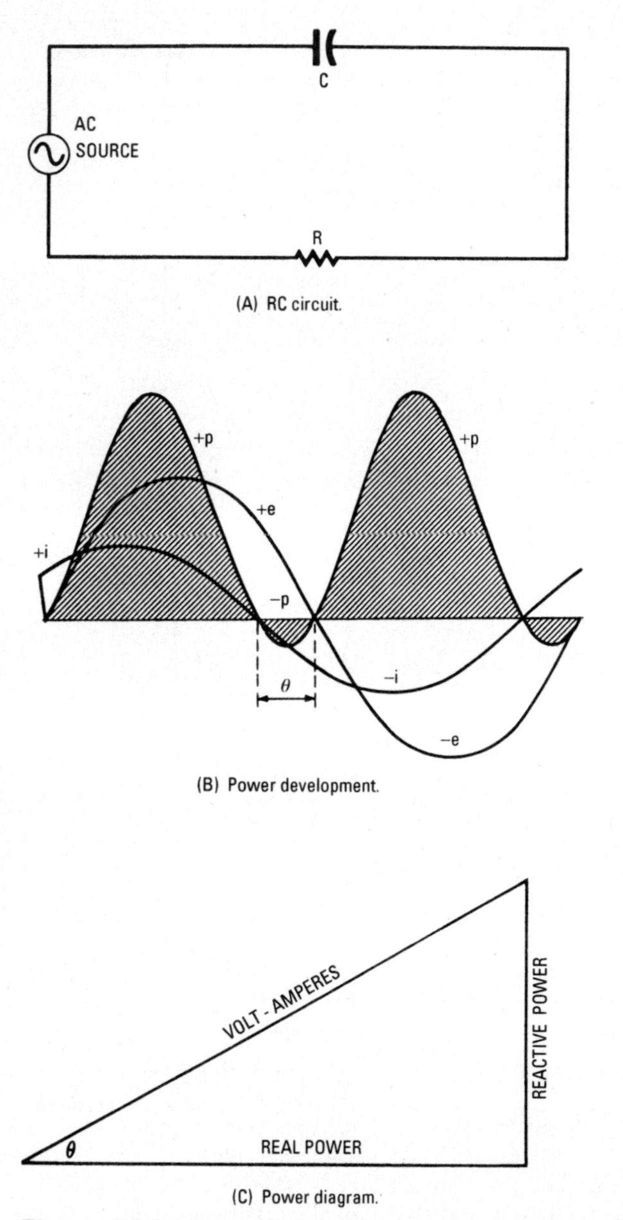

(A) RC circuit.

(B) Power development.

(C) Power diagram.

Figure 4-27 Power circuit with capacitance and resistance.

which contain both resistance (R) and capacitance (C). RL circuits are those which contain resistance (R) and inductance (L).

Inductance, Capacitance, and Resistance in Series

Many types of circuits have inductance, capacitance, and resistance connected in series with an AC voltage source, as shown in Figure 4-28. As we would expect, the voltage drop across the resistor is

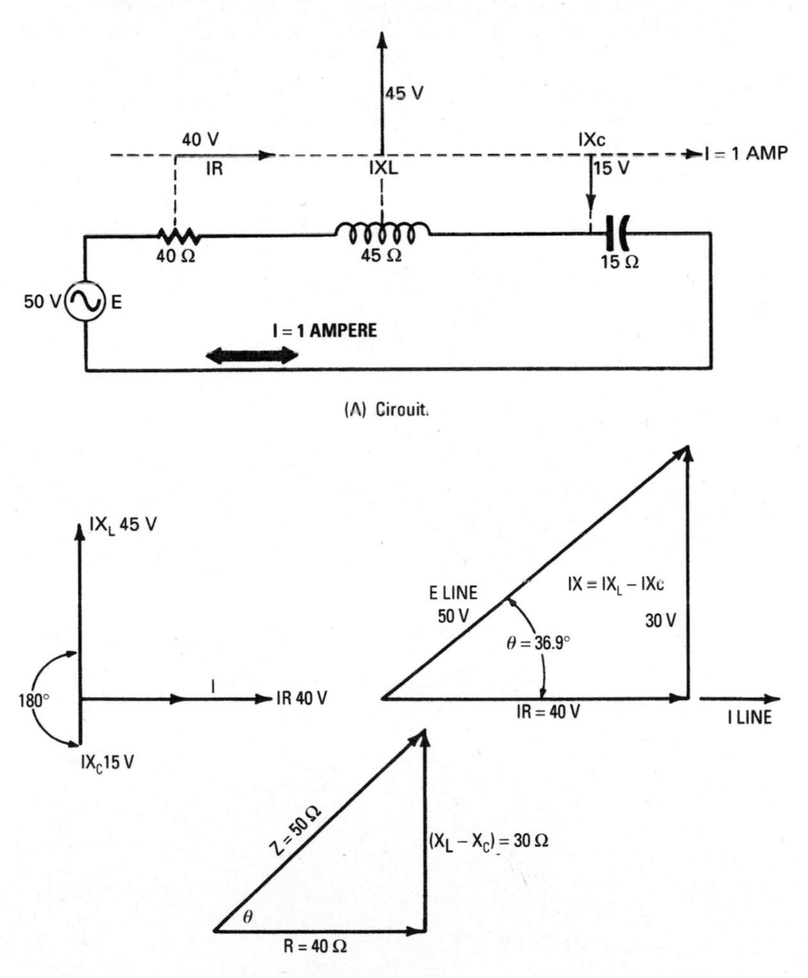

Figure 4-28 Resistance, inductance, and capacitance connected in series.

in phase with the circuit current, while the voltage drop across the inductor is 90 degrees out of phase with the circuit current, and the voltage drop across the capacitor is 90 degrees out of phase with the circuit current.

Since the current is the same at any point in the circuit of Figure 4-28, we take the current as a common reference in following the circuit action. In turn, the source voltage E is the vector sum of IR, IX_L and IX_C. In Figure 4-28B, we see the relations of IR, IX_L, and IX_C. Note that IX_L and IX_C are each 90 degrees away from I, and point in opposite directions (IX_L and IX_C have a phase angle of 180 degrees). Therefore, the total reactive voltage E_X is the difference between IX_L and IX_C. We write the following:

$$E_x = IX_L - IX_C = 45 - 15 = 30 \text{ volts}$$

Since X_L is greater than X_C in this example, the inductance dominates the circuit action, and the circuit current lags the applied voltage E. In other words, the phase angle in Figure 4-28B is lagging. The impedance diagram for the circuit shows the impedance is 50 ohms. To find the circuit impedance by arithmetic, we write the following:

$$Z = \sqrt{R^2 + (X_L - X_C)^2}$$

or,

$$Z = \sqrt{40^2 + (45 - 15)^2} = \sqrt{2500} = 50 \text{ ohms}$$

We know that the circuit current in Figure 4-28A is given by $I = E/Z$:

$$I = \frac{E}{Z} = \frac{50}{50} = 1 \text{ ampere}$$

In turn, the four power values in the circuit of Fig. 7.21 are as follows:

$$P_R = I^2 R = 40 \text{ watts}$$
$$P_L = I^2 X_L = 45 \text{ vars}$$
$$P_C = I^2 X_C = 15 \text{ vars}$$
$$P_{\text{apparent}} = EI = 50 \text{ volt-amperes}$$

Note that the difference between P_L and P_C is 30 vars. We know that apparent power is equal to the following:

$$P_{\text{app}} = \sqrt{\text{Real power}^2 + \text{Reactive power}^2}$$

or,

$$P_{\text{app}} = \sqrt{40^2 + 30^3} = 50 \text{ volts-amperes}$$

In large factories and power plants, electricians are particularly concerned with adjusting the capacitance in circuits such as shown in Figure 4-28 so that the inductive and capacitive reactances are equal. Figure 4-29 shows a circuit in which $X_L = X_C$. We recognize that X_L exactly cancels X_C. Therefore, *the circuit action is the same as though there were only resistance present in the circuit.* Since there is effectively no inductance or capacitance in this circuit, the phase angle between the source voltage and the circuit current is zero. Electricians say that this kind of a circuit has been *corrected for zero power factor.*

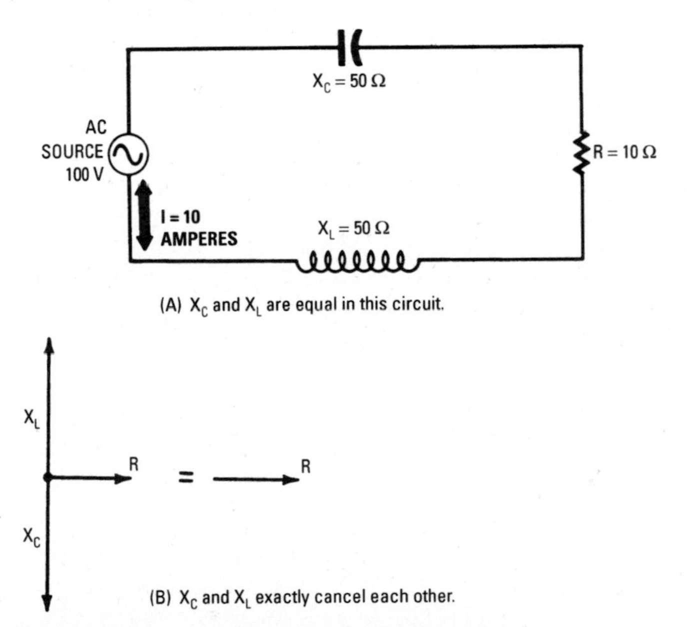

Figure 4-29 Conductive and reactance circuit.

It follows that the current in the circuit of Figure 4-29 is given by the following:

$$I = \frac{E}{R} = \frac{100}{10} = 10 \text{ amperes}$$

When the inductive reactance is equal to the capacitive reactance in a circuit, we see that the number of volt-amperes is the same as the resistive load power. This is a desirable condition in a power-distribution system because there is no reactive current surging in the line to increase the line losses. Therefore, the efficiency of a power-distribution system is greatest when the inductive reactance that may be present is exactly canceled by inserting an equal amount of capacitive reactance.

The circuit in Figure 4-29 is often called a *series-resonant circuit*. This simply means that the phase angle between the source voltage and circuit current is zero. It is important to note that there is a *voltage rise* across the capacitor and across the inductor in a series-resonant circuit. For example, in Figure 4-29, the voltage drop across the capacitor is equal to IX_C, or 500 volts. Similarly, the voltage drop across the inductor is equal to IX_L, or 500 volts. Although the source voltage is 100 volts in Figure 4-29, the voltage across L rises to 500 volts and the voltage across C rises to 500 volts. In turn, we must use inductors and capacitors that are rated for 500-volt operation.

This voltage rise across an inductor and across a capacitor in a resonant LCR circuit is sometimes called the *voltage magnification* of the circuit. The voltage magnification can be very great when the load resistance is small. For example, if we used a 1-ohm load resistor in Figure 4-30, the voltage across L would be 5000 volts, and across C would be 5000 volts. Figure 4-30 shows the instantaneous voltage, current, and power relations in a series-resonant circuit.

Inductance and Resistance in Parallel

Circuits that consist of inductance and resistance connected in parallel can be especially useful. Consider the impedance of an RL parallel circuit, such as shown in Figure 4-31. The inductive reactance has a value of $2\pi fL$ ohms. To find the circuit impedance, we may write:

$$Z = \frac{R X_L}{R^2 + X^2{}_L}$$

However, to simplify the calculation of Z, it is advisable to make use of the triangle relations shown in Figure 4-31. We lay off the line R with a length proportional to the resistance, and lay off the line X_L with a length proportional to $2\pi fL$. Then, we draw the line Z perpendicular to the hypotenuse of the triangle. The length of Z

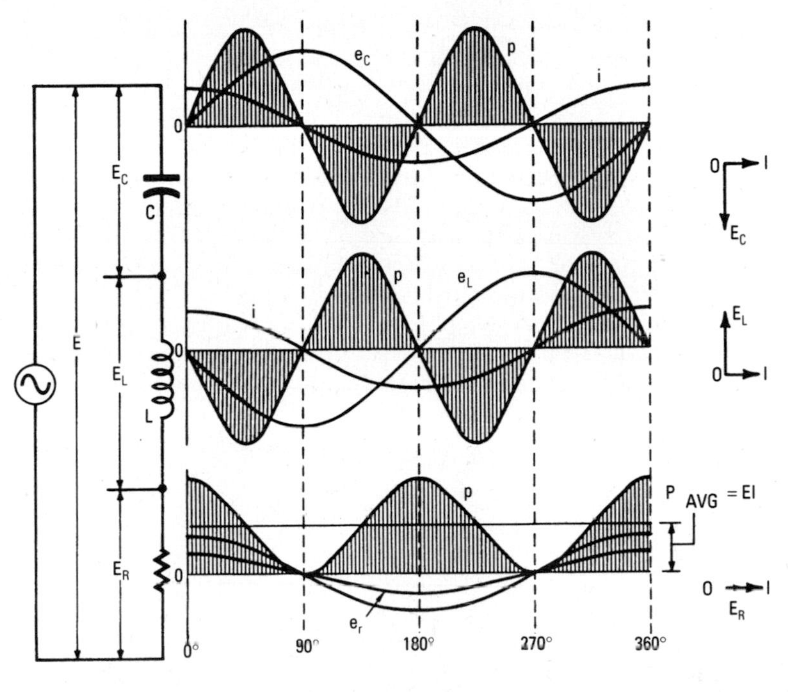

Figure 4-30 Instantaneous voltage, current, and power relations in a series-resonant circuit.

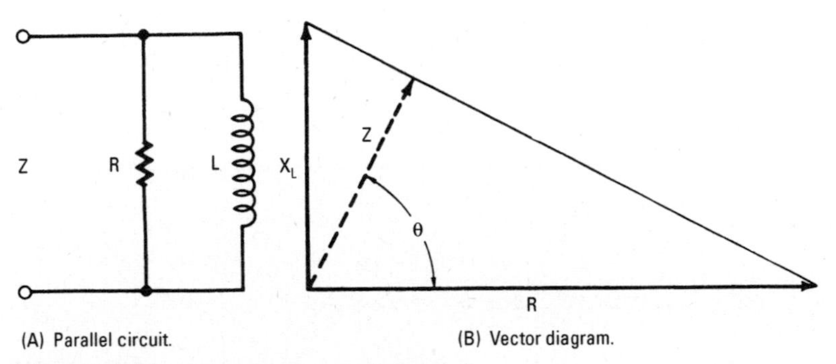

Figure 4-31 Impedance parallel circuit.

then gives the number of ohms of impedance. Note also that the angle θ is the power-factor angle of the RL parallel circuit.

When working with power-factor correction, as will be explained, electricians often find it useful to change a parallel RL circuit into an equivalent series RL circuit, as shown in Figure 4-32. In other words, when an inductance L_p is connected in parallel with a resistance R_p (see Figure 4-32A), we can make up an equivalent circuit by connecting an inductance L_S in series with a resistance R_S (see Figure 4-32B). The relations of inductance and resistance values in the series and parallel circuits are easily found by drawing the triangle relations shown in Figure 4-32C. Note that smaller values of series inductance and resistance have the same circuit action as larger values of parallel inductance and resistance.

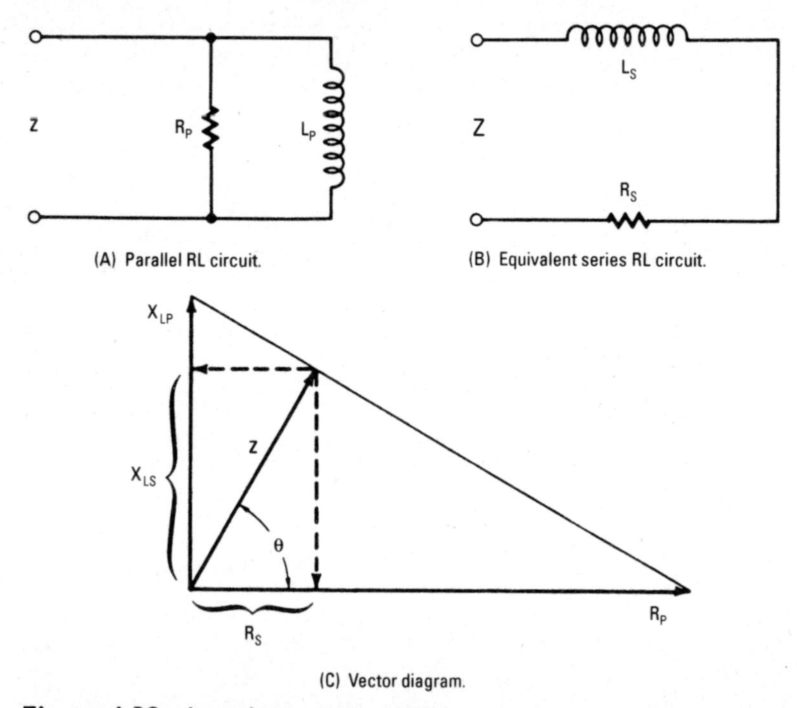

Figure 4-32 Impedance series circuit.

Capacitance and Resistance in Parallel

You will also see circuits that consist of capacitance and resistance connected in parallel. Figure 4-33 shows an RC parallel circuit. The

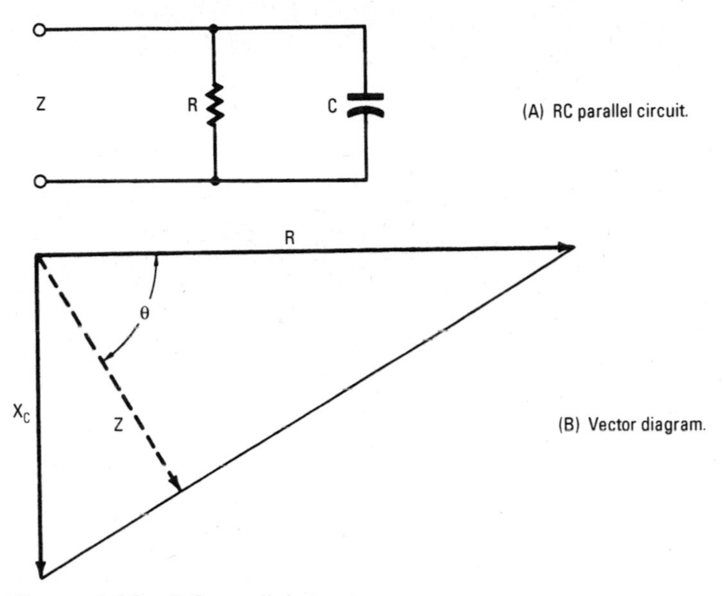

Figure 4-33 RC parallel circuit.

capacitive reactance has a value of $1/(2\pi fC)$ ohms. To find the circuit impedance, we may write the following:

$$Z = \frac{R\,X_C}{\sqrt{R^2 + X^2{}_C}}$$

However, to simplify the calculation of Z, it is advisable to make use of the triangle relations shown in Figure 4-33B. We lay off the line R with a length proportional to the resistance, and lay of the line X_C with a length proportional to $1/(2\pi fC)$. Then, we draw the line Z perpendicular to the hypotenuse of the triangle. The length of Z then gives the number of ohms of impedance. Note also that the angle θ is the power-factor angle of the RC parallel circuit.

It can be useful to change a parallel RC circuit into an equivalent series RC circuit, as shown in Figure 4-34. In other words, when a capacitance C_p is connected in parallel with a resistance R_p, we can make up an equivalent circuit by connecting a capacitance C_s in series with a resistance R_s. The relations of capacitance and resistance values in the series and parallel circuits are easily found by drawing the triangle relations shown in Figure 4-34C.

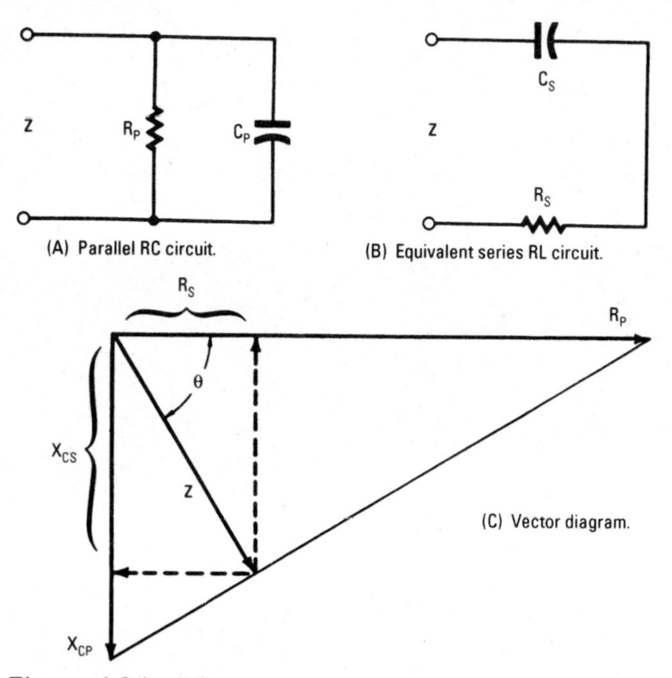

Figure 4-34 RC series circuit.

Inductance, Capacitance, and Resistance in Parallel

There are many circuits that contain resistance, capacitance, and inductance in parallel. A basic RCL parallel circuit is shown in Figure 4-35. The R, X_L, and X_C vectors are drawn as shown in Figure 4-35B. Note that X_L and X_C extend in opposite directions. Therefore, X_L and X_C will tend to cancel each other, and if $X_L = X_C$, the reactances cancel completely, leaving resistance only.

When the values of X_L and X_C are equal, the circuit in Figure 4-35A is a *parallel-resonant circuit*. This means that L and C are effectively not present, inasmuch as they cancel each other. Therefore, current is drawn by R only.

In power applications, when a circuit is parallel-resonant, it is also said to be *power-factor corrected*, since the power-factor angle is obviously zero in the power-factor corrected circuit, particularly when the power demand is heavy. This is because a power-factor corrected circuit has no surging line currents, and, therefore, operates at maximum efficiency. (The I^2R line loss caused by surging line currents has been eliminated.)

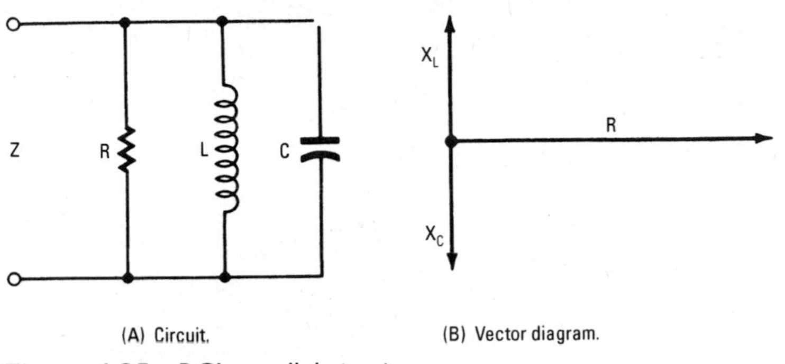

Figure 4-35 RCL parallel circuit.

Principles of Electromagnetic Induction

As mentioned in Chapter 1, electricity and magnetism are inseparable. Electronic devices work with electromagnetic induction in a myriad of applications. Various devices operate on the principle of electromagnetic induction, and circuits often have responses based on this principle. For example, we may note the following typical devices:

- Choke reactors
- Compensators
- Generators
- Ignition coils
- Induction motors
- Induction regulators
- Magnetos
- Transformers
- Watt-hour meters

When a permanent magnet is inserted into a coil, as shown in Figure 4-36, the voltmeter deflects. In other words, the changing magnetic field in the coil *induces* a voltage in the coil winding. When the permanent magnet is withdrawn from the coil, a voltage of opposite polarity is induced in the coiled winding. Note that if the permanent magnet is motionless in the coil, there is no induced voltage. Note also that when the magnet is moved faster, a greater voltage is induced.

It is the *relative* motion of a magnetic field and a wire (or coil) that induces a voltage in the wire. For example, if a coil is moved through

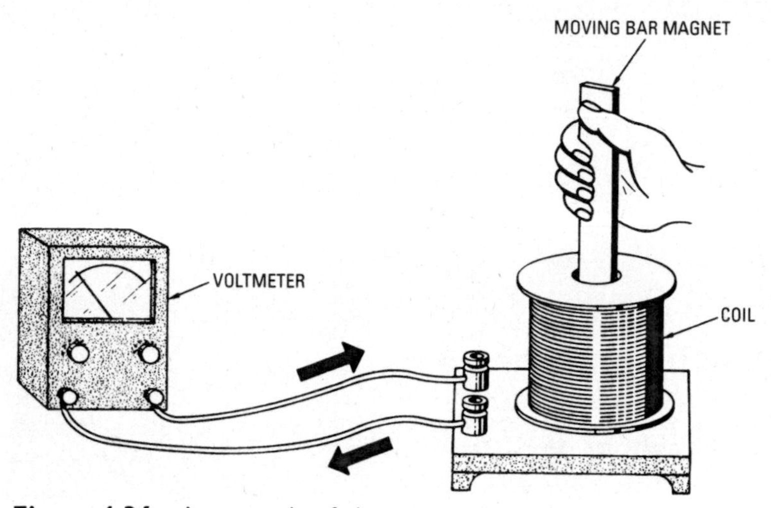

Figure 4-36 An example of electromagnetic induction.

a magnetic field, a voltage is induced in the coil. This principle is used in an electrical instrument called a *fluxmeter* (or *gaussmeter*) to measure the strength of a magnetic field. The instrument consists of a *flip coil* connected to a *galvanometer*, as shown in Figure 4-37. A flip coil consists of a few turns of wire, usually less than an inch in diameter. A galvanometer is simply a sensitive current meter. To use a fluxmeter, the electrician places the flip coil in the magnetic field to be measured and then flips the coil out of the field. In turn, the galvanometer reads the strength of the magnetic field in lines of force per square inch.

The meter in Figure 4-37 is constructed so that the pointer remains at the point on the scale to which it is deflected when the flip coil is moved out of the magnetic field that is to be measured. Therefore, it makes no difference how fast or how slow the electrician moves the flip coil out of the magnetic field. Before another measurement of magnetic field strength is made, the pointer is adjusted back to zero on the scale.

Laws of Electromagnetic Induction

When a conductor moves with respect to a magnetic field, a voltage is induced in the conductor only if the conductor *cuts* the lines of magnetic force. To cut lines of force means to move at right angles to the lines, as shown in Figure 4-38. In this example, conductor *AB* is

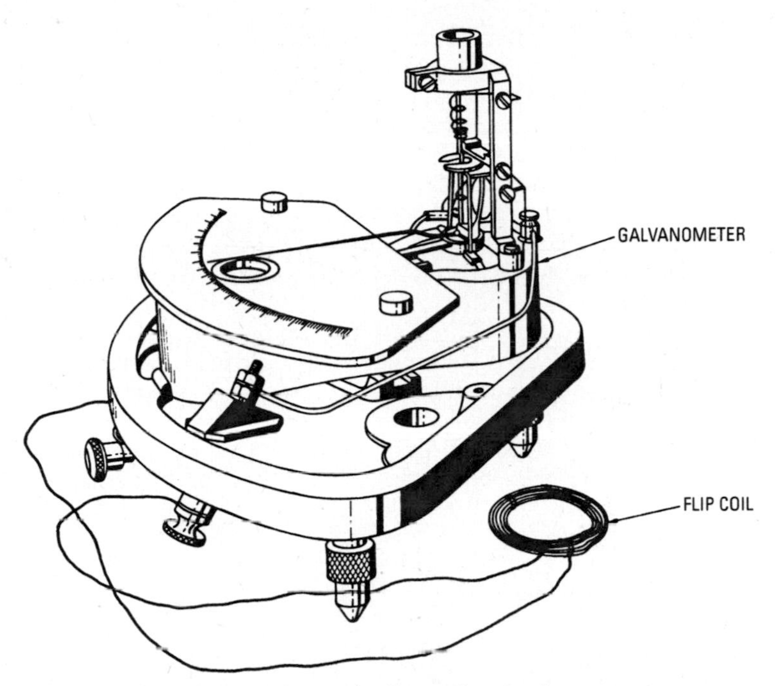

Figure 4-37 Construction of a flip coil and galvanometer.

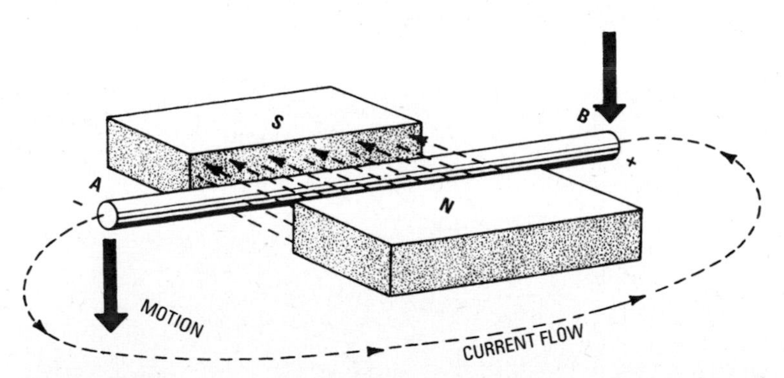

Figure 4-38 Induction of voltage in a conductor moving through a magnetic field.

moving downward through the magnetic field, and this motion is at right angles to the lines of force. The lines of force are directed from the North pole (N) to the South pole (S) of the magnet. A voltage is induced in the conductor. A negative potential appears at A, and a positive potential appears at B. If we connect a wire from A to B, electrons flow in the wire as shown.

Technicians need to know the polarity of the voltage induced in a conductor, or, what amounts to the same thing, the direction of induced current in the conductor. This is found by means of *Lenz's law.*

Heinrich Lenz was the Russian physicist who, in 1834, discovered the relationships between induced magnetic fields, voltage, and conductors. Figure 4-39 illustrates Lenz's law. If you point in the direction of the magnetic flux with the index finger of your left hand, and point your thumb in the direction that the conductor moves, your middle finger then points in the direction of electron flow

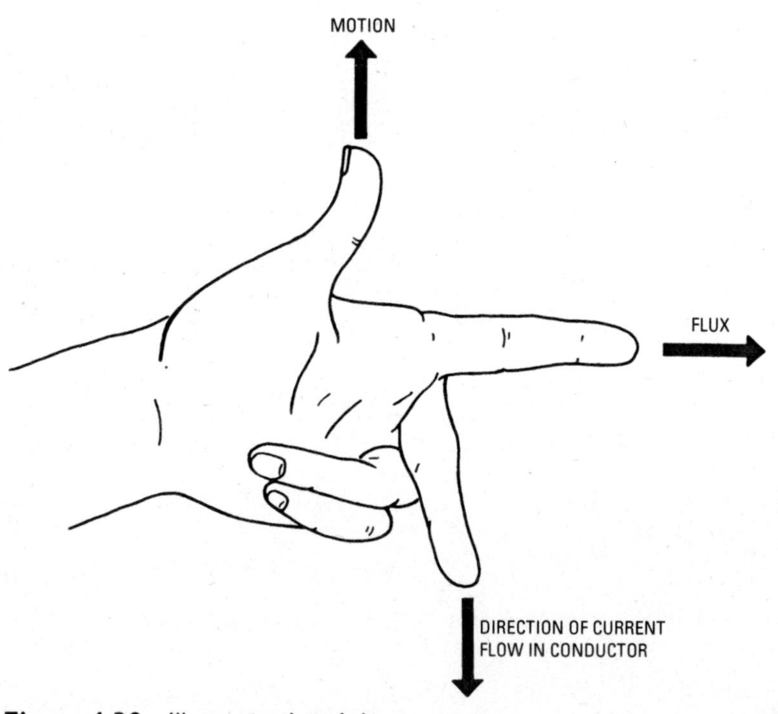

Figure 4-39 Illustrating Lenz's law.

in the conductor. Note that we hold the index finger, thumb, and middle finger at right angles to one another when we apply Lenz's law.

The amount of voltage that is induced in a moving conductor depends on how fast the conductor cuts lines of magnetic force. If a conductor moves at a speed such that it cuts 10^8 lines of force per second, there will be 1 volt induced in the conductor. Of course, if we use a number of conductors, the induced voltage will be multiplied by the number of conductors. For example, if the coil in Figure 4-36 has 100 turns, the induced voltage will be 100 times as great as if only one turn were used.

Since the amount of voltage induced in a conductor depends on how fast the conductor cuts lines of force, it might seem that the reading obtained on a galvanometer, such as the one shown in Figure 4-37, would depend on how fast the flip coil is removed from a magnetic field. However, we will find that the speed with which the flip coil is moved has no effect on the meter reading. The reason that speed makes no difference is because the galvanometer indicates the *total quantity of electricity* induced in the flip coil. This total quantity of electricity is the same, whether a small current is induced over a long period of time, or a large current is induced over a short period of time. A small current is produced by a small induced voltage, and a large current is produced by a large induced voltage.

Self-Induction of a Coil

A large flaming spark, or arc, frequently appears across switch contacts when the circuit to an electromagnet is opened. Figure 4-40 shows this circuit action. When a larger electromagnet is used, the arc is hotter and larger. This circuit action is explained as follows:

1. When the switch is closed (see Figure 4-40), current flows through the coil turns and build up a magnetic field in the space surrounding the coil.
2. A magnetic field contains *magnetic energy*; the larger the coil, the greater is the amount of energy in its magnetic field.
3. At the instant the switch is opened, current flow stops in the coil, and the magnetic force lines quickly collapse into the coil. This is a moving magnetic field that cuts the coil turns and induces a voltage in the turns—the *voltage of self-induction.*
4. While the magnetic flux lines are collapsing, the energy in the magnetic field becomes changed into heat and light energy in the form of a momentary arc between the switch contacts.

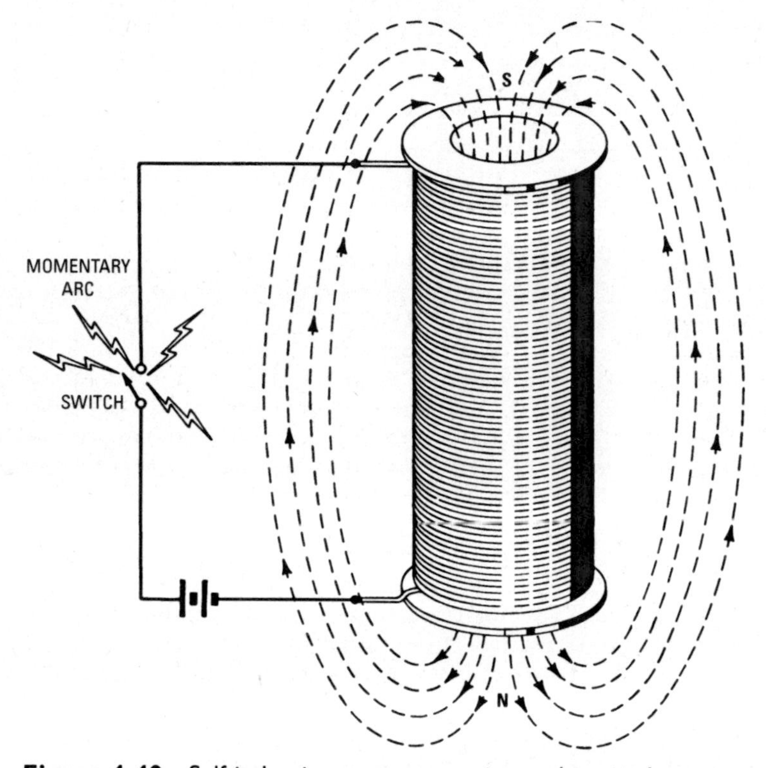

Figure 4-40 Self-induction causes an arcing at the switch contacts.

There are some important facts concerning self-induction that we should keep in mind. The voltage of self-induction depends on the resistance of the arc path. If the switch is opened quickly, the arc path is longer and the self-induced voltage rises to a higher potential than if the switch is opened slowly. If the switch is opened with extreme rapidity, it may be easier for sparks to jump between turns of the coil than between the switch contacts. The high voltage of self-induction can break down the insulation on the coil wire and damage the magnet. Therefore, electrical equipment that uses large electromagnets carrying heavy current is often provided with a protective device to prevent the voltage of self-induction rising to dangerous potentials.

We will find that the polarity of self-induced voltage is such that it tends to maintain current flow in the same direction as before the switch was opened. The direction of current in the arc between the switch contacts is the same as was flowing in the circuit before the switch was opened. This is often called the *flywheel effect* of an electromagnet. Just as a flywheel continues to turn in the same direction after engine is turned off, so does an electromagnet continue to supply current in the same direction to a circuit after the switch is opened.

Figure 4-41 shows a lamp connected across an electromagnet. When the switch is closed, the lamp glows dimly because the coils have comparatively low resistance and are connected in parallel with the lamp. However, when the switch is opened, the energy in the collapsing magnetic field causes the coil to apply a large momentary voltage across the lamp. Therefore, the lamp flashes brightly. If the coil is sufficiently large, the bulb will burn out when the switch is opened.

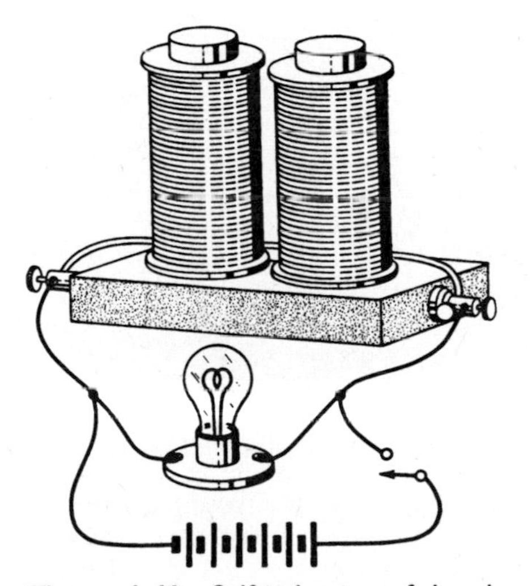

Figure 4-41 Self-induction of the electromagnet.

Figure 4-42 shows an electric bell. The contact breaker automatically opens the circuit to the electromagnet when the armature is attracted. Then, the armature springs back, and the circuit is again closed through the contacts. This opening and closing action repeats rapidly. Each time the contacts are opened, a spark jumps between the contacts because of the voltage of self-induction. To prevent the contacts from being burned away in a short time, special metals such as platinum are used. If a soft metal such as copper were

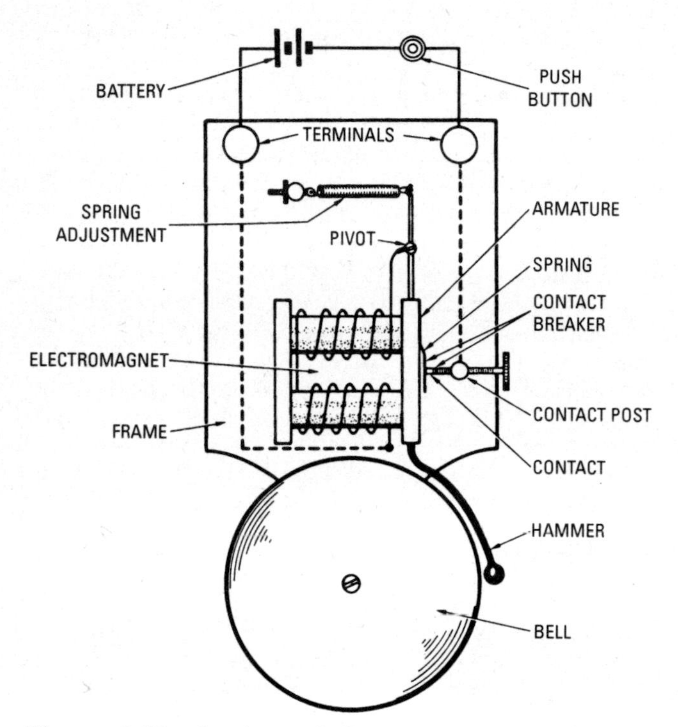

Figure 4-42 An electric bell.

used for the contacts, the life of the bell would be comparatively short.

Transformers

Transformers are widely used in electrical and electronic equipment. The only differences between transformers for power and electronics are mainly in size and in core types. Obviously, power transformers handle much more current than electronic transformers and are, therefore, built with heavier wire and better insulating methods. They are much larger and heavier. Electronic transformers operating at higher frequencies tend not to have iron cores, as these cores tend to overheat at higher frequencies. Such high-frequency transformers are commonly called *air-core* transformers. This section explains some very common and basic types of transformers called *ignition coils*, or *spark coils*.

Since a bar magnet produces an induced voltage when it is moved into or out of a coil, as shown in Figure 4-36, it follows that an

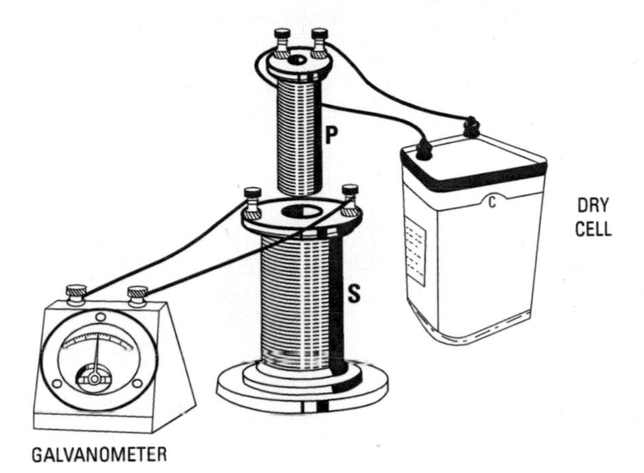

Figure 4-43 A moving electromagnet generates a current.

electromagnet can be substituted for the bar magnet, as depicted in Figure 4-43. It also follows that we can place the electromagnet into the coil and induce a current in the coil by opening and closing the circuit of the electromagnet. *This is an example of transformer action.* A transformer has two windings, called the *primary* (P) and the *secondary* (S).

When the primary circuit of a transformer is closed, as shown in Figure 4-44A, magnetic flux lines build up in the space surrounding each primary wire *B*. In turn, these expanding flux lines cut each secondary wire *A* and induce a voltage in the secondary. When the primary circuit is opened, the flux lines collapse, and again cut the secondary wire; therefore, a voltage is again induced in the secondary. Note that when the switch is closed, the galvanometer deflects in one direction, but when the switch is opened, the galvanometer deflects in one direction. However, when the switch is opened, the galvanometer deflects in the other direction. The flux lines cut the secondary wire in opposite directions when the switch is opened and closed.

If the switch is held closed in Figure 4-44A, the galvanometer does not deflect. In other words, the pointer "jumps" on the galvanometer scale only during the instant that the magnetic field is building up or collapsing. It is easy to see that if the flux lines from a primary wire cut more than one secondary wire, more voltage will be induced in the secondary, as shown in Figure 4-44C.

Since an ignition coil in an automobile must produce a very high voltage to jump the gap of the spark plug, the secondary is

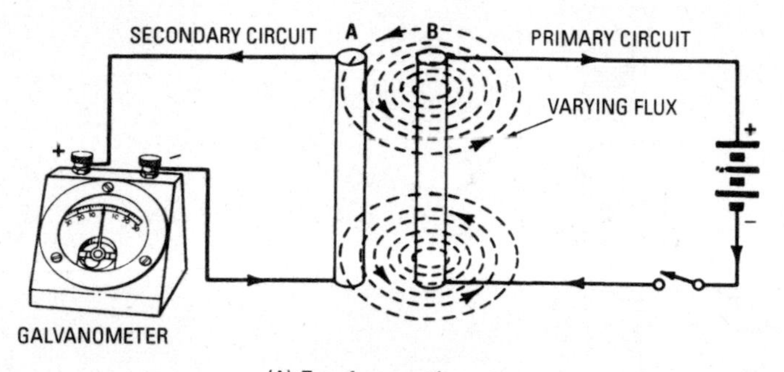

(A) Transformer action.

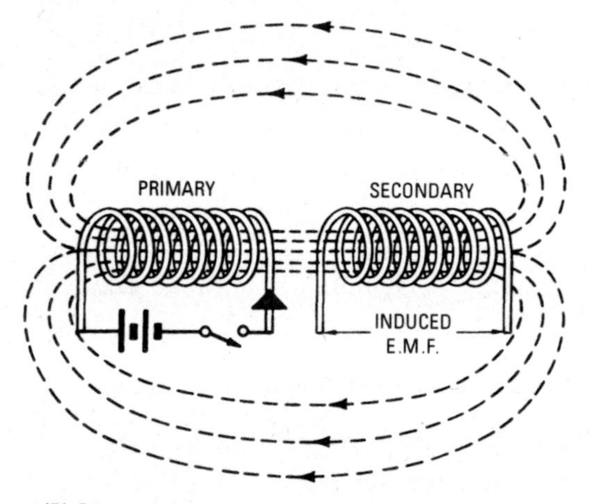

(B) Primary and secondary coils, showing magnetic flux lines.

Figure 4-44 Basic power transformer action.

wound with a large number of turns as compared with the primary. Figure 4-45 shows the arrangement of an ignition coil and its circuit. The primary is often called the *low-tension winding*, and the secondary the *high-tension winding*. High tension equates to high voltage, and low tension equates to low voltage.

The ground circuit, shown by dotted lines in Figure 4-45, is familiar to every automotive electrician. The engine block and car frame serve as part of the ignition circuit. We observe that both the primary and secondary currents travel through ground circuits. If the metalwork of the car were not used as a ground path for the ignition system, the wiring would be more complicated. Note that a pair of contacts in the distributor are opened and closed by a

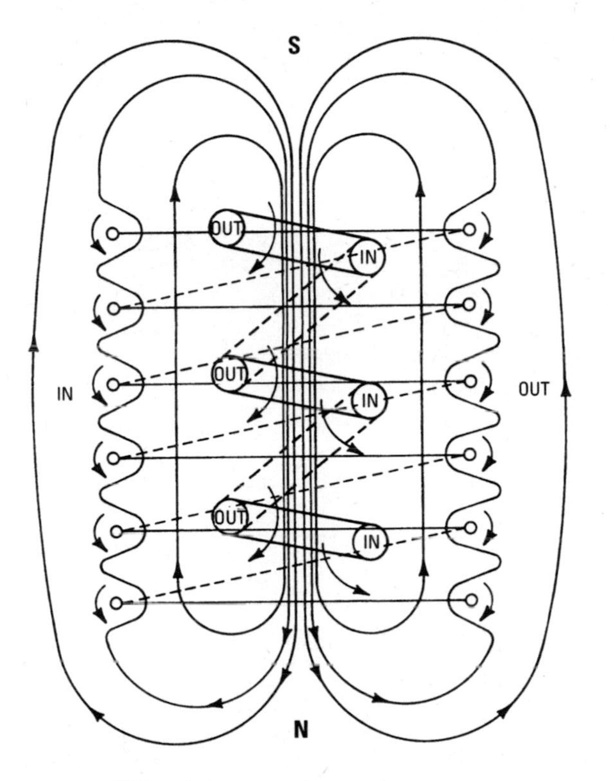

(C) Primary flux lines cutting secondary turns.

Figure 4-44 *(continued)*

revolving cam. These contacts operate as a switch to open and close the primary circuit.

A *spark coil* (or *vibrator coil*) is similar to an ignition coil except that the contacts are opened and closed by electromagnetic action, as seen in Figure 4-46. Note the *capacitor*, which is connected across the vibrator contacts. In actual practice, we will also find a capacitor connected across the contacts in Figure 4-45. Carefully observe the action of a capacitor. A basic capacitor consists of a pair of metal plates spaced near each other, with a sheet of insulating material (such as waxed paper, mica, or plastic) between the plates. (Also note that the old term for capacitor—*condenser*—is still occasionally used in the automotive industry.)

With reference to Figure 4-47, a momentary current flows when the switch is closed. The battery forces electrons into the negative

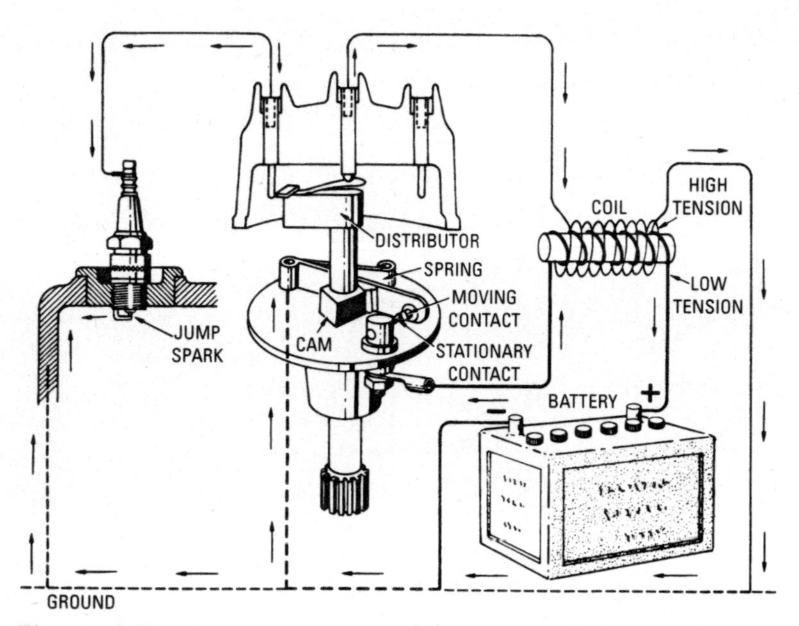

Figure 4-45 Diagram of automobile ignition system using battery, ignition coil, and high-tension distributor.

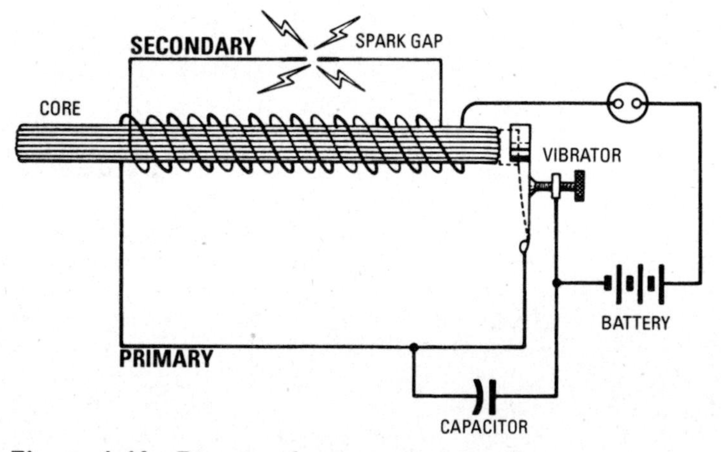

Figure 4-46 Diagram of a vibrator spark coil.

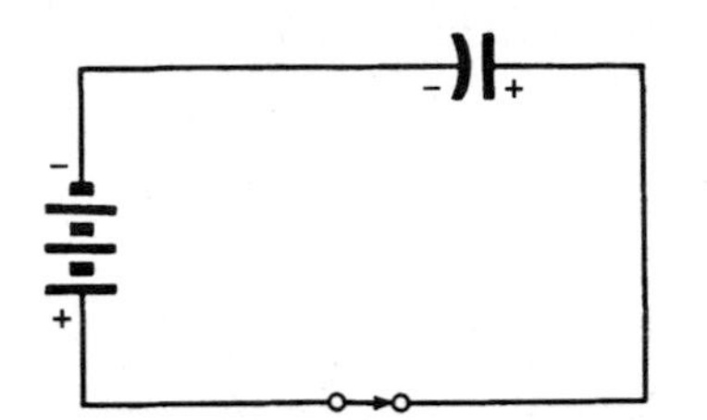

Figure 4-47 Capacitor action in a circuit.

plate and takes electrons away from the positive plate. As soon as the capacitor *charges* to the same voltage as the battery, current flow stops.

In its charged condition, a capacitor contains electrical energy that can be returned to the circuit. For example, suppose that we open the switch in Figure 4-47. The capacitor remains charged because there is no path for escape of electrons from the negative plate to the positive plate. However, if we short-circuit the capacitor terminals (as with a screwdriver), we will observe that a momentary current flows through the short circuit and produces a snapping spark at the point of contact. The larger the area of the plates in the capacitor, and the closer together the plates are mounted, the greater is the short circuit on *discharge*.

Now, let us return briefly to Figure 4-46. It is highly desirable to prevent the vibrator contacts from sparking or arcing when the contacts open. The reason for this is because a spark or arc wastes electrical energy that would otherwise appear as induced voltage in the secondary. The capacitor prevents sparking (or arcing) when the contacts open because the self-induced voltage from the primary is then used to charge the capacitor instead of jumping the air gap between contacts. The self-induced voltage is stored in the capacitor. A short time after the capacitor is charged, it proceeds to discharge back through the primary winding. During the charging process, a little time is provided for the vibrator contacts to separate enough that a spark does not jump between them. Therefore, sparking is practically eliminated at the contacts, and the electricity stored in the capacitor next produces a heavy momentary current flow through the primary. In turn, the secondary voltage is much greater than if a capacitor were not used.

Iron Cores

In Figures 4-45 and 4-46, the primary and secondary coils are wound on a single iron core. This core makes the ignition coil or spark coil

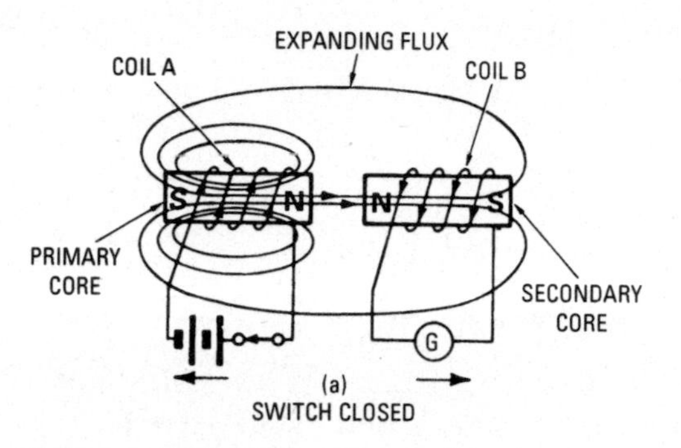

(A) The expanding flux fails to cut the secondary coil.

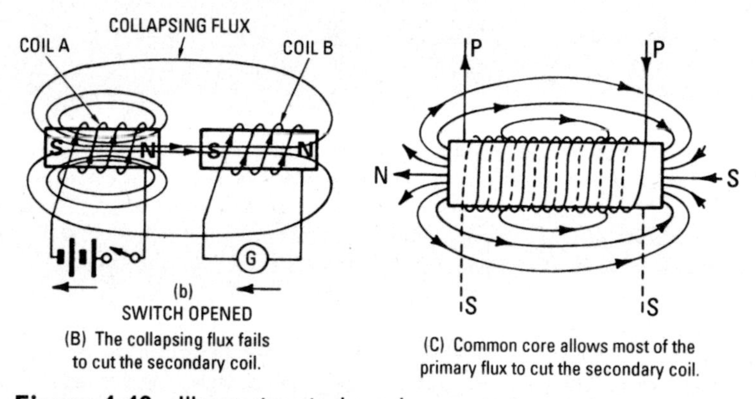

(B) The collapsing flux fails to cut the secondary coil.

(C) Common core allows most of the primary flux to cut the secondary coil.

Figure 4-48 Illustrating single and separate iron cores.

more efficient than if the primary and secondary were wound on separate iron cores with an air space between them. A single iron core also makes for much greater efficiency than an air core. The reason for this improvement in efficiency is seen in Figure 4-48. If an air core or if separate iron cores are used as shown in Figure 4-48A and 4-48B, most of the magnetic force line from the primary fail to cut the secondary coil. On the other hand, if a single iron core is used for both coils, as shown in Figure 4-48C, most of the primary magnetic force lines then cut the secondary coil.

Choke Coils

Choke coils are used in various types of equipment. For example, in Figure 4-49, we see how a choke coil is connected in a line conductor with a lightning arrester. If lightning strikes the line, a very large

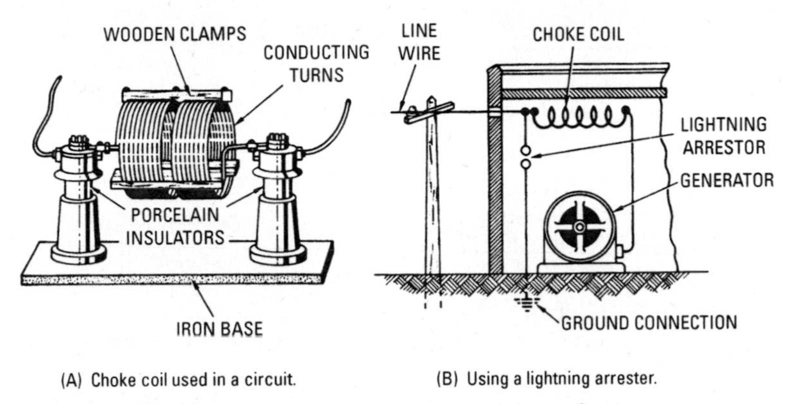

(A) Choke coil used in a circuit.

(B) Using a lightning arrester.

Figure 4-49 Illustrating the choke coil and its application.

surge of current travels along the line. This heavy current surge would damage equipment at the end of the line (such as the generator shown in Figure 4-49B) unless a protective device called a lightning arrester is used. A lightning arrester is basically a spark gap connected between the line and ground.

Since the high voltage produced by a lightning stroke will take the easiest path to ground, the strong electrical surge tends to jump the spark gap to ground instead of flowing through the generator in Figure 4-49B. We know that it takes quite a high voltage to break down a spark gap, even when the gap has a comparatively small spacing. Therefore, to make sure that the air gap breaks down before the high-voltage surge reaches the generator, a choke coil is connected between the gap and the generator.

A choke coil consists of a number of turns of large-diameter wire, as shown in Figure 4-49A. Since the resistance of the choke coil is very small, the steady current from the generator flows easily through the choke coil, and there is very little voltage drop across the coil. However, when a sudden electrical surge travels down the line, the choke coil blocks the surge and makes it jump the spark gap to ground. Let us see why a choke coil stops a sudden electrical surge.

When a voltage is suddenly applied to a coil, as shown in Figure 4-50, the magnetic flux lines build up, or expand. As the flux lines build up, they cut the turns of the coil and induce an emf. Note carefully that *the induced voltage opposes the applied voltage.* This is called *inductive opposition* to a sudden current change. (Electricians usually call this opposition *inductive reactance.*) Because the

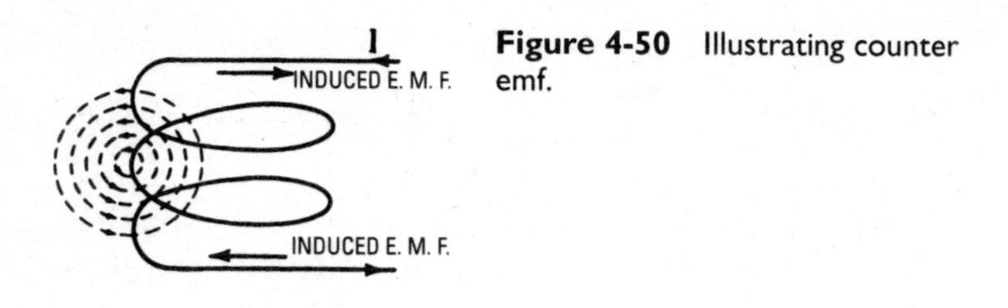

Figure 4-50 Illustrating counter emf.

sudden current surge is opposed by the induced emf, it takes a certain amount of time for the surge to overcome this opposition and get through the coil.

During the time that the current surge tries to build up a magnetic field in the choke coil, a very high voltage drop is produced across the spark gap, and the surge follows the easiest path to ground by jumping the gap. In other words, before a magnetic field can build up to any great extent in the choke coil, the lightning surge has been harmlessly passed to ground through the spark gap.

The induced emf in Figure 4-50 is called a *counter emf* (cemf), or *counter voltage.* Counter emf is found in any coil arrangement when voltage is suddenly applied to the coil. Therefore, it takes a certain amount of time for a surge voltage to produce current flow through a coil. On the other hand, after the switch has been closed for a short time in a circuit such as that shown in Figure 4-51, the current will have risen to its full value given by Ohm's law, and the current thereafter flows steadily as if the coil were not in the circuit.

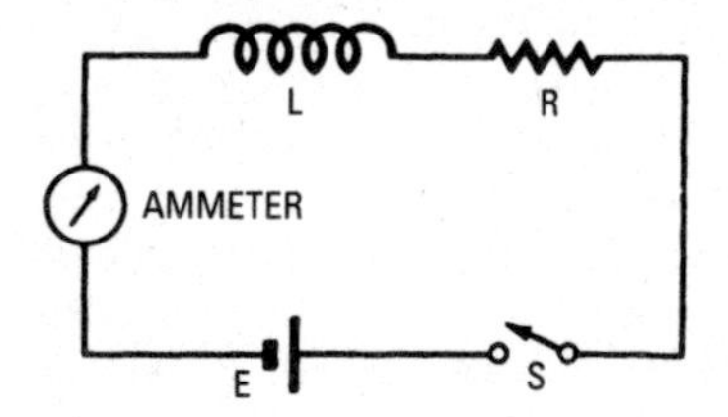

Figure 4-51 A time delay before the ammeter indicates a current flow.

Reversal of Induced Secondary Voltage

We will find that the secondary voltage reverses in polarity when the magnetic flux in a spark coil stops expanding and starts to collapse. For example, we see in Figure 4-48A that the expanding lines of force cut the secondary in a direction such that the right end of the secondary core has a South polarity. (This is shown by our left-hand rule.) Next, we see in Figure 4-48B that the collapsing lines of force cut the secondary in the opposite direction, and the right-hand end of the secondary core now has a North polarity. Therefore, the induced secondary voltage changes its polarity as the expanding magnetic field changes into a collapsing field.

Transformer Operation

Transformers are sometimes termed *static transformers* since they have no moving parts. The principle of a transformer is illustrated in Figure 4-52.

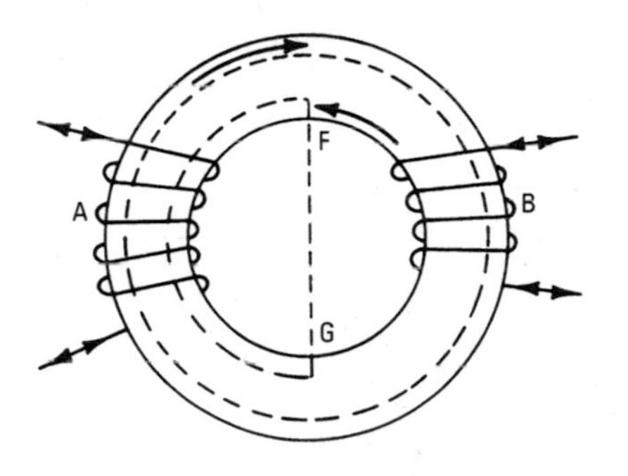

Figure 4-52 Principle of a transformer.

Assume that 100 volts AC are impressed on winding *A*, wound around a soft iron ring *C*, and the coil contains 300 turns. A magnetic flux is set up by the current in *A*, within the soft iron core *C*. This flux is set up around each turn or convolution of coil *A*. The flux lines are like rubber bands that expand outward, cutting each turn of winding *A*, expanding until they cut each turn of coil *B*, before they eventually occupy the cross-section of core *C*.

As the impressed voltage and the current in coil *A* rises and falls, the flux, in turn, rises and falls, cutting the turns of coil *B*. From the principle of induction, the rising and falling of the flux cutting coil *B* induces an emf in coil *B*. There is also a cemf induced in coil *A*. This emf and the emf induced in coil *B* are in opposition to the impressed emf in coil *A*.

If coil *B* is open (that is, has no load connected), there will be no flux generated by coil *B* in opposition to the flux generated by the impressed emf and current in coil *A*. The cemf in coil *A*, however, in opposition to the impressed emf on coil *A*, opposes the changes in current and tends to keep the flux oscillating with the frequency of the impressed emf.

Windings and Voltages

The winding that receives the impressed emf is always termed the *primary* winding, regardless of the impressed voltage. The winding that receives the induced emf (in this case, winding B) is the *secondary* winding. There is very often confusion in that the high-voltage winding is called the primary and the low-voltage winding is called the secondary. As may be seen, this is not the case. The input side is the primary, and the output side is the secondary.

The voltages of a transformer are proportional to the number of turns in the windings. Thus, if in Figure 4-52 winding A has 300 turns and an impressed voltage of 100 volts, and winding B has 150 turns, then winding B will have an induced voltage of 50 volts.

Each convolution or turn of transformer coils is cut by magnetic flux four times per cycle:

1. When the flux is rising
2. When the flux is falling to zero
3. When the flux is rising in the reverse direction
4. When the flux is falling to zero

The average voltage that is induced in each winding with a flux having a maximum value of φ will be:

$$E_{avg} = 4\varphi nT/10^8$$

where the following is true:

E_{avg} = average emf induced in each winding
φ = maximum magnetic flux
n = frequency in hertz
T = number of turns in coil
4 = constant: number of times each turn is cut by flux per cycle
10^8 = constant to reduce absolute lines of force for conversion of terms to practical volts

The rms voltage, or effective voltage, as read on the voltmeter, equals 1.11 times the average voltage. So, for the effective voltage instead of the average voltage, the 4 in the previous formula becomes 4×1.11, or 4.44. Thus,

$$E_{avg} = 4.4\varphi nT/10^8$$

It was shown that for the transformer in Figure 4-52, with no load on winding *B*, the current into winding *A* was limited to a low value by the counter emf induced in that winding.

If, however, AC voltage is impressed across the primary of an iron-core transformer, a current surge may occur. The reason for this is as follows. The voltage to be impressed has associated with it a flux curve, which is simply the flux that would occur were the AC voltage impressed on the primary. The flux curve, in its relation to its steady-state axis, is determined by two factors:

- There is no current at the instant of application of voltage, thus the instantaneous value of the flux is zero.
- The rate of change of flux must be that required to cause the desired voltage to be induced (approximately equal to the applied voltage).

Thus, the flux may reach a peak density of as much as twice the steady-state value. The result of this is complete core saturation (that is, a condition in which a great change in current is required to produce even a small change in flux). The current necessary to produce the required flux may therefore be very great.

The inrush of current only exists for the first half-cycle and then drops off rapidly during subsequent cycles and reaches the normal charging current in two to three cycles. The maximum peak of inrushing current can reach 30 to 100 times the normal line current. As an example, a transformer with 100 volts impressed and 0.01-ohm primary resistance could reach 10,000 amperes of inrush current.

If the switch is closed, and the voltage wave value is maximum, the inrush current would be very low. Thus the inrush current is dependent upon the point of the voltage wave that the switch is closed.

With three phases, the voltage waves are 120 degrees apart, so that there will always be an inrush of current upon closing the switch. Therefore, the inrush current must be considered when determining the primary overcurrent protection.

If a load is added to winding *B* of Figure 4-52, this load will draw current from winding *B*. Thus the current in *B* (the flux of which opposes the flux of the primary current) reduces the cemf in *A*, which allows winding *A* to take more current from the source of supply in direct proportion to the secondary current. The reverse action takes place if the current load on winding *B* is removed or reduced.

From this, it may be observed that the primary winding of a transformer draws current in proportion to the current load on the secondary.

Earlier it was stated that the primary voltage and turns were proportional to the secondary voltage and turns. Thus, there would be one winding of 100 volts and 300 turns as opposed to 50 volts and 150 turns on the other winding. This is fundamentally true of an ideal transformer with no losses, but, practically, windings must be added or subtracted from one coil or the other, because of losses in flux and the like.

So far, the current relationships between windings were merely mentioned as being proportional. They are inversely proportional. A transformer with a 100-volt primary and 5-amperes full load would draw 500 volt-amperes. The secondary has 50 volts and, neglecting efficiencies, the secondary would be carrying 500 volt-amperes also. So, $I = P/E = 500/50 = 10$ amperes.

Figure 4-53 recaps the analysis of the ratio of turns, voltage, current, and power. Here the high side has 100 volts and 300 turns. So 100 V/300 T = 1/3 volt per turn. The low side has 50 volts and 150 turns. So, 50 V/150 T = 1/3 volt per turn. Thus the voltage-and-turns ratios are equal.

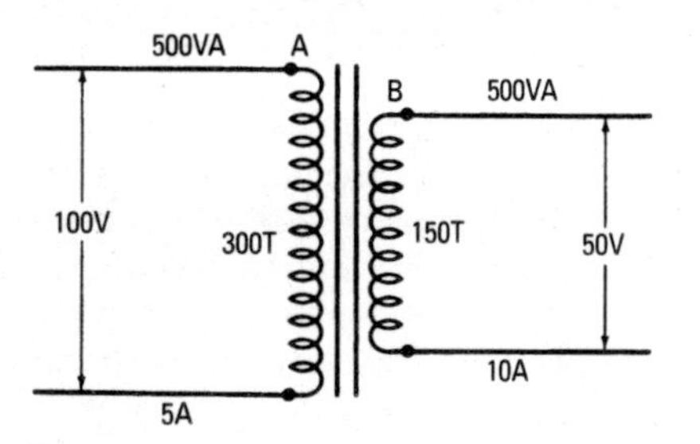

Figure 4-53 Ratio of turns, voltages, currents, and power in transformers.

Winding *A* carries 500 volt-amperes and so does winding *B*, from which it may be observed that primary and secondary windings carry the same power.

As to the current in the windings, for winding *A*:

$$I = P/E = 500\text{ VA}/100\text{ V} = 5\text{ amperes}$$

Winding *B* has

$$I = P/E = 500\text{ VA}/50\text{ V} = 10\text{ amperes}$$

From this it is found that the ratio of primary turns to secondary turns is in inverse proportion to the ratio of primary current to secondary current.

Recap and Formulas

The past few chapters have covered quite a bit of material. I want to restate most of the critical formulas and concepts again at the end of this chapter. This is important in the learning process. Please read this short section carefully. Review the formulas and be sure that you understand them. Think about which formula you would use in which certain situation.

Most importantly, try to understand *why* these formulas work a certain way. You should now have amassed enough information that some of these things are beginning to fit together. Review this recap material carefully and think about why all of these formulas are true. Remember, these formulas represent facts that people had to discover. They watched, guessed, and verified all of these formulas over long periods of time, until they really came to know how electricity operates. All of this took immense effort.

The Primaries

The three primary factors in the operation of electricity are voltage, current flow, and resistance. These are the fundamental forces that control every electrical circuit.

Voltage is the force that pushes the current through electrical circuits. The scientific name for voltage is *electromotive force* (*emf*). It is represented in formulas with the capital letter E and is measured in *volts*. The scientific definition of a volt is "the electromotive force necessary to force 1 ampere of current to flow through a resistance of 1 ohm."

In comparing electrical systems to water systems, voltage is comparable to water pressure. The more pressure there is, the faster the water will flow through the system. Likewise, with electricity, the higher the voltage (electrical pressure), the more current will flow through any electrical system.

Current (which is measured in *amperes*, or *amps* for short) is the rate of flow of electrical current. The scientific description for current is *intensity of current flow*. It is represented in formulas with the capital letter I. The scientific definition of an ampere is a flow of 6.25×10^{23} electrons (called 1 *coulomb*) per second.

I compares with the rate of flow in a water system, which is typically measured in gallons per minute. In simple terms, electricity is thought to be the flow of electrons through a conductor. Therefore, a circuit that has 9 amps flowing through it will have three times as many electrons flowing through it as does a circuit that has a current of 3 amps.

Resistance is the resistance to the flow of electricity. It is measured in ohms and is represented by the capital of the Greek letter omega Ω. The plastic covering of a typical electrical conductor has a very high resistance, whereas the copper conductor itself has a very low resistance. The scientific definition of an ohm is "the amount of resistance that will restrict 1 volt of potential to a current flow of 1 ampere."

In the example of the water system, you can compare resistance to the use of a very small pipe or a large pipe. If you have a water pressure on your system of 10 pounds per square inch, for example, you can expect that a large volume of water would flow through a 6-inch-diameter pipe. A much smaller amount of water would flow through a half-inch pipe, however. The half-inch pipe has a much higher resistance to the flow of water than does the 6-inch pipe.

Similarly, a circuit with a resistance of 10 ohms would let twice as much current flow as a circuit that has a resistance of 20 ohms. Likewise, a circuit with 4 ohms would allow only half as much current to flow as a circuit with a resistance of 2 ohms.

The term *resistance* is frequently used in a very general sense. Correctly, it is the direct current (DC) component of total resistance. The correct term for total resistance in alternating current (AC) circuits is *impedance*. Like DC resistance, impedance is measured in ohms, but is represented by the letter *Z*. Impedance includes not only DC resistance, but also *inductive reactance* and *capacitive reactance*. Both inductive reactance and capacitive reactance are also measured in ohms.

Ohm's Law

The three primaries have a relationship of one to another (more voltage, more current; less resistance, more current). These relationships are calculated by using Ohm's law.

Ohm's law states the relationships between voltage, current, and resistance. The law explains that in a DC circuit, current is directly proportional to voltage and inversely proportional to resistance. Accordingly, the amount of voltage is equal to the amount of current multiplied by the amount of resistance. Ohm's law goes on to say that current is equal to voltage divided by resistance and that resistance is equal to voltage divided by current.

These three formulas are shown in Figure 4-54, along with a diagram to help you remember Ohm's law. The Ohm's law circle can easily be used to obtain all three of these formulas.

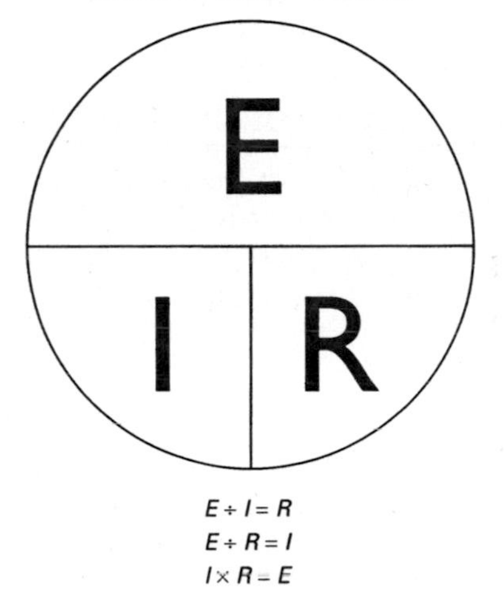

Figure 4-54 Ohm's law diagram and formulas.

The method is this. Place your finger over the value that you want to find (E for voltage, I for current, or R for resistance), and the other two values will make up the formula. For example, if you place your finger over the E in the circle, the remainder of the circle will show $I \times R$. If you then multiply the current by the resistance, you will get the value for voltage in the circuit. If you want to find the value for current, you will put your finger over the I in the circle, and then the remainder of the circle will show E/R. So, to find current, you divide voltage by resistance. Last, if you place your finger over the R in the circle, the remaining part of the circle shows E/I. Divide voltage by current to find the value for resistance. These formulas set up by Ohm's law apply to any electrical circuit, no matter how simple or how complex.

If there is one electrical formula to remember, it is certainly Ohm's law. The Ohm's law circle found in Figure 4-54 makes remembering the formula simple.

Watts

Another important electrical term is *watts*. A watt is the unit of electrical power, a measurement of the amount of work performed. For example, 1 horsepower equals 746 watts; 1 kilowatt (the measurement the power companies use on our bills) equals 1000 watts.

The most commonly used formula for power (or watts) is voltage times current ($E \times I$). For example, if a certain circuit has a voltage of 40 volts with 4 amps of current flowing through the circuit, the wattage of that circuit is 160 watts (40×4).

Figure 4-55 shows the Watt's law circle for figuring power, voltage, and current, which is similar to the Ohm's law circle that was used to calculate voltage, current, and resistance. For example, if you know that a certain appliance uses 200 watts and that it operates on 120 volts, you would find the formula P/E and calculate the current that flows through the appliance, which in this instance comes to 1.67 amps. In all, 12 formulas can be formed by combining Ohm's law and Watt's law. These are shown in Figure 4-56.

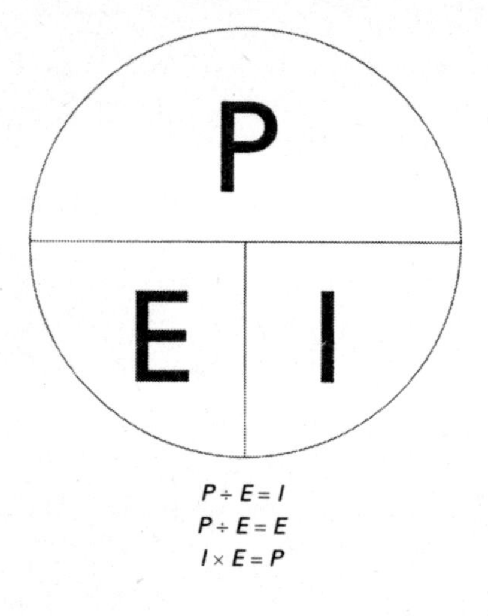

Figure 4-55 Watt's law circle.

Reactance

Reactance is the part of total resistance that appears in alternating-current circuits only. Like other types of resistance, it is measured in ohms. Reactance is represented by the letter X.

There are two types of reactance: inductive reactance and capacitive reactance. Inductive reactance is signified by X_L, and capacitive reactance is signified by X_C.

Inductive reactance (inductance) is the resistance to current flow in an AC circuit caused by the effects of inductors in the circuit. Inductors are coils of wire, especially those that are wound on an iron core. Transformers, motors, and fluorescent light ballasts are the most common types of inductors. The effect of inductance is to

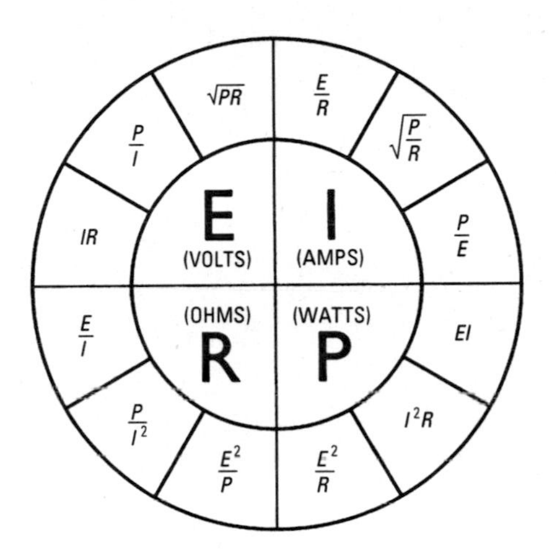

Figure 4-56 The 12 Watt's law formulas.

oppose a change in current in the circuit. Inductance tends to make the current lag behind the voltage in the circuit. In other words, when the voltage begins to rise in the circuit, the current does not begin to rise immediately, but lags behind the voltage a bit. The amount of lag depends on the amount of inductance in the circuit.

The formula for inductive reactance is as follows:

$$X_L = 2\pi FL$$

In this formula, F represents the *frequency* (measured in *hertz*) and L represents inductance, measured in *Henries*. You will notice that, according to this formula, the higher the frequency, the greater the inductive reactance. Accordingly, inductive reactance is much more of a problem at high frequencies than at the 60 Hz level.

In many ways, capacitive reactance (capacitance) is the opposite of inductive reactance. It is the resistance to current flow in an AC circuit caused by the effects of capacitors in the circuit. The unit for measuring capacitance is the *farad (F)*. Technically, one farad is the amount of capacitance that would allow you to store 1 *coulomb* (6.25×10^{23}) of electrons under a pressure of 1 volt. Because the storage of 1 coulomb under a pressure of 1 volt is a tremendous amount of capacitance, the capacitors you commonly use are rated in *microfarads* (millionths of a farad).

Capacitance tends to make current lead voltage in a circuit. Note that this is the opposite of inductance, which tends to make current lag. Capacitors are made of two conducting surfaces (generally some

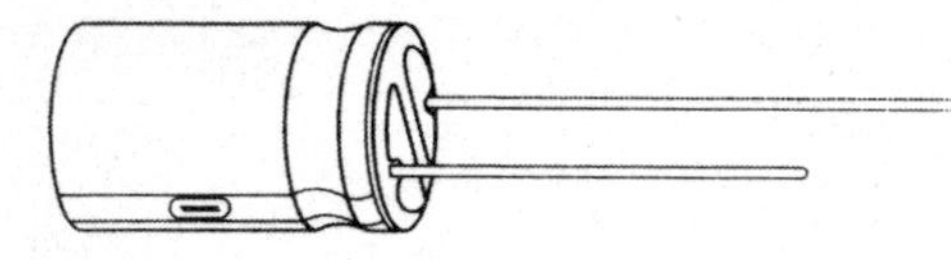

Figure 4-57 Capacitor.

type of metal plate or metal foil) that are just slightly separated from each other (see Figure 4-57). They are not electrically connected. Thus, capacitors can store electrons, but cannot allow them to flow from one plate to the other.

In a DC circuit, a capacitor gives almost the same effect as an open circuit. For the first fraction of a second, the capacitor will store electrons, allowing a small current to flow. But after the capacitor is full, no further current can flow because the circuit is incomplete. If the same capacitor is used in an AC circuit, though, it will store electrons for part of the first alternation, and then release its electrons and store others when the current reverses direction. Because of this, a capacitor (even though it physically interrupts a circuit) can store enough electrons to keep current moving in the circuit. It acts as a sort of storage buffer in the circuit.

In the following formula for capacitive reactance, F is frequency and C is capacitance, measured in farads.

$$X_C = \frac{1}{2\pi FC}$$

Impedance

As explained earlier, impedance is very similar to resistance at lower frequencies and is measured in ohms. Impedance is the total resistance in an alternating-current circuit. An alternating-current circuit contains normal resistance, but may also contain certain other types of resistance called *reactance*, which are found only in AC circuits. This reactance comes mainly from the use of magnetic coils (*inductive reactance*) and from the use of capacitors (*capacitive reactance*). The general formula for impedance is as follows:

$$Z = \sqrt{R^2 + (X_L - X_C)^2}$$

This formula applies to all circuits, but specifically to those in which DC resistance, capacitance, and inductance are present.

The general formula for impedance when only DC resistance and inductance are present is this:

$$Z = \sqrt{R^2 + X_L{}^2}$$

The general formula for impedance when only DC resistance and capacitance are present is this:

$$Z = \sqrt{R^2 + X_C{}^2}$$

Series Circuits

The simplest circuits are series circuits—circuits that have only one path in which current can flow, as shown in Figure 4-58. Notice that all of the components in this circuit are connected end-to-end in a series.

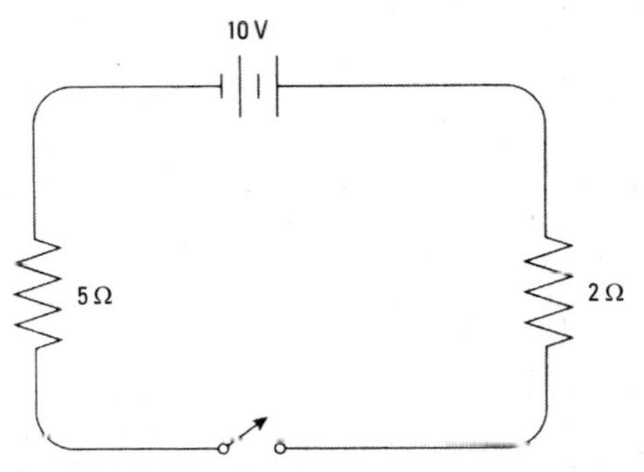

Figure 4-58 Series circuit.

Voltage

The most important and basic law of series circuits is *Kirchhoff's law.* It states that the sum of all voltages in a series circuit equals zero. This means that the voltage of a source will be equal to the total of voltage drops (which are of opposite polarity) in the circuit. In simple and practical terms, the sum of voltage drops in the circuit will always equal the voltage of the source.

Current

The second law for series circuits is really just common sense—that the current is the same in all parts of the circuit. If the circuit has only one path, what flows through one part will flow through all parts.

Resistance

In series circuits, DC resistances are additive, as shown in Figure 4-59. The formula is this:

$$R_T = R_1 + R_2 + R_3 + R_4 + R_5$$

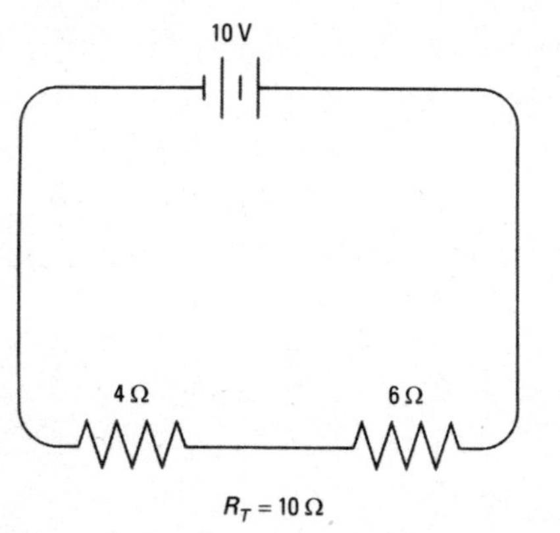

Figure 4-59 DC resistances in a series circuit.

Capacitive Reactance

To calculate the value of capacitive reactance for capacitors connected in series, use the product-over-sum method (for two capacitances only) or the reciprocal-of-the-reciprocals method (for any number of capacitances). The formula for the *product-over-sum method* is as follows:

$$X_T = \frac{X_1 \times X_2}{X_1 + X_2}$$

The formula for the *reciprocal-of-the-reciprocals method* is this:

$$X_T = \frac{1}{\frac{1}{X_1} + \frac{1}{X_2} + \frac{1}{X_3} + \frac{1}{X_4} + \frac{1}{X_5}}$$

Inductive Reactance

In series circuits, inductive reactance is additive. Thus, in a series circuit, the following formula is used:

$$X_T = X_1 + X_2 + X_3 + X_4 + X_5$$

Parallel Circuits

A parallel circuit is one that has more than one path through which current will flow. A typical parallel circuit is shown in Figure 4-60.

Voltage

In parallel circuits with only one power source (as shown in Figure 4-60), the voltage is the same in every branch of the circuit.

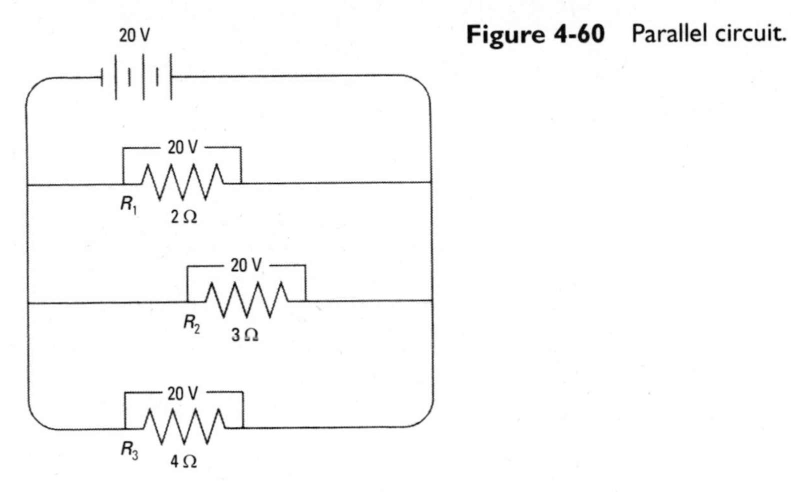

Figure 4-60 Parallel circuit.

Current

In parallel circuits, the amperage (level of current flow) in the branches adds to equal the total current level seen by the power source. Figure 4-61 shows this in diagrammatic form.

Resistance

In parallel circuits, resistance is calculated by either the product-over-sum method (for two resistances):

$$R_T = \frac{R_1 \times R_2}{R_1 + R_2}$$

or by the reciprocal-of-the-reciprocals method (for any number of resistances):

$$R_T = \frac{1}{\frac{1}{\frac{1}{R_1}} + \frac{1}{\frac{1}{R_2}} + \frac{1}{\frac{1}{R_3}} + \frac{1}{\frac{1}{R_4}} + \frac{1}{\frac{1}{R_5}}}$$

If the circuit has only branches with equal resistances, the following formula is used:

$$R_T = R_{\text{branch}} \div \text{number of equal branches}$$

The result of these calculations is that the resistance of a parallel circuit is always less than the resistance of any one branch.

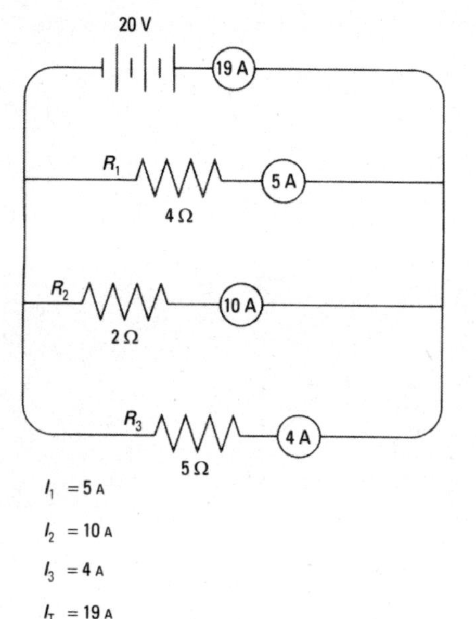

Figure 4-61 Parallel circuit, showing current values.

Capacitive Reactance

In series circuits, capacitances are additive. For an example, refer to Figure 4-62. Notice that each branch has a capacitance of 100 microfarad (μfd, meaning *micro*). If the circuit has four branches, each of 100 μfd, the total capacitance is 400 μfd.

Inductive Reactance

In parallel circuits, inductances are calculated by the product-over-sum or the reciprocal-of-the-reciprocals methods.

Series-Parallel Circuits

Circuits that combine both series and parallel paths are obviously more complex than either series or parallel circuits. In general, the rules for series circuits apply to the parts of these circuits that are in series; the parallel rules apply to the parts of the circuits that are in parallel.

A few clarifications follow.

Voltage

Although all branches of a parallel circuit are exposed to the same source voltage, the voltage drops in each branch will always equal the source voltage (see Figure 4-63).

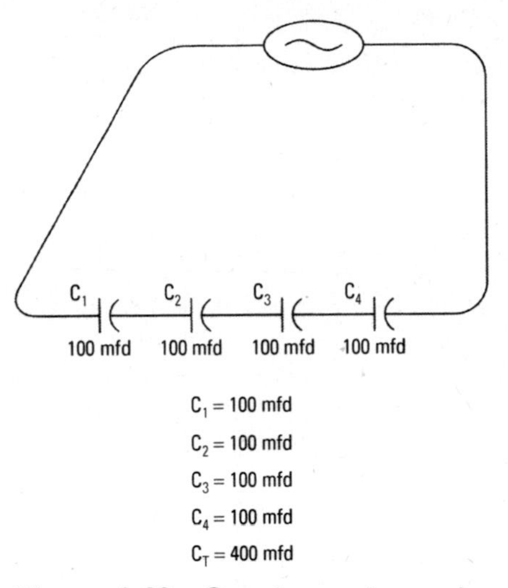

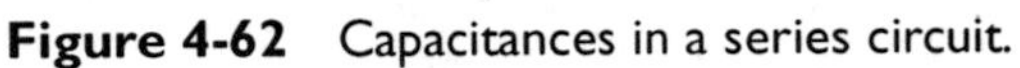

Figure 4-62 Capacitances in a series circuit.

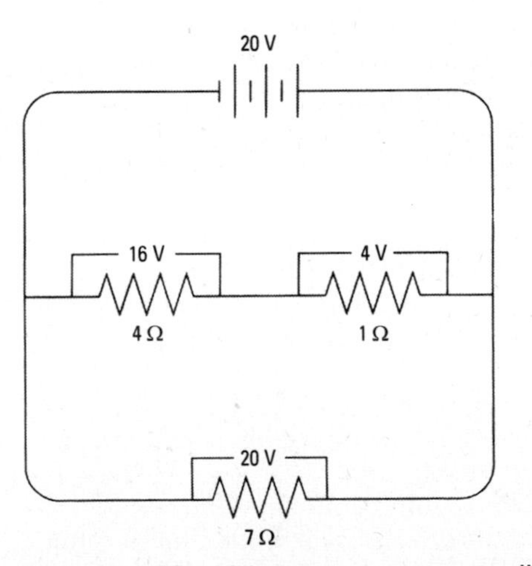

Figure 4-63 Voltages in a series-parallel circuit.

Current

Current is uniform within each series branch, whereas the total of all branches equals the total current of the source.

Resistance

Resistance is additive in the series branches, with the total resistance less than that of any one branch.

Capacitive Reactance

X_C is calculated by the reciprocal-of-the-reciprocal method within a series branch, and the total X_C of the branches is additive.

Inductive Reactance

X_L is additive in the series branches, with the total inductive reactance less than that of any one branch.

Formulas

The following formulas are important to keep in mind:

$$E_{avg} = 4\varphi nT/10^8$$

$$E_{avg} = 4.4\varphi nT/10^8$$

Summary

An alternating current (AC) is usually defined as a current that charges in strength according to a sine curve. An alternating voltage reverses its polarity on each alternation, and an alternating current reverses its direction of flow on each alternation.

The frequency of an AC voltage of current is its number of cycles per second. For example, electricity supplied by public utility companies in the United States has a frequency of 60 cycles per second. Cycles per second (cps) are also called hertz (Hz).

An instantaneous voltage is the value of an AC voltage at a particular instant. The electrical industry generally considers that the most important instantaneous AC voltages are the maximum (peak) voltage, and the effective voltage. In most practical electrical work with AC circuits, we use effective values of voltage and current in Ohm's law. The only thing that we must watch for is to use the same kind of values in Ohm's law. We would get incorrect answers if we used a peak voltage value and an effective current value in Ohm's law.

Induction is a magnetic effect. A magnetic field forms around a conductor any time a current flows through it and a magnetic field moving through a conductor causes a current. With alternating current, the magnetic fields around wires never stop moving, which gives rise to many circuit effects that we call induction. Inductive AC circuits are used in many, many types of equipment. For example, induction coils are used in theaters and motion-picture studios for lighting control equipment.

Inductance opposes any change in current flow. If we increase the voltage across an inductor, an increase in current flow builds up gradually because of the self-induced voltage that opposes current change. On the other hand, if we decrease the voltage across an inductor, the current flow falls gradually because of the self-induced voltage. In other words, the most noticeable difference between a resistive circuit and an inductive circuit is in their response speeds. The current flow in a resistive circuit changes immediately if the applied voltage is changed. On the other hand, the current flow in an inductive circuit is delayed with respect to a change in applied voltage.

The power in a DC resistive circuit is equal to volts multiplied by amperes. This is also true in an AC resistive circuit. On the other hand, in a circuit that has inductance only, the true power is zero. The current flow through an inductor is delayed when voltage is applied due to the self-induced voltage of the inductor, which opposes the applied voltage. If we multiply instantaneous voltages by instantaneous currents, we will find the instantaneous power. The average power in an inductor is half positive and half negative; no power is consumed by the inductor. So, we can simply say that an inductor takes power from the source during the positive alteration and returns this power to the source during the negative alternation.

We know that in a circuit containing only resistance, the current and voltage are in phase. Therefore, there is true power in the resistor, which appears as heat. True power is also called real power.

The total opposition to AC current flow is called impedance. Impedance is measured in ohms, just as resistance and reactance are measured in ohms. If an AC circuit contains reactance only, the circuit impedance is the same as the circuit reactance. Or, if an AC circuit contains resistance only, the circuit impedance is the same as the circuit resistance.

When inductance and resistance are connected in series with an AC source, there are three power values to be considered: a true power in watts, a reactive power in vars, and an apparent power in volt-amperes. Only the true power does useful work in the circuit. The reactive power merely surges back and forth in the circuit. The apparent power provides reactive power and true power.

Electricians often work with capacitive reactance, as well as with inductive reactance. Capacitance is measured in farads with a capacitor tester. A 1-farad capacitor will store 1 coulomb of charge when 1 volt is applied across the capacitor. A coulomb is the amount of charge produced by 1 ampere flowing for 1 second. Since the farad is a very large unit of capacitance, we generally work with microfarads; a microfarad is one millionth of a farad (10^{-6} farad). The symbol μ or the letter *m* is used to represent micro.

Many types of circuits have inductance, capacitance, and resistance connected in series with an AC voltage source. Circuits that consist of inductance and resistance connected in parallel can be especially useful. You will also see circuits that consist of capacitance and resistance connected in parallel. There are many circuits as well that contain resistance, capacitance, and inductance in parallel.

Electronic devices work with electromagnetic induction in a myriad of applications. Various devices operate on the principle of electromagnetic induction, and circuits often have responses based on this principle.

This chapter concludes with a recap of principles and formulas presented in the first few chapters of this book.

Review Questions

1. What is meant by the frequency of an AC voltage.
2. What does effective voltage mean?
3. What is the numerical difference between an effective voltage and a peak voltage?
4. What is a vector? Why is it helpful to represent AC voltages and currents by vectors?
5. Show AC voltage sources connected in series-aiding.
6. Show AC voltage sources connected in series-opposing.
7. How can the brightness of a lamp be controlled with a variable inductance?
8. State Ohm's law for a purely inductive circuit.
9. In what way does true power differ from reactive power?
10. How is impedance defined?
11. Define a henry.
12. Define a farad.
13. Explain the principle of transformer action.
14. What is inrush current?
15. What is the principle of electromagnetic induction?

16. Explain Lenz' law.
17. If 1 volt is induced in a moving conductor, how many magnetic lines of force does the conductor cut each second?
18. What is meant by the self-induction of a coil?
19. Explain how a capacitor stores electric charge when it is connected to a battery.
20. Define counter electromotive force (cemf).
21. Reactance appears only in ________ circuits.
22. Name the two types of reactance.
23. What is the term for total resistance in an alternating current circuit?
24. What kind of power is measured in watts?
25. What kind of power is measured in volt-amperes?

Exercises

1. Sketch an induction coil and explain its operation. Label the parts and show interaction.
2. Draw a parallel circuit with inductive reactance, capacitive reactance, and resistance. Show all values.
3. Draw a circuit with a 100-volt DC voltage source, a resistor, an inductor, and a capacitor. Then, draw the same circuit with a 100 volt, AC, 400 Hz voltage source. Show all values for both circuits.

Chapter 5

Resonance

Resonance is the condition that occurs when the inductive reactance and capacitive reactance in a circuit are equal. When this happens, the two reactances cancel each other out, leaving the circuit with no resistance except for whatever DC resistance exists in the circuit.

Remember that both induction and capacitance are dependent upon frequency. If the frequency of a circuit goes up or down, inductance and capacitance change. So, more or less inductive reactance and capacitive reactance are present in the circuit at different frequencies. With careful design, we can arrange the circuit so that the inductive and capactive reactance are equal at a certain frequency, and cancel each other.

Figure 5-1 shows a graph of impedance in a specific circuit. Notice that the impedance varies with frequency. At 0.4 kHz Impedance is 400 ohms. It is also 400 ohms at about 1.6 kHz. However, at 1 kHz, it dips almost to zero. It is at this frequency that inductive and capacitive reactance are canceling out each other.

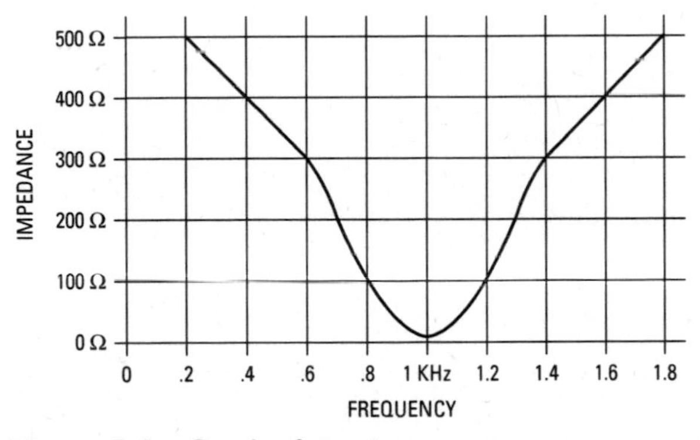

Figure 5-1 Graph of circuit resonance.

Resonance is commonly used for filter circuits or for tuned circuits. By designing a circuit that will be resonant at a certain frequency, only the current of that frequency will flow freely in the circuit. Currents of all other frequencies will be subjected to much higher impedances and will thus be greatly reduced or essentially eliminated. This is how a radio receiver can tune in one station at a time. The capacitance or inductance is adjusted until the circuit is resonant at the desired frequency. Thus, the desired frequency flows through the circuit and all others are shunned. Parallel resonances occur at the same frequencies and values as do series resonances.

In the following formula for resonances, F_R is the frequency of resonance, L is inductance measured in henrys (H), and C is capacitance measured in farads.

$$X_C = \frac{1}{2\pi FC}$$

To fully understand resonance, it is necessary to understand RC circuits (circuits with resistance and capacitance) and RCL circuits (circuits with resistance, capacitance, and inductance). We will go through examples of working with these circuits in some detail.

RC Circuits

Figure 5-2 shows a 100-ohm noninductive resistor in series with a 4 μF capacitor, connected to a 130 Hz AC source of 2000 volts.

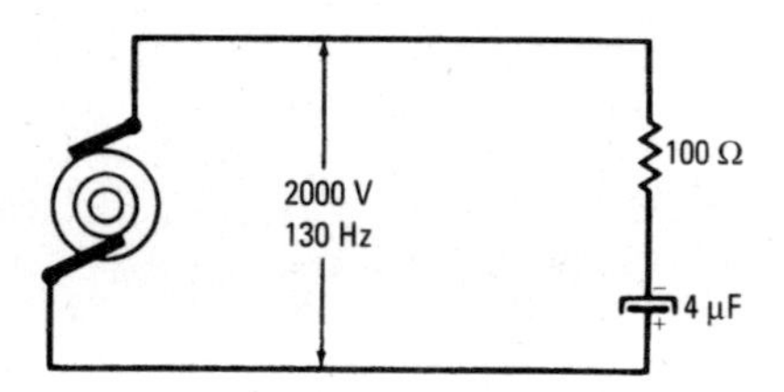

Figure 5-2 Resistance and capacitance in series.

Let us find the following:

1. Capacitive reactance in the circuit
2. Total impedance of the load
3. Current in the circuit
4. Power factor of the circuit
5. Angle of lead in degrees

$$X_C = \frac{1}{2\pi f C} = \frac{1}{6.28 \times 130 \times 0.0000004} = 322 \text{ ohms}$$

$$Z = \sqrt{R^2 + X_C^2} = \sqrt{100^2 + 306^2} = 322 \text{ ohms}$$

$$I = E/Z - 200/322 = 6.21 \text{ amperes}$$

$$\cos\varphi = R/Z = 100/322 = 0.311$$

$$\text{PF} = \cos\varphi \times 100 = 0.311 \times 100 = 31.1 \text{ percent}$$

From the table of cosines that appears in Appendix A, we find the following:

$$0.311 = 72 \text{ degrees}$$

This may be plotted vectorially, as shown in Figure 5-3. Draw a line, AB, 100 units long for R. Then at 90 degrees from AB, draw BC 306 units long for X_C. Connect A and C by a line, measure the units of length in this line, and it will be found to be 322 units in length, or 322 ohms impedance.

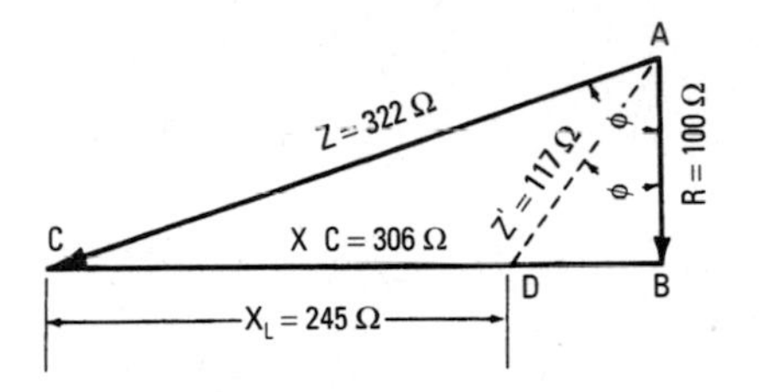

Figure 5-3 Vector value of R; X_C and X_L in series as shown in Figure 5-2.

With a protractor, measure angle N and it will be found to be 72 degrees. Since this is a capacitive circuit, the power factor will be leading, and $PF = (R/Z) \times 100 = (100/322) \times 100 = 31.1$ percent.

RCL Circuit

Next, we will add a coil into the circuit, as shown in Figure 5-4. This coil is 0.3 H.

Let us find the following:

1. The inductive reactance of the coil
2. The total impedance of the circuit

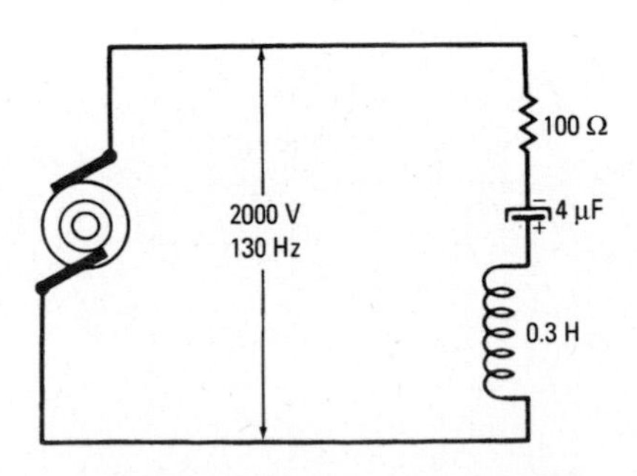

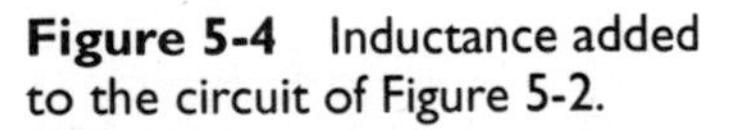
Figure 5-4 Inductance added to the circuit of Figure 5-2.

3. The current in the circuit
4. The power factor
5. The angle N in degrees.

$$X_L = 6.28fL = 6.28 \times 130 \times 0.3 = 245 \text{ ohms}$$

$$Z = \sqrt{R^2 + (X_C - X_L)^2} = \sqrt{100^2 + (306 - 245)^2}$$

$$= \sqrt{100^2 + 61^2} = 117 \text{ ohms}$$

$$I = E/Z = 2000/117 = 17.09 \text{ amperes}$$

$$\cos\varphi = R/Z = 100/117 = 0.855$$

$$PF = \cos\varphi \times 100 = 0.855 \times 100 = 85.5 \text{ percent}$$

Cosine φ is 0.855, and (from the table of cosines shown in Appendix A) is 31 degrees.

The capacitive reactance is greater than the inductive reactance, so the power factor is leading.

Now, in Figure 5-4 measure *CD* up from *C* as 245 units of length (inductive reactance), which must be subtracted from the capacitive reactance, *BC*. This leaves the effective capacitive reactance as *BD*, or 61 ohms X_C. Angle N can be measured by a protractor as 31 degrees leading.

No circuit actually can have inductance without some resistance. So there are core losses and copper losses. With a capacitor, there is also a very slight loss involved. The two problems just covered, therefore, did not include these losses. Figure 5-5 shows the values for a problem covering such losses.

Find the following:

1. Total current
2. PF of the entire circuit

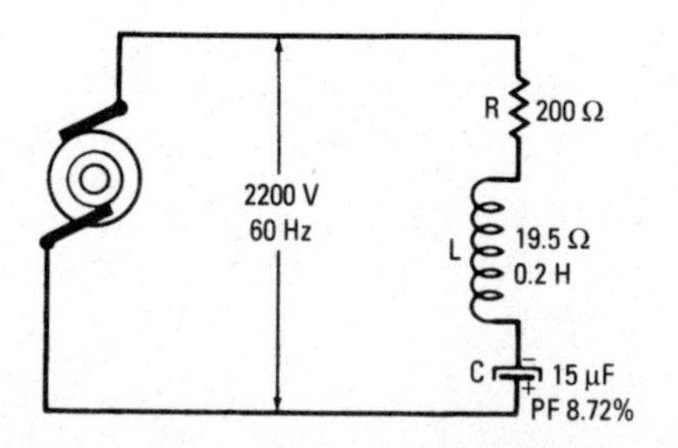

Figure 5-5 Resistance, inductive reactance, and capacitive reactance in series.

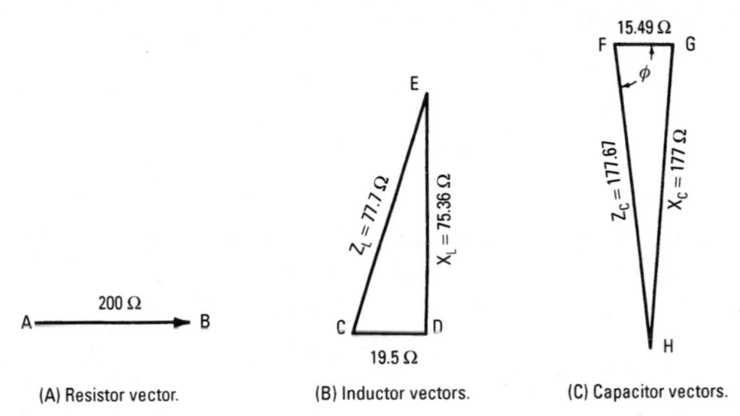

(A) Resistor vector. (B) Inductor vectors. (C) Capacitor vectors.

Figure 5-6 Vector diagram for impedance of R, L, and C of Figure 5-4.

Figure 5-6 will be used in conjunction with this problem. Line AB, 200 units long, represents the resistance of R. The inductive reactance of coil L is $X_L = 6.28 \times 60 \times 0.2 = 75.36$ ohms. The impedance of coil L is calculated as follows:

$$Z_L = \sqrt{R_L^2 + X_L^2} = \sqrt{19.5^2 + 75.36^2} = 77.8 \text{ ohms}$$

The impedance vector for inductor L is given by CE. The capacitive reactance of C is calculated as follows:

$$X_C = \frac{1}{6.28fC} = \frac{1}{6.28 \times 60 \times 0.000015} = 177 \text{ ohms}$$

There is no ohmic resistance in a capacitor, so the energy component of a capacitor cannot be expressed as such. It is, however, possible to express the energy component in equivalent ohms of the total impedance of the capacitor. To do this, the power factor of the capacitor must be obtained. This may be done with a voltmeter, ammeter, and a wattmeter:

$$W/VA = \cos \text{N}$$

$$\cos \text{N} \times 100 = \text{PF}$$

From these figures, vector diagram FGH may be constructed. Line GH was found to be 177 ohms. If the power factor is known, the energy component, FG, and the total impedance may be found, as follows:

$$FH = GH/\sin \text{N}_2$$

Since neither angle N_2 or its sine are known, this cannot be found directly. The power factor of the capacitor is cos N_C, so if PF = 0.0872 = cos N_C, the sine of N_C may be found by using the cosine table in Appendix A. Then, applying the previous formula, $FH = GH/\sin N_C = 177/0.9962 = 177.67$ or Z_C. $Z_C \cos N_C = 177.67$ $0.0872 = 15.49$, which is line *FG* or the energy component of the capacitor.

To get the combined impedance of these three devices, use the following equation:

$$R = 200 + 15.49 + 19.5 = 234.99 \text{ ohms}$$

$$Z = \sqrt{R^2 + (X_C - X_L)^2} = \sqrt{234.99^2 + (177 - 75.36)^2}$$
$$= 256.22 \text{ ohms}$$

The current is $I = E/Z = 2200/256.2 = 8.5$ amperes

$$PF = (R/Z) \times 100 = 234.99/256.22 \times 100 = 91.7 \text{ percent}$$

Figure 5-7 shows the combined vectors displayed in Figure 5-6 for the impedance of *R*, *L*, and *C* in Figure 5-5. With vectors, the

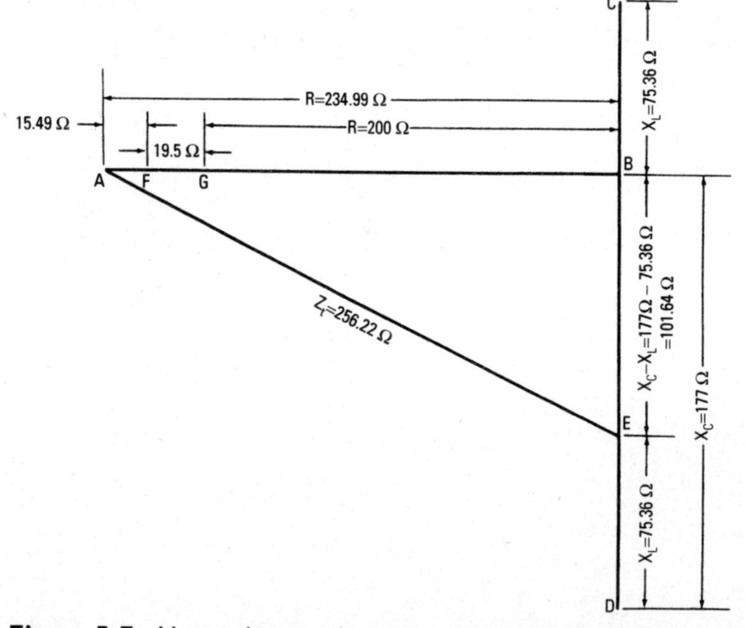

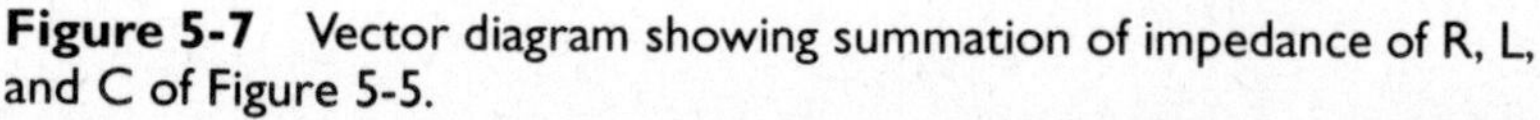

Figure 5-7 Vector diagram showing summation of impedance of R, L, and C of Figure 5-5.

wattless component of capacitors is vertical and below the true component (that is, *BD* is the wattless component of the capacitor). Line *AF* is the true-power component of the capacitor, *FG* the true component of the inductance, and *GB* is the resistance of *R*. The wattless component of inductance is vertical (*BC*) and above the true component, *FG*.

The common expressions wattless power or wattless current are used with AC. There is actually no wattless current, because wherever a current flows, there must be an expenditure of energy. These currents are out of phase with the applied voltage, and the consideration of power involved is based on energy components and wattless components of these currents. Figure 5-8 illustrates the relationship of the three components in an inductive circuit.

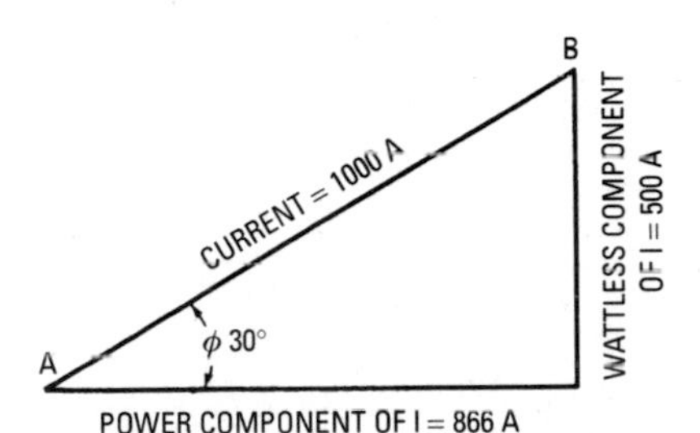

Figure 5-8 Vector diagram showing typical relationships of current components in an AC circuit.

Taking the values of *I* in Figure 5-8, and using 1000 volts, the line *AB* (1000 A) times 1000 volts equals 1,000,000 volt-amperes apparent power, or 1000 kilo-volt-amperes (kVA). *BC* (500 A) times 1000 volts equals 500,000 volt-amperes as the wattless power component, and *AC* (866A) times 1000 volts equals 866,000 watts or 866 kilowatts (kW) of true power.

This tells us that the vector diagram in Figure 5-8 is an inductive circuit with an angle lag of 30 degrees or a power factor of 86.6 percent. Thus, the circuit must be designed for 1000 amperes because of the 86.6 percent PF, while 866 amperes would be all the actual current that would be required to do the work if we had unity (100 percent) power factor.

Parallel LC Circuit

Inductance and capacitance are not often found in series; they are usually in parallel. There is a capacitance between conductors of circuits (the conductors are the plates and the insulation, and air is the dielectric).

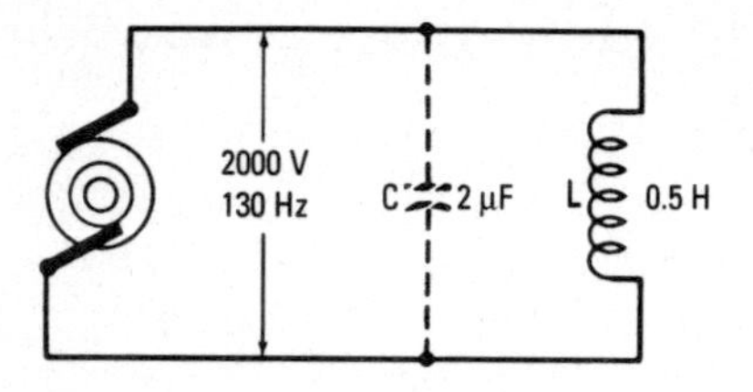

Figure 5-9 Capacitive effect of conductors.

As a practical example, consider the circuit shown in Figure 5-9. The system has a 2 μF capacitor and the load is 0.5 henry, voltage 2000 volts, and frequency 130 hertz.

$$X_C = \frac{1}{6.28\,fC} = \frac{1}{6.28 \times 130 \times 0.000002} = 613 \text{ ohms}$$

The energy components will be neglected in this example, so C is all X_C or Z_C, and L is all X_L or Z_L. Thus, we can deduce the following:

$I_C = E/Z_C = 2000/613 = 3.26$ amperes in C (leading)

$I_L = E/X_C = 2000/408 = 4.9$ amperes in L (lagging)

I_C is thus out of phase 90 degrees so is all wattless component

I_L is also out of phase 90 degrees so is likewise wattless component

$I_L - I_C = 4.9 - 3.26 = 1.64$ amperes is the total current supplied by the alternator. Now, $Z = E/I = 2000/1.64 = 1219$ ohms combined impedance.

Since we have neglected the energy losses, the power factor is zero.

The current required is considerably less than that required by either device, because C and L are diametrically opposed to each other. The capacitor acts as sort of a generator, storing energy in one alternation and releasing it in the next alternation.

Series Resonant Circuits

With capacitance and inductance in series, X_C and X_L are in series; and as $X_L = 6.28fL$ approaches the value of $X_C = 1/6.28fC$, the line voltage and current will increase. When X_C and X_L are in parallel (as shown in Figure 5-10) the line current diminishes as X_C approaches X_L.

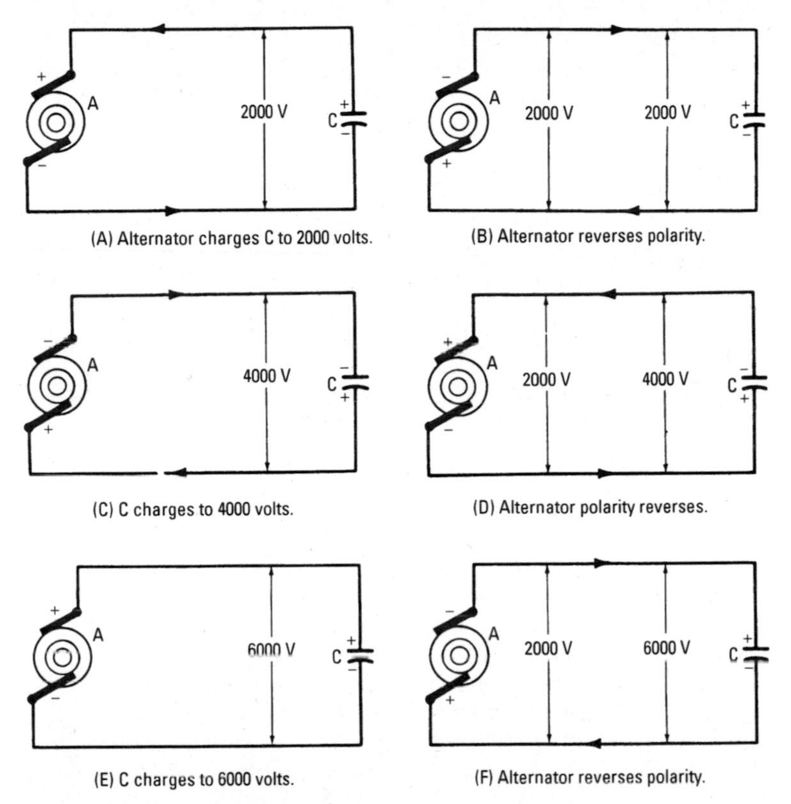

(A) Alternator charges C to 2000 volts.

(B) Alternator reverses polarity.

(C) C charges to 4000 volts.

(D) Alternator polarity reverses.

(E) C charges to 6000 volts.

(F) Alternator reverses polarity.

Figure 5-10 Voltage cumulation.

Circuits possessing X_C and X_L result in resonance. Resonance cannot exist without inductance and capacitance. Perfect resonance results when $X_C = X_L$.

To illustrate, the alternators in Figure 5-10 and the capacitors are such that voltage builds up as it does when X_C and X_L balance. In Figure 5-10A alternator *A* charges C to 2000 volts. In Figure 5-10B, the alternator has reversed polarity. Capacitor *C*, in series with *A*, sends its emf back so that at the end of that alternation the two voltages have added, and C is now charged with 4000 volts, in Figure 5-10C. In Figure 5-10D the polarity of the alternator reverses, and the action in Figure 5-10B is repeated, except the capacitor receives a charge of 6000 volts, and so on. (In reality, this circuit would have to be modified to behave as shown in Figure 5-10.)

Resonance occurs at a certain frequency. Should the frequency be increased, the value of X_L will rise. At the same time that an increase in frequency increases X_L, however, it reduces X_C. Where resonance exists, altering the frequency will check it. The smaller the resistance in a series circuit, the greater is the local voltage set up across the capacitor and inductance.

Voltage at resonance is very dangerous and undesirable, because of the damage and possible breakdown of insulation caused by the increase in voltage. Resonance in power systems must be avoided by proper designs.

Current resonance is highly desirable because it relieves the alternator of the necessity of furnishing the wattless components of the current and tends to correct the power factor toward unity, or 100 percent.

Parallel Resonant Circuits

Parallel resonant circuits (which you may hear called *tank circuits*) are interesting in that they experience what is called a *flywheel effect.* Flywheels are used in many applications to store kinetic energy and to keep something moving long after the primary source of motion is removed. A flywheel on a motor, for example, can keep the motor turning for a long time after power has been disconnected.

Figure 5-11A shows a parallel resonance circuit. Figure 5-11B shows the same circuit with the power source removed. Imagine for a moment that we turn on the power source long enough to charge the capacitor and set up a magnetic field around the inductor.

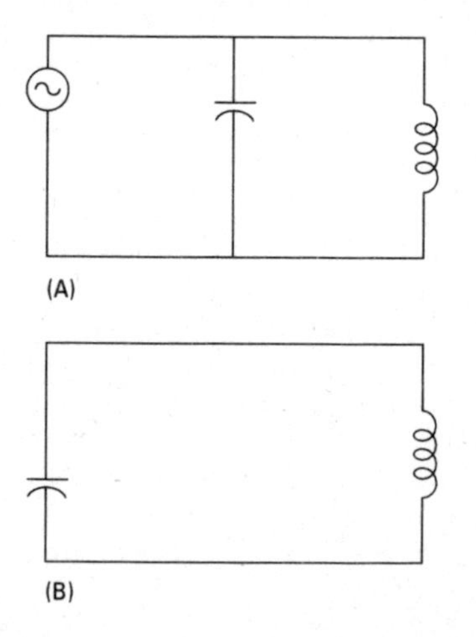

Figure 5-11 Parallel resonance ciruit.

Then we remove it from the circuit. What happens next is a long-term oscillation of current back and forth through this circuit.

At first, current will remain flowing because of the effect of the inductor, then the magnetic field around the inductor will collapse and the capacitor will begin to discharge. This will send current moving in the opposite direction until the other side of the capacitor is charged and the reversed magnetic field collapses. Then current will run in the original direction again.

Depending on the amount of resistance in the circuit, this oscillation of current can last for a considerable length of time. This is especially true at resonant frequencies, when circuit impedance is very low. (There is never zero impedance, simply because wires and connections have a small amount of resistance.) Figure 5-12 shows the oscillation currents that would result from the situation just described. This is called a *damped oscillation.*

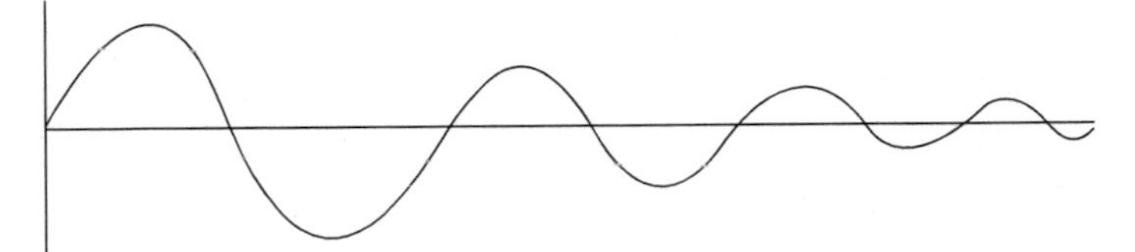

Figure 5-12 Damped oscillation.

Uses of Series and Parallel Resonant Circuits

With their differing characteristics, series resonant circuits and parallel resonant circuits are used in quite different applications. The central characteristics of these circuits are the following:

- Series circuits present a high impedance to all frequencies but the resonant frequency.
- Parallel circuits present a high impedance only to its resonant frequency.

So, series circuits are used when we want to accept one frequency only, and parallel circuits are used when we want to exclude one frequency only. In both cases, the formula for calculating F_R is the same.

Figure 5-13 shows one way this is done. The parallel resonant circuit shown in Figure 5-13A will present a high resistance to one frequency (or small range of frequencies) only. So, to all other frequencies but the F_R, the circuit looks like Figure 5-13B (a direct, nearly zero-impedance path to ground). In this case, almost no

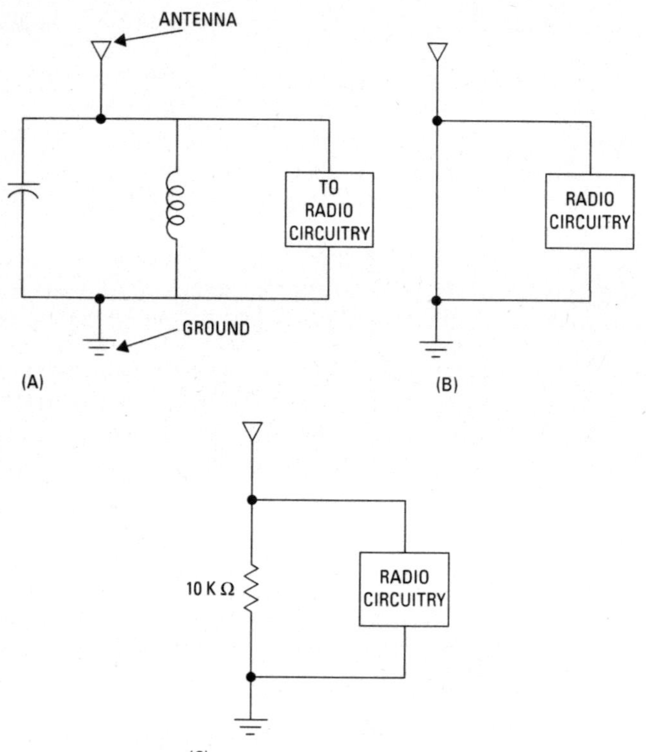

Figure 5-13 Parallel resonant circuit.

current will flow through the higher-impedance radio circuitry. However, to the resonant frequency, the circuit looks like Figure 5-13C (the path to ground having a very high impedance). In this case, the available voltage will push most of the current through the radio circuitry, rather than sending current through a much higher impedance to ground.

You can see from this example how we can use a parallel resonant frequency to separate out one frequency for a specific use. Actually, this is exactly how radio tuners work. If you take apart an older style of radio tuner, you will find a variable capacitor in it, which changes the resonant frequency of the parallel tuning circuit to separate out the frequency of each specific radio station.

This is an example of a basic filter circuit. We will be covering filters in more depth in Chapter 10, but you can see from this discussion how filters operate. Note that is not just the basic resonant circuit that can be useful, but the relation of other circuit elements to the basic characteristics of resonance. In the previous example, we acknowledged the fact that resonance shows a high voltage to F_R and used that phenomenum to shunt off circuit voltage in a specific way.

Series resonant circuits present a low resistance to the resonant frequency and allow for higher voltages to form at each circuit element. Parallel resonant circuits present a high resistance to the resonant frequency and allow for higher currents to flow through each circuit element.

Summary

Resonance is a characteristic of AC circuits that allows us to isolate specific frequencies. This is accomplished by balancing inductive and capacitive reactance. Since the levels of both types of reactance are determined by frequency, we can modify them purposefully, and to attain specific situations.

Since resonance is determined by balancing inductance and capacitance, they are generally identified as RCL circuits, with the initials representing resistance, capacitance, and inductance. It is by balancing these that we can obtain the resonance characteristics we want.

Parallel circuits are series circuits that reach resonance at the same values of reactance but behave differently. Series circuits present a high impedance to all frequencies *except* the resonant frequency. Parallel circuits present a high impedance *only* to its resonant frequency.

In the next chapter, we will examine the various types of semiconductors.

Review Questions

1. A coil has 5 ohms resistance and 7.5 ohms inductive reactance. A capacitor with capacitive reactance of 20 ohms and a power factor of 0.06 is connected in parallel with the coil on a 220-volt circuit.

(a) What is their combined impedance?

(b) What is the current in each device?

(c) What is the total current?

(d) What is the overall power factor?

2. What causes resonance?

3. With capacitive reactance and inductive reactance in parallel, what happens to the line current as X_C approaches X_L?
4. With X_L and X_C in series, what happens to the line current and voltage as resonance is approached?
5. Does the frequency affect resonance?
6. Is current resonance desirable?
7. What effect does current resonance have on power factor?
8. The cosine of N is 0.866. What is the sine of N?
9. Which type of resonant circuit presents a high impedance to the resonant frequency?

Exercises

1. Construct a circuit that is resonant at 4 kHz. Show calculations to prove that you are correct.

Chapter 6

Semiconductors

Semiconductor substances can be made into amazingly useful devices. The name *semi*conductor indicates that these materials are partway between being insulators and conductors. In fact, semiconductors are composed of elements (silicon and germanium) that have four electrons in the outer electron shell, out of a maximum of eight.

This chapter discusses the basic definitions and operations regarding semiconductors. It covers a few semiconductor devices for the sake of illustration, but not many. The key thing to gain from this chapter is to understand what semiconductors are, and how they operate. This will not be a long chapter, but it will be a critical one. Go slowly and make sure you understand it well.

Doping and Molecular Structure

The raw silicon and germanium used to create semiconductors are by themselves not that useful. It is when we modify these elements that the really interesting things start happening.

To make semiconductors do useful jobs, we modify them slightly. In the trade, this is called *doping* the silicon or germanium. And by modify I mean that we add a small amount of other elements (called *dopants*) to the silicon or germanium.

In specific, we add other substances that have either three or five electrons in their outer electron shells. By mixing the materials this way, we create semiconductors that are a little more positive than the original silicon, or a little more negative that the original silicon.

If we take the basic four-electron silicon (or germanium) and add some three-electron substance (such as indium, boron, or gallium), we obtain a compound material that is slightly more positive (lacking in electrons) than the original silicon.

If we take the basic four-electron silicon (or germanium) and add some five-electron substance (such as arsenic, antimony, or bismuth), we obtain a compound material that is slightly more negative (having extra electrons) than the original silicon.

Where this gets very interesting is when you consider that these materials are essentially crystalline in structure. That is, they have a very tight, rigid, and regular molecular structure, as is shown in Figure 6-1. Because of this rigid structure, the doping atoms and their electrons are held in place. This means that the positive material and negative materials stay where they are, and the differences in potential from this electron imbalance don't even themselves out very much.

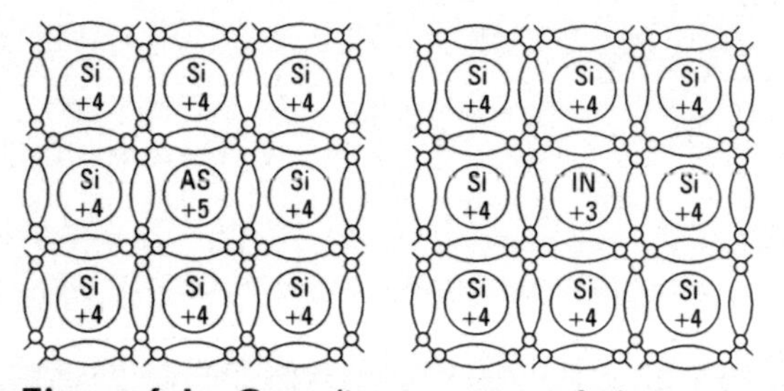

Figure 6-1 Crystaline structure of silicon, showing dopants.

P and N Types

As you might infer from the previous section, the actual semiconductor substances we use in circuits are called *N-Type* semiconductors and *P-Type* semiconductors. N-types would be those that are doped to have an excess of electrons (such as when you add some arsenic to silicon). P-types would be those that are doped to have a deficit of electrons (such as when you add some indium to silicon).

Note again that, while I am using silicon in these explanations, germanium would work equally well. In fact, both silicon and germanium are very commonly used in semiconductor devices.

Also note that the amount of dopant added to the base silicon or germanium is very small—on the order of 0.00001 percent. But this is enough to accomplish our goals.

Vacuum Tube Devices

The original electronic devices were vacuum tubes. Even though tubes are very seldom used today, it is worth spending a few paragraphs explaining their operation. Understanding how tubes work makes understanding semiconductor operation easier.

Most of the things we now do with semiconductors were possible with vacuum tubes, long before anyone had heard of semiconductors. Without vacuum tubes, radio, television, X-rays, and a host of other things would have been impossible. These tubes were the first electronic devices. They took time to heat up before they could operate, they often burnt out, and they were relatively expensive. Nevertheless, they could do things that no electrical device could do, and thus they were very widely used. Even the first computers were composed of vacuum tubes.

Vacuum tubes are similar to standard light bulbs in structure. They contain heated elements inside of a glass tube that has air evacuated from it. As you probably know, vacuum tubes are usually more cylindrical in shape than most light bulbs and are generally made from clear glass. But beyond this, they differ a great deal in operation from a light bulb.

The first tubes were developed following an observation (accredited to Thomas Edison) that a heated metal plate will give up electrons (current flow) and that a cold plate will not. Based upon this, John Ambrose Flemming (who was scientific advisor to the Marconi Company) invented the first diode. This device, as is shown in Figure 6-2, allowed current to flow in one direction only. Current will flow from the hot plate to the cold plate, but not from the cold plate to the hot plate.

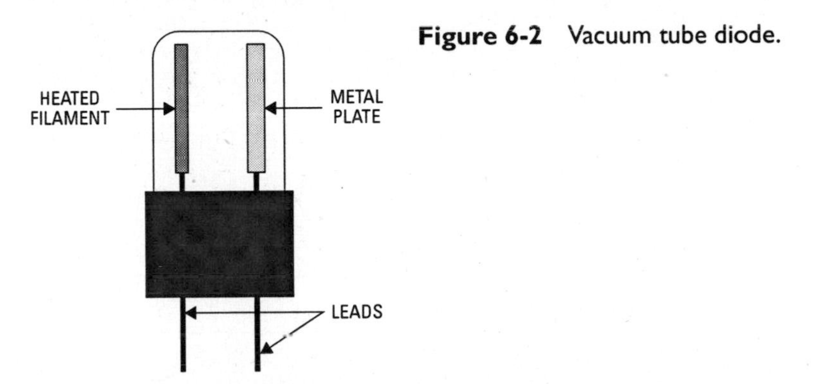

Figure 6-2 Vacuum tube diode.

This device works very well to turn alternating current (AC) into a pulsating direct current (DC), as shown in Figure 6-3. Notice that the diode allows one-half of the AC sine wave to flow, and does not

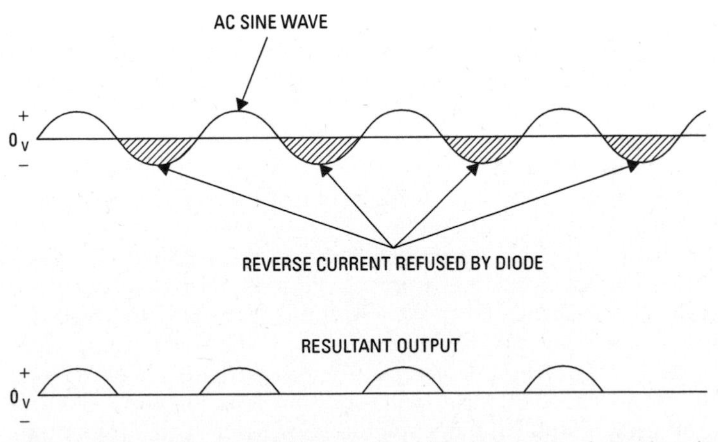

Figure 6-3 AC current before and after being run through a diode.

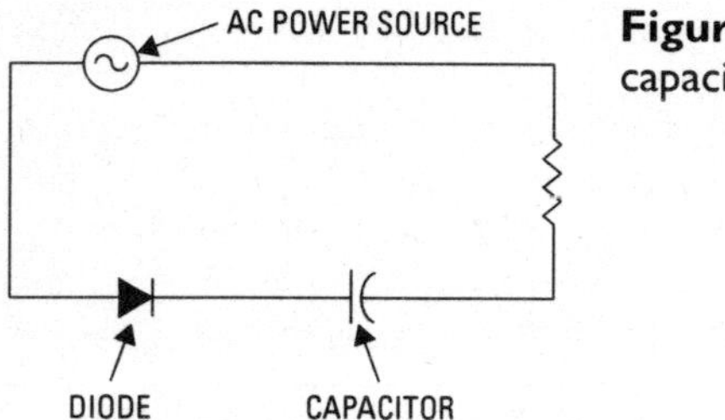

Figure 6-4 Diode in circuit with capacitor.

allow the other side, resulting in a pulsed DC. Figure 6-4 shows how a capacitor is added to the circuit to smooth out the wave, and make it more usable in practical applications. (A smooth current allows most device to operate more efficiently than does a rough pulse.) Figure 6-5 shows the smoother output current from the circuit of Figure 6-3.

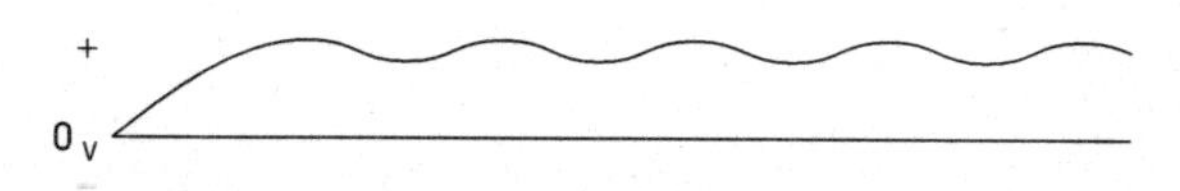

Figure 6-5 Resultant output from Figure 6-4.

Another advance was made in 1907 by Lee DeForest, who used a wire grid in between the two plates of a diode to control the flow of current from plate to plate. By varying the voltage applied to the grid, overall current could be changed continually and very quickly. This is the tube that made radio a practical possibility, since it allowed for the amplification of small audio signals to be increased to levels high enough to operate a loudspeaker. In fact, the original name of DeForest's tube was an *audion*, though we now call it a *triode*. This is shown in Figure 6-6.

While other types of tubes were made and used, these two were the first and most fundamental. The proper name for the vacuum tube is *thermionic valve*, and, for a long time, most people referred to tubes as *valves*.

Tubes had to be applied very carefully, and increasingly complex tubes had to be developed to get improved sensitivity and consistent operation. Besides, they were very hot and fragile. So when semiconductors (also called *solid-state devices*) came along, tubes were phased out of use. Why use a technology as difficult as tubes when you can use a small solid piece of crystal that does the same jobs and more, and that is smaller, more reliable, and safer at the same time?

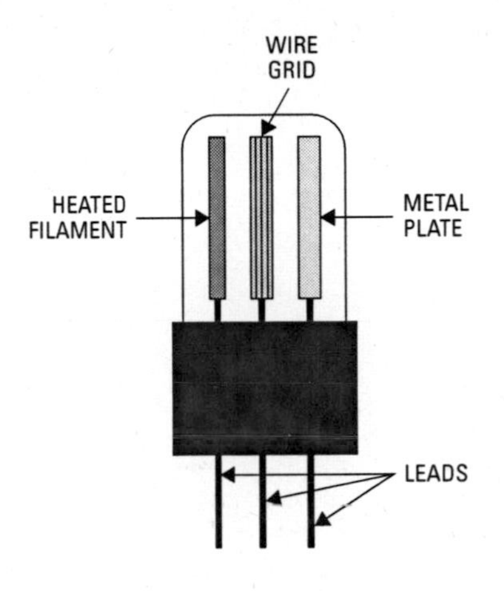

Figure 6-6 Audion or triode tube.

A few types of vacuum tubes are still used, because of their high reliability in special applications. For example, microwave ovens use magnetron tubes. But this is the exception rather than the rule. Semiconductors are used in almost every electronic application.

Semiconductors do virtually all the jobs that tubes do—plus a few extra jobs—and they operate more efficiently. They don't need to warm up before they can operate, and they are very small. The first computer filled up a space the size of a large garage because of the size of the tubes. With the small size of semiconductor devices today, you can fit a far, far more powerful computer on a desktop. In the case of the computer, the tubes and semiconductors primarily did the same jobs, but the size difference was extremely important.

One more step was critical: developing a method of putting hundreds of semiconductor devices on one small piece of silicon. This device, the *integrated circuit (IC) chip*, is merely a large number of semiconductor devices squeezed into a very small area. Needless to say, the IC chip has had a major impact on the modern world.

The invention of the electronic tube was crucial to many of the most important developments of the first half of the twentieth century. Likewise, semiconductors and IC chips were critical to developments in the last half of the twentieth century.

The PN Junction

Almost all semiconductor devices are based on junctions between P-type semiconductors and N-type semiconductors. What happens at these junctions is the central event of semiconductor operations.

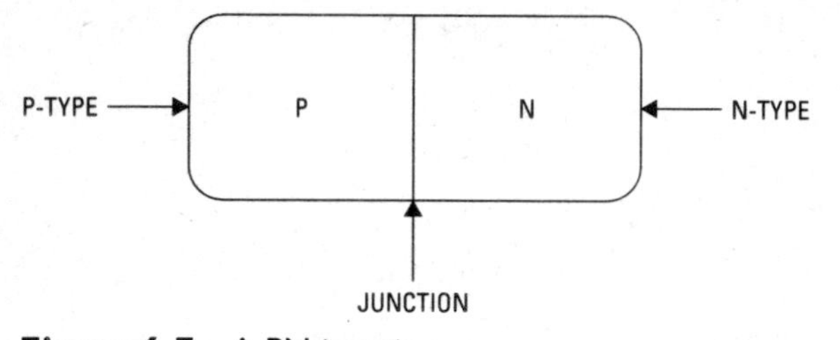

Figure 6-7 A PN junction.

Figure 6-7 shows a simple PN junction. This is simply a piece of P-type semiconductor and a piece of N-type semiconductor formed adjacently. The junction is where the two types of material meet.

When the two types of semiconductors are placed together this way, a small voltage will appear. Figure 6-8 shows why this is so. On the left, you see the P-type material, which is doped with a material that has three electrons in its outer shell, giving it a small deficit in electrons as compared to pure silicon or germanium. On the right, you see the N-type material, which is doped with a material that has four electrons in its outer shell, giving it a small excess of electrons as compared to pure silicon or germanium. This variance in electrons creates a small difference in potential between the two halves. If you were to connect a sensitive voltmeter as shown in Figure 6-8, it would show a small voltage.

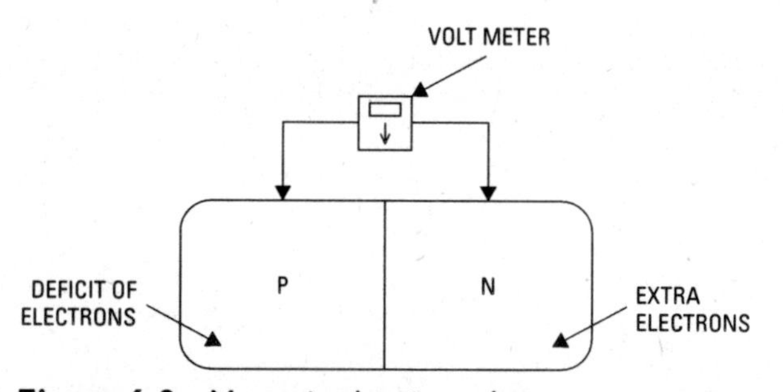

Figure 6-8 Measuring barrier voltage.

Because of this voltage, a few free electrons will cross from the N side to the P side. This creates what is called a *barrier layer* on each side of the junction, as shown in Figure 6-9. The P material on the left side of the junction gains a few electrons, and the N material on the right side of the junction loses a few electrons. (But remember, since the atoms on both sides are a crystalline structure, the doping material remains locked in place, so this effect is very much a local thing.)

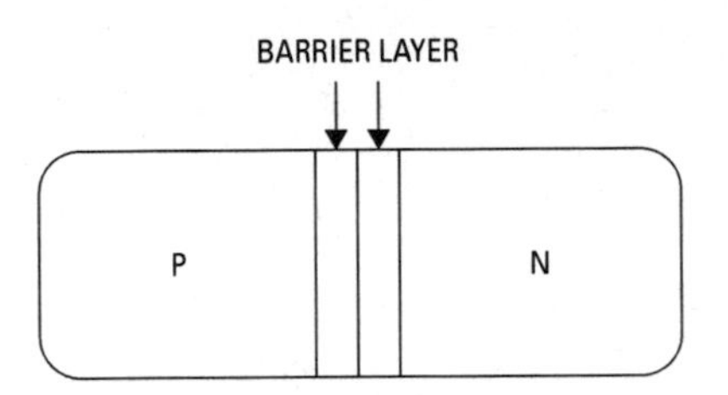

Figure 6-9 The barrier layer.

On either side of the junction, you have materials that are less polarized than they are in their natural state. That is, the P material next to the junction on the left side will become less positive than the rest of the P material. Likewise, the N material next to the junction on the right side will become less positive than the rest of the N material. This set creates something of a barrier to current flow, called the *barrier layer* or the *depletion layer*. In this text, we will generally use *barrier layer*, though either term can be used.

Holes

The *hole* is a not-so-elegant term that refers to a spot where an electron would normally be, but is presently empty. That is, a hole is an electron's spot that is vacant at the moment.

In a semiconductor, holes are said to flow in the opposite directions of electrons. But it is important to understand that holes are not physical things, and electrons *are* physical things. Holes are openings. But, when these holes open and are refilled in a sequence, we can say that the hole moves in a specific direction. Think of this as eggs in a container; if you leave one spot open in the container, then move the eggs one at a time, always filling the spot made open by the previous move, this would move the hole from end to end.

The concept of moving holes is critical to understanding how semiconductor devices work. We will talk about the flow of holes in addition to the flow of electrons.

Figure 6-10 shows some of the symbols used to explain actions inside semiconductors. We are deviating from tradition a bit to use symbols that are as simple and instinctive as possible to make this clear. Notice that we are not using the usual battery symbol here but rather a drawing of a battery.

In Figure 6-10, the battery terminals are noted to explain not only the movement of electrons but the movement of holes as well. The rules are these:

- Positive terminals push holes and pull electrons.
- Negative terminals pull holes and push electrons.

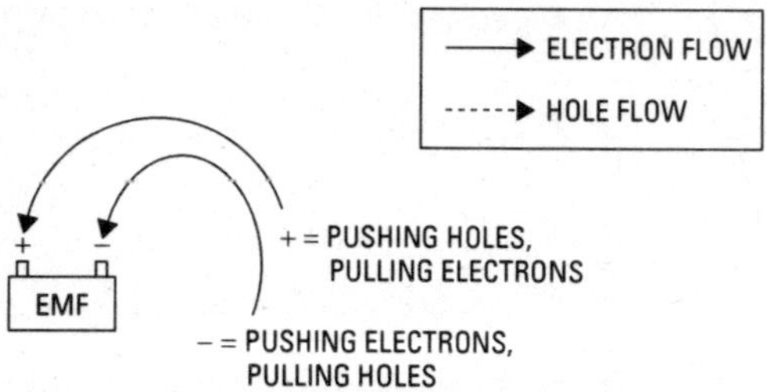

Figure 6-10 Representing electron flow and hole flow.

This makes sense, since like charges repel and unlike charges attract. A negative terminal would attract positives (holes) and repel negatives (electrons). In the reverse, a positive terminal would attract negatives (electrons) and repel positives (holes).

Be sure you understand the concepts presented in the previous few paragraphs. The concepts are not especially difficult but may be strange to many people and difficult to grasp at first.

Note also in Figure 6-10 that we will be representing electron flow with solid lines and hole flow with broken lines.

How the PN Junction Works

Figure 6-11 shows the state of a PN junction before outside voltages are applied. Notice the barrier layer. On the P side, near the junction, the semiconductor material is more negative that the rest of the P material (represented by the lowercase "n"). On the N side, near the junction, the semiconductor material is more positive that the rest of the N material (represented by the lowercase "p").

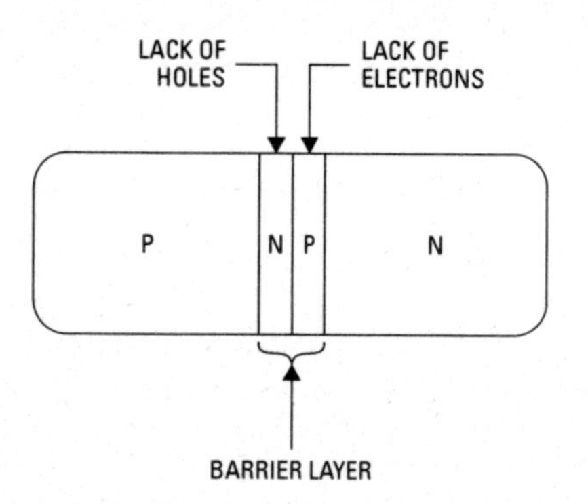

Figure 6-11 PN junction with no outside voltage applied.

The thing that makes these PN junctions work is the barrier layer and the way it reacts to external voltages applied to each side of the junction.

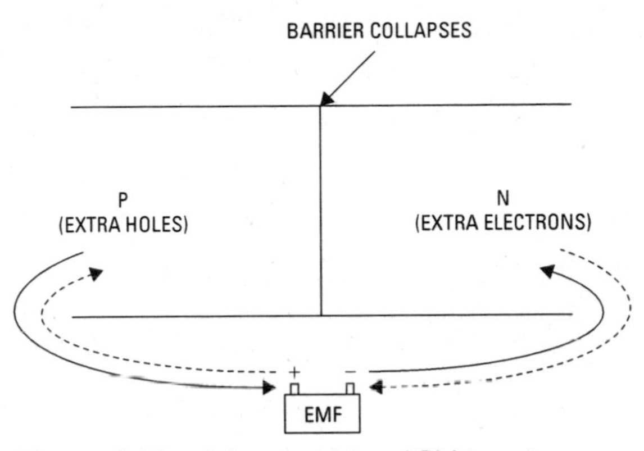

Figure 6-12 A forward-biased PN junction.

Forward Bias

Figure 6-12 shows what happens when a voltage of a particular polarity is connected. On the right side, notice that the battery is giving the N-type semiconductor more electrons. Refer back to Fgure 6-11 to see that the N side of the barrier layer had a lack of electrons. By pushing extra electrons into the N side, we are collapsing the barrier layer. By giving it the electrons it needs, we are shrinking the layer that is depleted of electrons.

The same thing happens on the P side of the barrier layer. Note that in Figure 6-11 we showed this area as having a lack of holes. And remember that the positive terminal of a battery pushes holes. So, the positive terminal of the battery is pushing holes into the P side of the barrier layer that was depleted of holes.

When connected this way, the barrier layer collapses, and current will flow from end to end. We then say that the PN junction is connected to the battery in the *forward-biased* configuration. Forward bias simply means that the device will allow current to flow.

Reverse Bias

Figure 6-13 shows the same PN junction, but with the polarity of the battery reversed. Notice carefully what happens this time.

The N side is connected to the positive terminal of the battery. This pulls electrons from the N side. Notice that the N side next to the barrier layer is already showing a lack of electrons. Now, we are using the battery's electrical force to pull more electrons away. This will cause the N side of the barrier to expand.

The P side is connected to the negative terminal of the battery. This pulls holes from the P side. Potice that the P side next to the

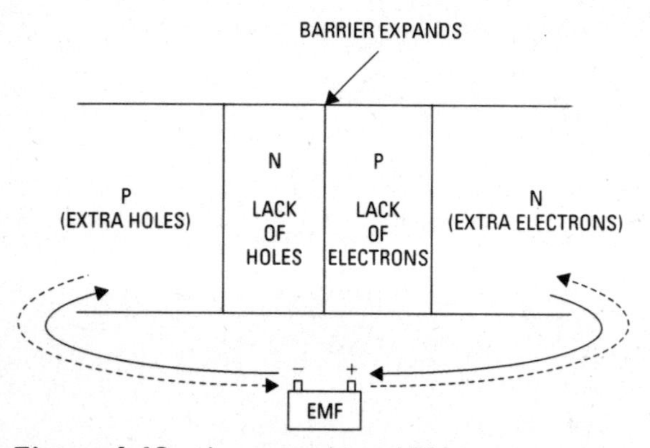

Figure 6-13 A reverse-biased PN junction.

barrier layer is already showing a lack of holes. Now, we are using the battery's electrical force to pull more holes away. This will cause the P side of the barrier to expand.

In Figure 6-13, the battery's emf cause the barrier layer to expand, making it much harder for current to flow through the device. We call this condition *reverse bias*. When connected this way, current will not flow through the PN junction.

As you may have guessed, this PN junction device is commonly called a *diode*. It works in a circuit much the same as the vacuum tube diode shown in Figure 6-2, but is entirely solid and requires no external heat source to make it work. In addition, it is much more reliable, tougher, and has a longer life.

It should be noted that if you raise the voltage high enough, the barrier layer can be broken through, but this is similar to standard electrical devices arcing over, or sparking. The level at which the semiconductor fails to operate properly is called the *breakdown voltage*. But, operated within its normal limits, the PN junction will allow current to flow in the forward-biased configuration and will not allow current to flow in the reverse-baised configuration.

Please make sure you understand this section well. Remember, whether or not electricity will flow through the junction depends on how the emf of the battery affects the barrier layer. If the polarity of the battery causes the barrier layer to grown, no current will flow. If the polarity of the battery causes the barrier layer to collapse, current will flow easily.

PNP Junctions

The PN junctions discussed so far have been two-part semiconductor devices. The PNP junction is a three-part device, as shown in

Figure 6-14. In this device, there are two (not one) junctions and two (not one) barrier zones.

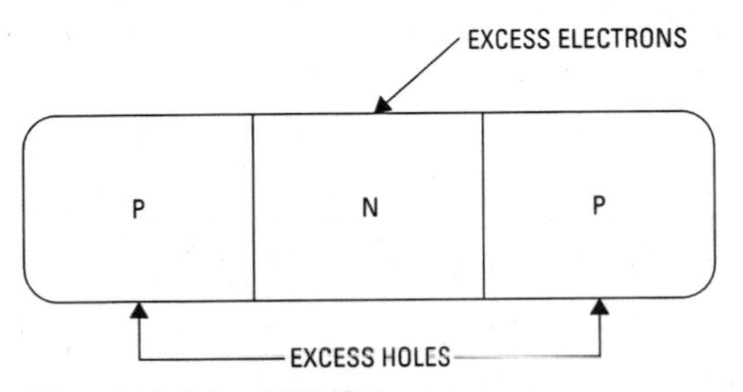

Figure 6-14 A PNP junction.

Figure 6-15 shows how the barrier zones develop in the PNP junction. Figure 6-16 shows how the PNP junction is connected to two battery power sources, as well as the flow of electrons and holes.

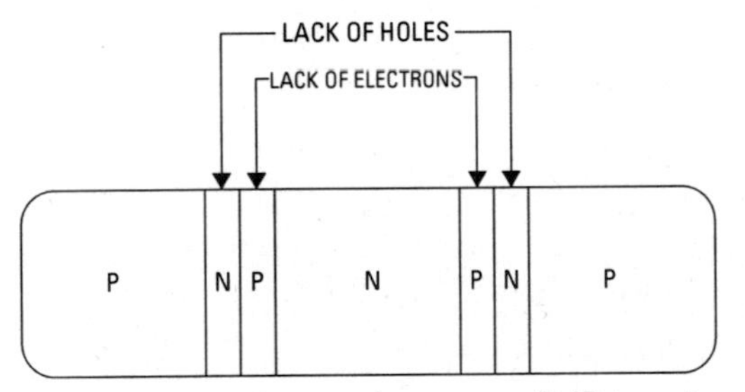

Figure 6-15 Barrier layers in a PNP junction.

Figure 6-17 shows the PNP junction split in half to illustrate what happens at each of the PN (or NP) junctions. If you look carefully and compare this figure to Figures 6-12 6-13, you will see that the

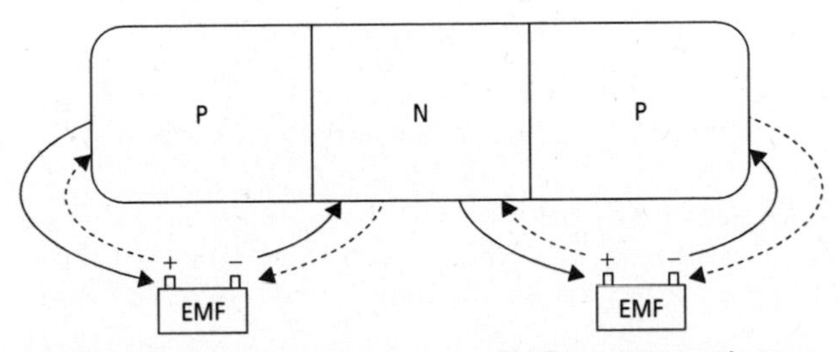

Figure 6-16 Connection of a PNP junction, electron flow and hole flow.

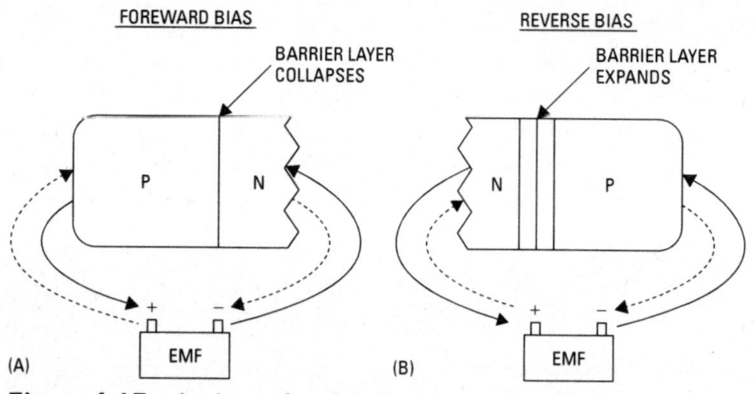

Figure 6-17 Analysis of each half of the PNP junction.

left-hand side shown in Figure 6-17A is forward-biased, and the right-hand side in Figure 6-17B is reverse-biased. This is important.

So, from Figure 6-17, we would conclude that current would flow in the left side of this PNP junction but not in the right side. And you would be correct, *except* for one thing: the attraction of oppositely charged particles.

Figure 6-18 shows how the PNP junction works. Look at the left side of the drawing, and notice that holes flow from the positive terminal into the P material. Then, since this junction is forward-biased, the holes will continue into the N material.

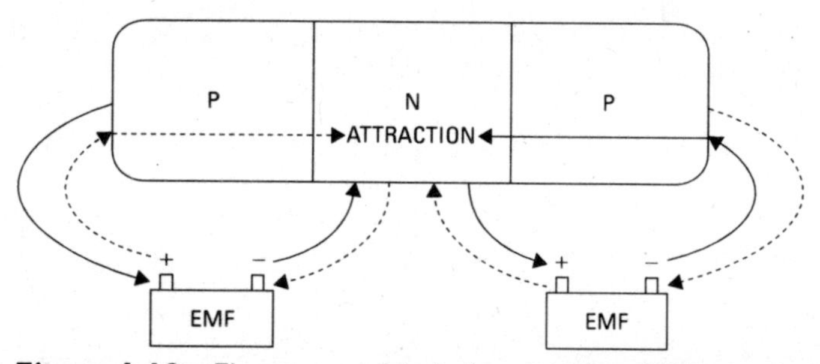

Figure 6-18 Electrons and hole functions in a PNP junction.

Staying with Figure 6-18, notice the flow of electrons on the right side of the illustration. They flow from the battery to the P material. But once there, they are stuck, since this is a reverse-biased connection.

Notice carefully that there are now free holes flowing in the N material at the center of the PNP junction. There are also electrons being fed into to right-hand side P material with nowhere to go. Finally, recall that unlike charges attract.

The holes in the N material attract the electrons in the P material. This attraction is strong enough to draw the electrons across the NP junction of the right side of the device.

The holes flowing in the left-hand circuit attract the electrons in the right-hand circuit and assist them in crossing the reverse-biased NP junction.

Again, be sure you understand this concept. This is the secret of how transistors work. The PNP junction is a transistor.

Transistors

The name *transistor* is a shortening of the original name, *transfer resistor*. The transitor allows us to control one side of the device by altering the current in the other side. In other words, it transfers resistance from one side to the other. We use this effect to amplify currents, or to simply turn them on and off.

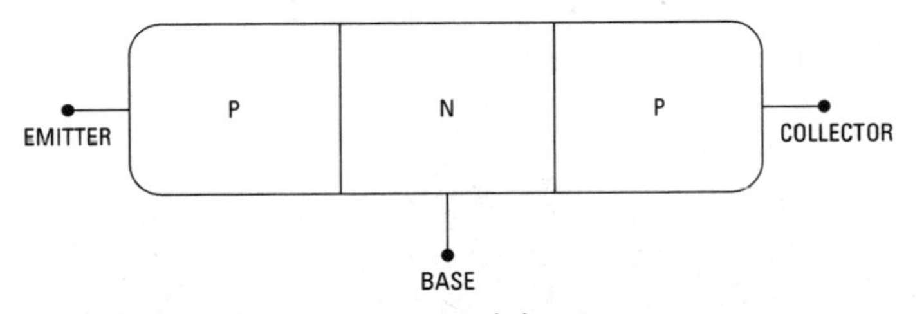

Figure 6-19 Transistor terminal designations.

Figure 6-19 shows the basic PNP transistor with its terminals identified. The terminals are *base*, *collector*, and *emitter*. Figure 6-20 shows a typical transistor. From this drawing, you can see how the name base came into use. It is actually much easier to manufacture the transistor shown in Figure 6-20 than the bar shape we have been using for illustration (though both types are made). In Figure 6-20, the N material would have been formed into the base, then the smaller P sections formed afterward. Each of the three sections would have a small wire lead attached to it.

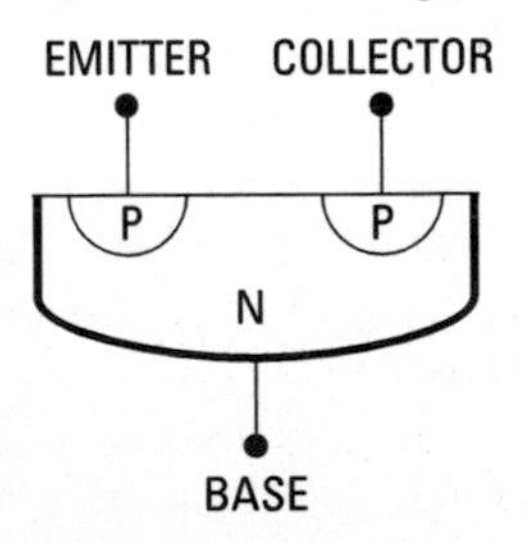

Figure 6-20 Typical transistor construction.

Figure 6-21 shows the PNP transistor connected in a circuit. This is the same device and connections that we have been using

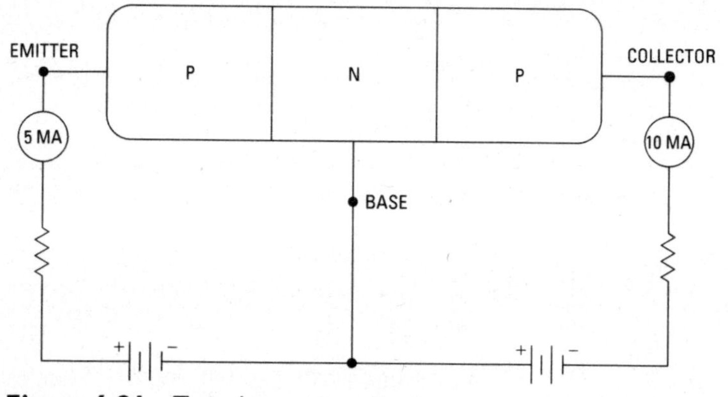

Figure 6-21 Typical transistor circuit.

throughout this discussion. But this time, we are also showing current values. In the left-hand side of the device, 5 milliamps (mA) are flowing. But, in the right-hand side of the device, 10 mA are flowing. This doubling of current from one side to the other is typical. Note that the collector circuit (base-collector) has twice as much current flowing as the emitter circuit (base-emitter). This also is typical.

Another very important characteristic of the transistor is that varying the current on the base-emitter circuit will automatically vary the current on the base-collector side. The current can vary continuously, and both sides will move up and/or down in unison. Can you see from this why transitors are so useful? Without a device that was finely variable like this one, we could not amplify audio signals, which vary thousands of times per second. The same is true for video signals and many other types as well.

In our examples we have always used PNP transistors. However, NPN junction transistors work in precisely the same way, just with the polarities reversed.

Summary

Semiconductors are composed of materials with four valence electrons (such as silicon or germanium). To these substances are added small amounts of doping, which are materials that have either three or five valence electrons. This gives the doped materials either a positive or negative charge. We call these P-type and N-type semiconductors.

Vacuum tubes were the first electronic devices. These devices used evacuated glass tubes and various electrodes and grids to finely control the flow of electrical currents. Effective tubes were developed after observing that hot plates give up electrons easily, and that cold plates do not. Vacuum tubes (especially the audion or triode) made radio and many other technologies possible and affordable. Tubes were rather quickly replaced with solid-state devices, once semiconductor devices entered the market at low prices.

PN junctions are P-type and N-type semiconductors that are formed together. They have a small voltage where the two types of materials meet. This is because of the small negative charge of one side and the small positive charge of the other side. Because of the localized movement of charges near the junction, a barrier layer forms.

When analyzing the operation of semiconductor devices, we use the concepts of holes and hole flow. A hole is an empty opening for an electron. Holes move in the reverse direction that electrons do.

A PN junction can be connected to a power source in such a way as to make its barrier layer either expand or collapse. When the barrier layer is made to collapse, current can easily flow from one end of the device to the other. We call this forward-biased. When the voltage source is connected in such a way as to make the barrier layer expand, no current will flow through the device, and this is called reverse-biased.

A PNP junction is essentially a sandwich of three layers of semiconductor types. Barrier layers form between the layers of this device. To operate this device, we set up a forward bias on one side of the device and a reverse bias on the other side. As electrons and holes begin to flow through one side of the device, they attract unlike charges (electrons or holes) from the other (reverse-biased) side. This helps the particles in the reverse-biased side across the barrier layer, allowing current to flow in both sides of the device.

The PNP device (and the NPN device, which is identical, although in the reverse) is a transistor. Current in the reverse-biased side of the transistor will be several times that of the forward-biased side. Furthermore, the side with the higher current is controlled by the side with the lower current level. This variation of current level can be very very fast and continual, allowing this device to be used for amplification of very intricate signals. Transistors may also be used for simply turning circuits on or off.

In addition, there are other types of transistors besides this basic type that we have illustrated here. We will explain many of them in Chapter 7.

Review Questions

1. Why are silicon and germanium used as semiconductors?
2. Name two three-electron doping substances.
3. Name two five-electron doping substances.
4. What was the observation that led to the development of vacuum tubes?
5. Who invented the tube that made radio possible?
6. Why does a barrier layer form at a PN junction?
7. Why does the barrier layer collapse in a forward-biased diode?
8. Why does the barrier layer expand in a reverse-biased diode?
9. What is a hole?
10. How many barriers layers are in a PNP junction?
11. How is a PNP junction biased?
12. How are electrons assisted across the base-collector barrier layer?
13. What are the two transistor circuits?
14. Which transistor circuit carries more current?

Exercises

1. Illustrate the flow of holes.
2. Draw an NPN transistor. Show electron and hole flow.

Chapter 7

Semiconductor Devices

You should now have a good understanding of how semiconductor devices operate. This chapter describes several semiconductor devices. The discussion does not go into the intricacies of hole flow and charge attraction that Chapter 6 did, but it does cover many more devices, explaining what they are, what they do, and how they are applied.

Diodes

As explained in Chapter 6, a diode is simply a PN junction: a piece of type N semiconductor joined to a piece of type P semiconductor. This is shown in Figure 7-1. If a battery is connected to the diode as shown (positive terminal to N, negative terminal to P), no current will flow through the diode with the exception of a very small leakage current. Figure 7-1 shows the diode connected to the battery in a way that makes it *reverse-biased*. This means that it is connected so that it opposes current flow—its bias is reversed.

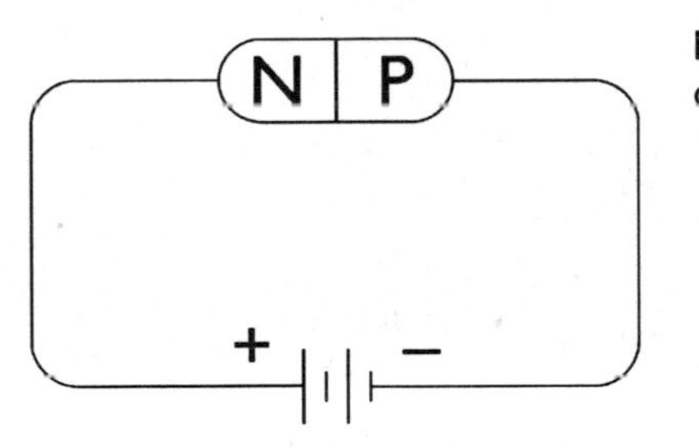

Figure 7-1 Reverse-biased diode.

In Figure 7-2, you see the same diode connected the opposite way—with the positive terminal to P and the negative terminal to N. When connected this way, current will flow with very little

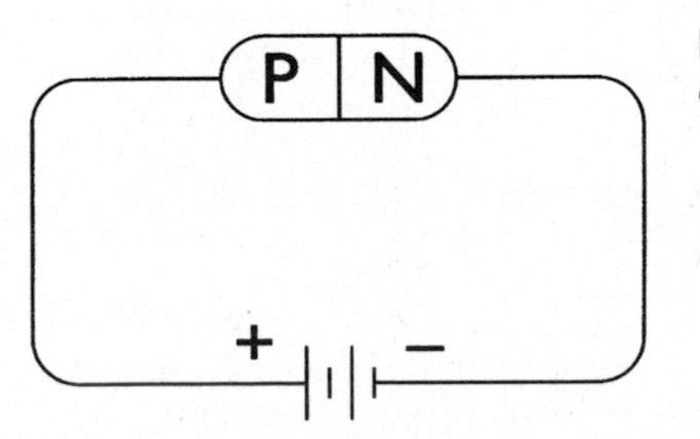

Figure 7-2 Forward-biased diode.

resistance. Figure 7-2 shows the diode connected to the battery in such a way as to make it *forward-biased*. This connection allows the current to move forward through it.

Diode Applications

Diodes are commonly used to convert alternating current (AC) into direct current (DC). By simply connecting the diode in series with a circuit, you allow current to flow in only one direction—it won't flow in reverse. Thus, the current can no longer alternate; it can flow in only one direction.

Diodes come in all sizes and ratings, and must be applied correctly. If you connect a diode rated for 5 volts to a 24-volt circuit, it will fail. Usually, diodes look like resistors, but they can come in varied sizes and shapes. Figure 7-3 shows several types.

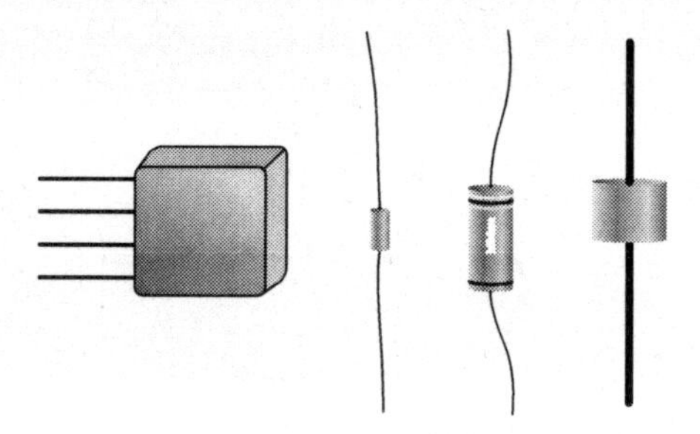

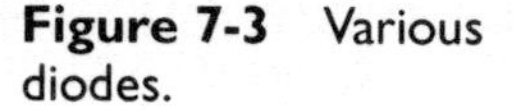

Figure 7-3 Various diodes.

Diodes are used anywhere a one-way valve for electricity is required and are frequently used in power-conversion equipment, where AC current must be changed to DC current. AM radios use diodes to remove the negative portion of the AM signal and to use the positive portion.

Diodes are used in overvoltage protection devices. In this application, they are connected so that they are reverse-biased normally, and go to a forward bias if the voltage rises too high. The diode is connected between the circuit being protected and ground, so that if a high-voltage situation occurs, the circuit will connect to ground, giving the current a path away from the circuit that must be protected.

Diodes are used to construct logic gates, which are discussed in Chapter 14.

Types of Diodes

There are many types of diodes. The common types are discussed here, but be aware that you may come across additional types of diodes as well.

Normal Diodes

These are the diodes described to this point. They are almost always made of silicon, although a few germanium diodes are used.

One of the most common types of normal diodes is the *point contact* diode. The name point contact refers merely to the way the device is constructed, not some special operating mechanism. Point contact diodes are simple to manufacture, and, as a result, are inexpensive and common. Figure 7-4 shows a point contact diode, and Figure 7-5 shows the usual symbol for a diode. Figure 7-6 shows a graph of the current flow in a diode.

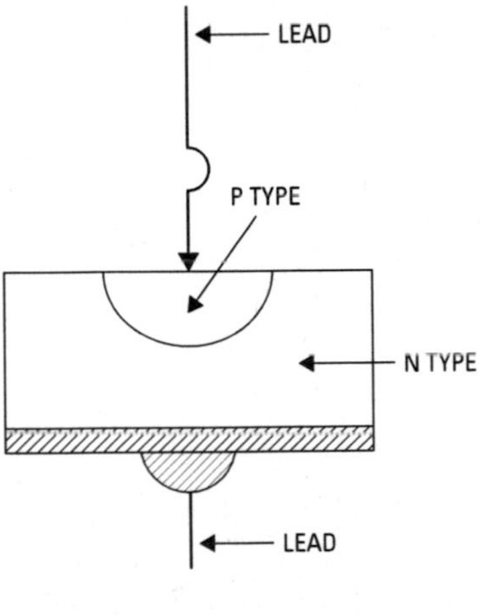

Figure 7-4 Point contact diode.

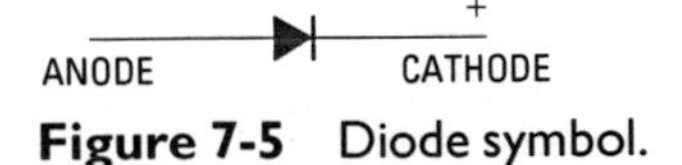

Figure 7-5 Diode symbol.

Zener Diodes

The Zener diode is a PN diode that has been specially doped. Zener diodes are usually connected in circuits in the reverse-biased position and are used as surge protectors.

Zeners are named for Dr. Clarence Melvin Zener, who invented the device at Southern Illinois University.

Typically, Zener diodes are installed in parallel with a load that is to be protected. They are connected to oppose current flow (the definition of reverse-biased). But when the voltage applied to them reaches a certain level (called the *breakdown voltage*), they will conduct a current backward. This has the effect of shunting the voltage away from the load being protected and sending it to ground instead. Obviously, the device must be properly constructed to handle breakdown voltages and still survive.

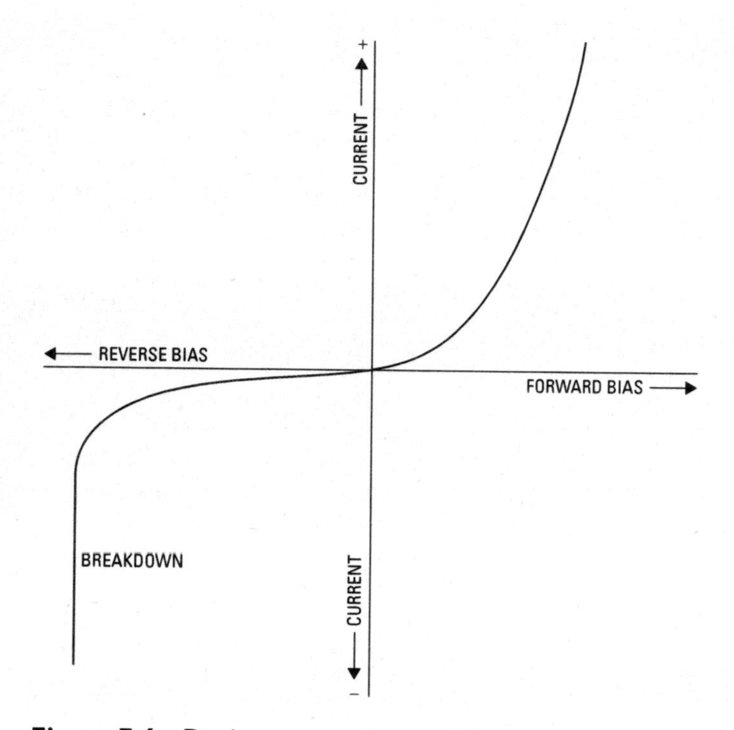

Figure 7-6 Diode current value graph.

When properly sized Zener diodes are used this way, they provide a high level of overvoltage protection for sensitive circuits. They are especially useful because they have a very fast response time. The Zener diode will respond to an overvoltage within a few nanoseconds, rather than taking the many milliseconds of response time required by other types of surge suppressors before they can protect the circuit.

Figure 7-7 shows a current graph for a Zener diode and Figure 7-8 shows a typical circuit containing a Zener diode. Figure 7-9 shows the symbol for a Zener diode.

Gold-Doped Diodes

Normal diodes add a fair amount of capacitance to a circuit. At lower frequencies, this is not a problem, except in the most sensitive circuits. At higher frequencies, however, this extra capacitance can cause performance problems. By adding some gold doping, this capacitance is significantly reduced. So, for circuits that operate in the megahertz (MHz) ranges (millions of cycles per second), gold-doped capacitors are common.

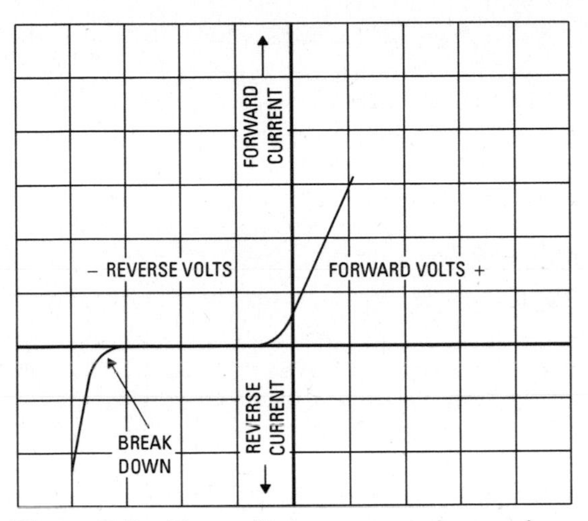

Figure 7-7 Zener diode current value graph.

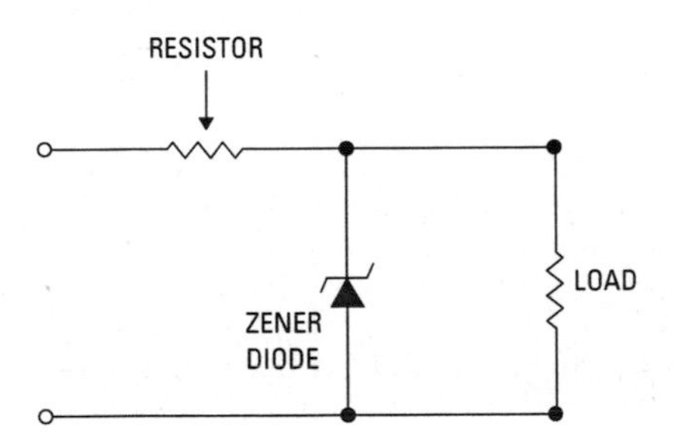

Figure 7-8 Zener diode circuit.

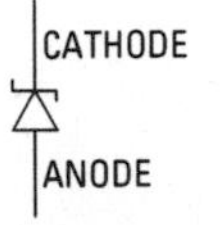

Figure 7-9 Zener diode symbol.

Avalanche Diodes

Avalanche diodes are made so that they will conduct in reverse if the reverse-bias voltage is greater than the diode's breakdown voltage. (Again, *breakdown voltage* is the value at which the circuit overpowers the semiconductor device's bias and its barrier zone, pushing electricity through the device by brute force.)

Avalanche diodes function much the same as do Zener diodes, but the mechanism by which they work is different. People will sometimes confuse the two, so be specific in your use of these terms and verify the terms when others use them.

The avalanche effect occurs when a reverse electrical field across the PN junction causes the ionization inside the device, and, as a result, a large current is permitted to flow. Note that avalanche diodes are designed to break down this way and to not be destroyed in the process. This involves specific materials and types of construction. The reverse breakdown voltage of these devices is 6.2 volts or higher.

A specialized type of avalanche device is called a *transient voltage suppression diode*, or *TVS diode*. These are avalanche diodes that have a much larger cross-sectional area at their PN junction than do normal avalanche diodes. This allows the device to conduct large currents to ground without being damaged.

Light-Emitting Diodes (LEDs)

We will cover light-emitting diodes (LEDs) and photodiodes briefly here, because they are semiconductor devices. But since they handle light, they are also classified as optoelectronic devices, which we'll cover in Chapter 8.

When an electron crosses the PN junction of a diode, it emits a photon (a particle of light). In standard diodes, these photons are at infrared frequencies, which are not visible to the human eye, and are also reabsorbed by the semiconductor material. However, if the certain materials are formed in specific shapes, these photons are emitted as visible light. We call these devices light-emitting diodes, or LEDs.

Special polishing and manufacturing techniques allow LEDs to function as a *laser diode*. This is a very useful device for fiber-optic technologies, which are discussed in Chapter 14.

Photodiodes

Photodiodes are PN junction devices that have transparent junctions. When photons enter the device, they add enough energy to push electrons across the junction. Thus, in these devices, light causes the flow of electricity.

Photodiodes are used to make solar cells, which generate electricity from sunlight. They are also used in light meters.

LEDs are sometimes paired with photodiodes to create an *optical isolator* or *opto-isolator*. This allows signals to be transferred from one electronic circuit to another without electrons passing through the junction. Photons (particles of light) are used to pass the signal at the junction instead. This is useful in many highly sensitive circuits

Schottky Diodes

Schottky diodes are made with materials that provide for a very low voltage drop inside the device.

All semiconductors have a certain internal resistance, which leads to a voltage drop. Thus, the semiconductor device also acts as a small resistor in addition to its desired actions. In most cases, this voltage drop is negligible, much as the small votage drop in copper wire is negligible.

Schottky diodes are used where power is provided by batteries and conservation of power must be carefully obtained. With an internal votage drop of less than half of a volt, Schottky diodes are very useful in conserving power.

Tunnel Diodes

Tunnel diodes are some of the more exotic types. By using an effect called *quantum tunneling*, they are able to create a region in the semiconductor having a negative resistance that allows signals to be amplified.

Quantum tunneling and the overall field of quantum mechanics are beyond the scope of this text. We will make mention of them from time to time (as they do impact electronics), but to explain them properly requires high-level training in physics.

Gunn Diodes

Gunn diodes use the same quantum tunneling effect as do tunnel diodes, but they are made differently and operate differently (they use traveling dipole domains, which, again, are beyond the scope of this text). Gunn diodes are useful in constructing high-frequency microwave oscillators.

Transistors

As explained in Chapter 6, transistors are remarkable devices that generally function as switches or amplifiers. Considering that they can be produced in extremely small sizes, they become much more useful, especially in IC chips.

The basic transistor is an NPN junction in which one side is more heavily doped than the other. In other words, one of the N sides is more negative than the other N side. The more heavily doped side is called the *emitter*, and the less heavily doped side is called the *collector*. The P section that is sandwiched in between is called the *base*. This is shown in Figure 7-10.

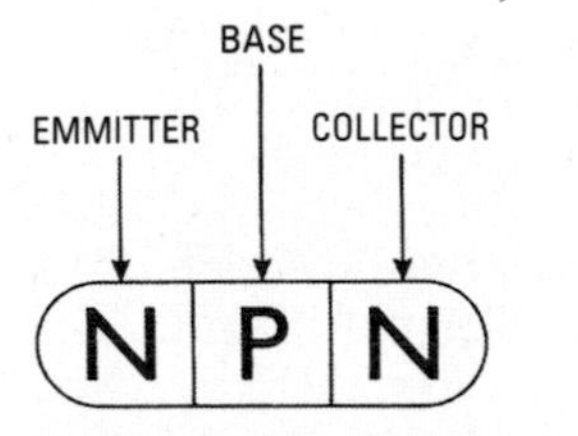

Figure 7-10 NPN transistor.

To understand how this device works as an automatic switch, look at Figure 7-11. As shown, you will connect the same transistor in a circuit. Looking at the right side of Figure 7-11, you see that the collector-to-base NP junction is reverse-biased. (Refer to

Figure 7-1 again.) Therefore, except for a very small leakage current, no current flows through this junction.

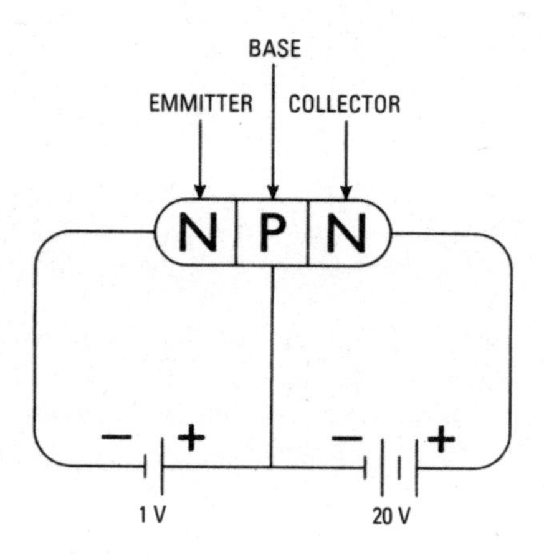

Figure 7-11 NPN transistor connected in a circuit.

So far, so good. But when you connect a voltage to the left side of the transistor, something unique happens. As more current flows through the base-to-emitter NP junction (on the left side of Figure 7-11), it changes the charges in the other NP junction and allows current to flow through it, too. If current flows in the left side of the circuit (base-to-emitter), current will flow through the right side also (collector-to-emitter). If no current flows in the left (base-to-emitter) side, none will flow in the right (collector-to-emitter) side either.

If you look at Figure 7-11 again, you see that the voltage of the battery supplying power on the left side of the diagram is only 1 volt, but the voltage on the right side is 20 volts. So, with this circuit, you can use a 1-volt circuit to control a 20-volt circuit. This is a basic amplifier.

Now, to make it really interesting, here is one last thing that the transistor does: It keeps the current that flows through the right side of our circuit proportional to the current level in the left side of the circuit. In other words, if 5 milliamps flow through the left side, allowing 100 milliamps to flow through the right side, then increasing the current in the left side to 10 milliamps will automatically increase the current in the right side to 200 milliamps. (You are assuming here that all other things remain unchanged.) Figure 7-12 shows this relationship.

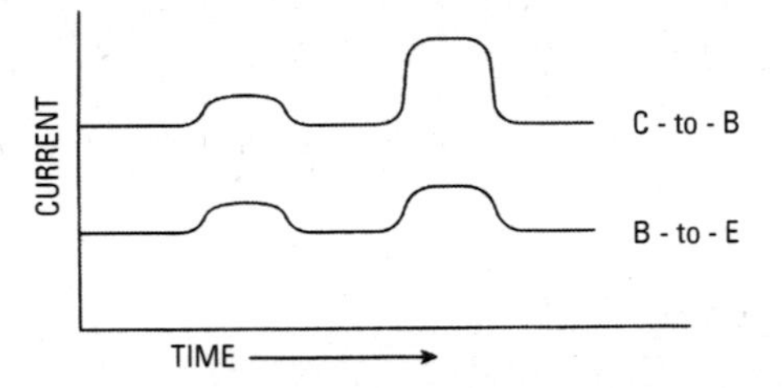

Figure 7-12 Current relationships in transistor circuit.

You can see from this description how useful transistors are. Considering that they can be produced in extremely small sizes, they become much more important.

Transistor Applications

Transistors are used in an immense number of applications. Probably the most common application is in microprocessors, where transistors are used as on/off switches. This use makes modern computers possible. Computers use binary code; they interpret the presence of a voltage as a 1 and the absense of voltage as a 0. In this application, transistors are used to create the zeros and ones that the computer processor can read as data.

Another application of transistors is in the amplification of signals. Transistors can frequently produce amplified signals that are 100 times larger than the original signal. Such amplification is necessary for almost all audio and video equipment, as well as for many other devices.

Types of Transistors

Many types of transistors have been developed; many more than can be discussed in this text. But we will cover the common types of transistors. Several common transistors are illustrated in Figure 7-13.

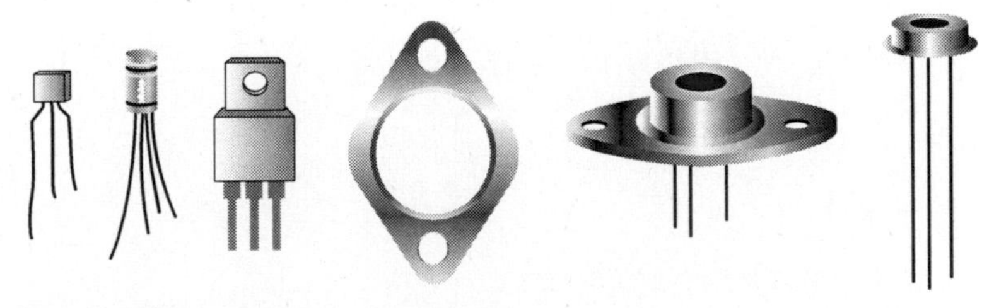

Figure 7-13 Common transistors.

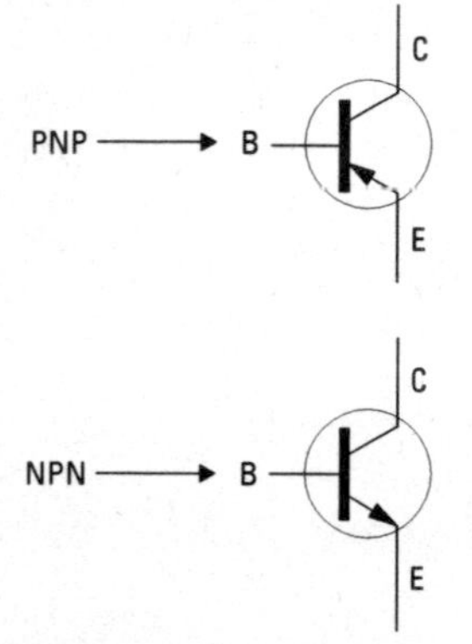

Figure 7-14 Symbols for the bipolar junction transistor.

Bipolar Junction Transistors (BJTs)

Bipolar junction transistors are the most common type, and the type that we have described thus far in this text. Figure 7-14 shows some common symbols for the bipolar junction transistor. Figure 7-15 shows some common (point contact) transistors, with circuits.

Field-Effect Transistors (FETs)

The most common field-effect transistor is one made with metal oxides. Thus it is called the metal-oxide semiconductor field-effect transistor, or MOSFET. MOSFET devices control current through a transistor differently than do bipolar junction devices. In the MOSFET, current flows from end to end only, and is controlled by a *gate*. Figure 7-16 shows a diagram of a MOSFET.

When a voltage is connected between the source terminal and the gate, an electrical field is generated, and this field creates an *inversion channel* in the semiconductor material in the channel below the

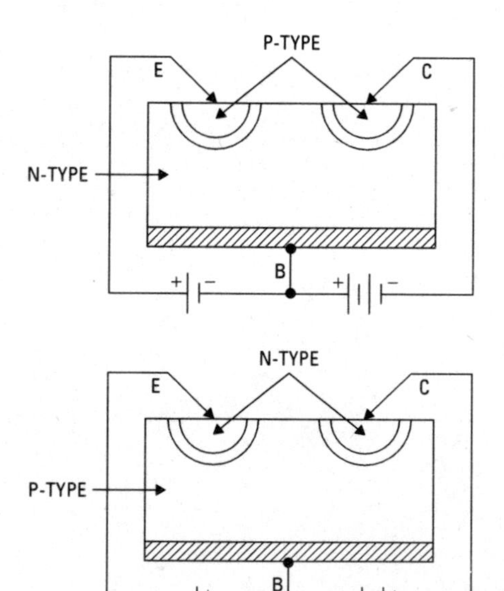

Figure 7-15 Point contact transistors.

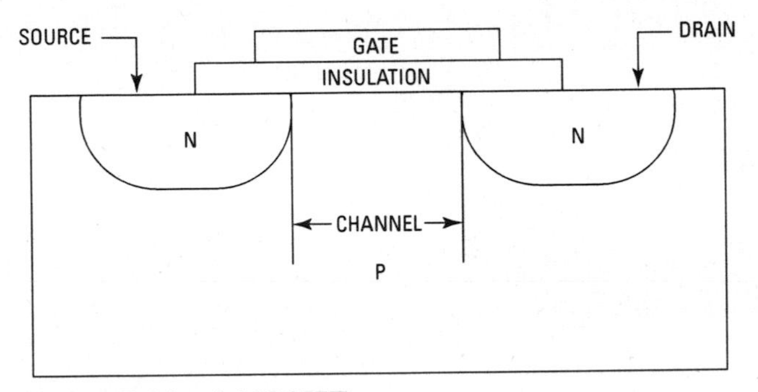

Figure 7-16 A MOSFET.

gate. This means that the semiconductor material in the channel effectively inverts its polarity, as shown in Figure 7-17. Note that once the voltage is applied, the P material near the gate becomes negatively charged. This creates a single N-type channel between the source and the drain parts of the device, and current flows quite easily from end to end.

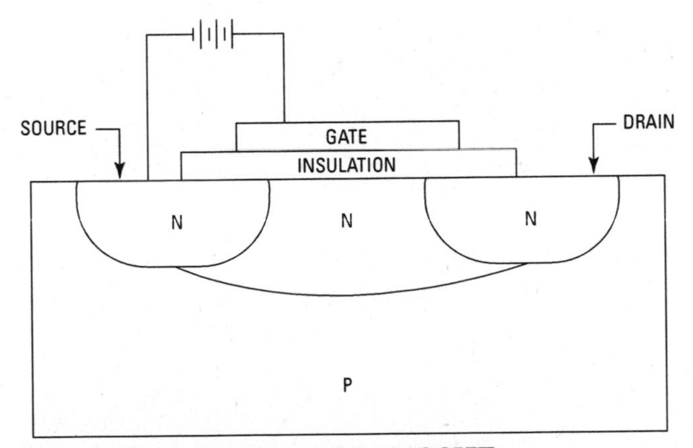

Figure 7-17 Operation of the MOSFET.

Using differing levels of source-gate voltage will invert more or less of the channel material, and allow more or less current to flow from source to drain. So the MOSFET is commonly used as an on/off switch, but it can function as a variable resistor as well. Figure 7-18 shows a diagram of the construction of a MOSFET.

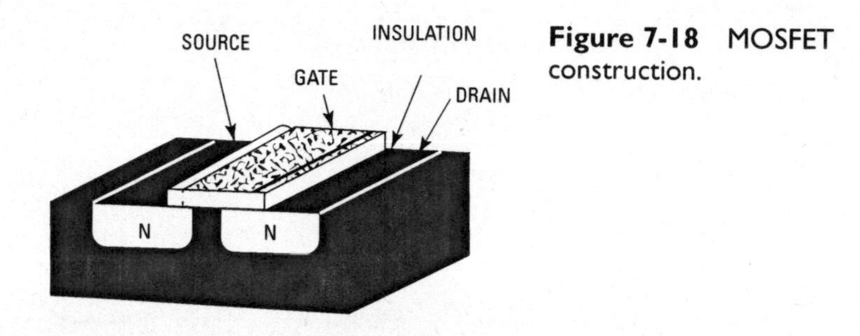

Figure 7-18 MOSFET construction.

The primary advantage of the MOSFET is that it acts very quickly. That is, it can turn on and off at very high speeds, which makes it perfect for use in microprocessors, where speed is critically important.

Additionally, MOSFETs are very efficient. Since no current flows between the gate and the source (insulation being present between the two), very little current is wasted, and very little heat generated. This is important when trying to put millions of transistors into a very small area.

There are many variants on the MOSFET, most of which are specially designed for specific uses. By altering the materials and manufacturing techniques, specific operating characteristics (operation at different voltages, maximum speed, minimum heat generation, and so on) can be obtained. However, the basic operating mechanism of all these devices are the same as described here.

Figure 7-19 shows symbols for MOSFETs.

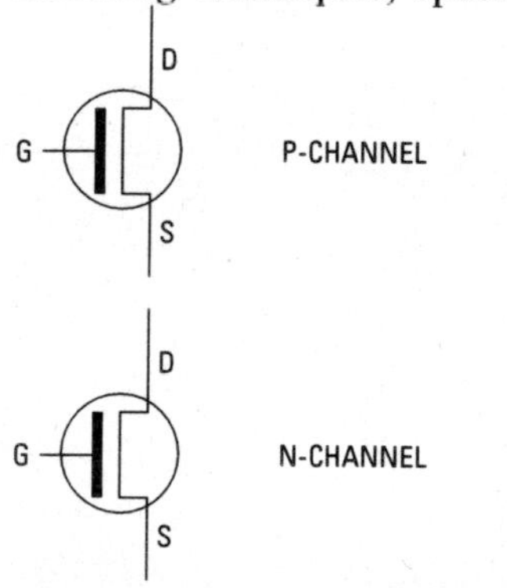

Figure 7-19 MOSFET symbols.

Another type of FET is the *junction gate field-effect transistor (JFET)*. The JFET is different from the MOSFET in that in its de-energized state current will flow through it from end to end. In other words, it acts as a normally closed switch; energized, it stops the flow of current through it.

Figure 7-20 shows the JFET. Notice the *depletion zone*. This is the same area as the barrier zone, but by a different name. In the depletion zone, the normal carriers (electrons in N material, holes in P materials) are depleted. For FETs, this terminology is more commonly used than barrier zone.

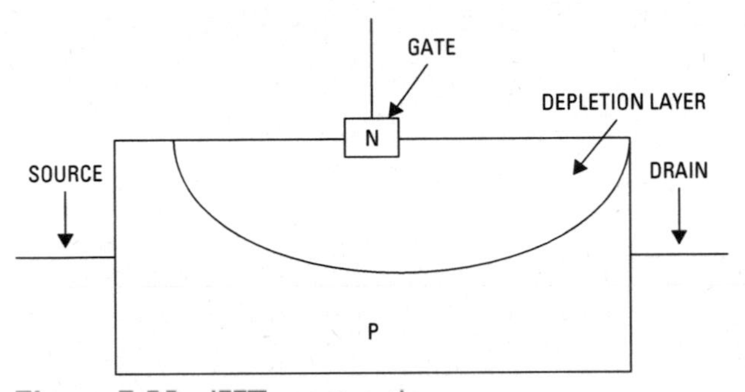

Figure 7-20 JFET construction.

Figure 7-21 shows how the JFET operates. When a voltage is applied to the gate, the depletion zone expands to the edge of the P-type material, blocking current flow from source to drain.

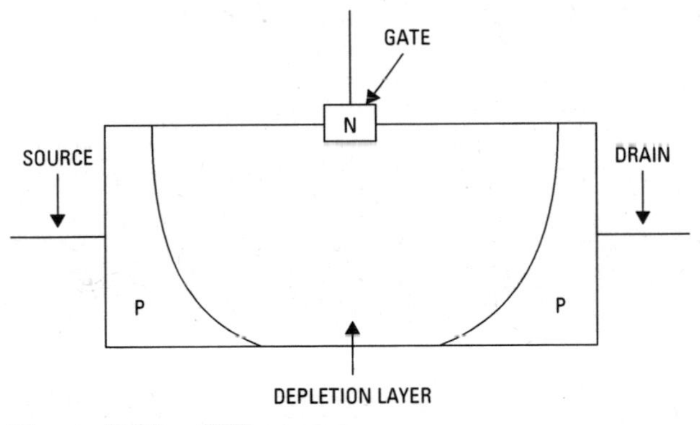

Figure 7-21 JFET operation.

Thyristors

Thyristors, also called *silicon-controlled rectifiers (SCRs)*, are composed of four layers of silicon P and N semiconductors. In effect, they are two transistors connected together, as shown in Figure 7-22.

Unless current is put through the gate lead of the device, no current will flow from the anode to the cathode. If there is a gate current, the resistance between the anode and the cathode drops to almost zero, allowing current to flow freely. Thus, the gate current is necessary to start the rest of the thyristor conducting. Unlike the transistor, however, the current will continue to flow from the anode

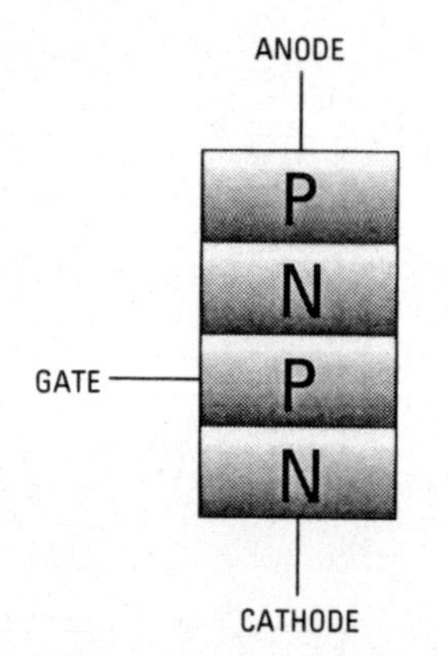

Figure 7-22 A thyristor, or silicon-controlled rectifier.

to the cathode, even when the gate current ceases. Once started, the anode-to-cathode current will flow until it stops on its own; it won't be stopped by the thyristor.

Thyristors can be connected so that current will be turned on at a very precise voltage level. Thus, the thyristor can be designed to cut off (or *clip*) part of an AC sine wave, then turn on at a specified level. After this, it will remain on until zero voltage is reached again. However, the thyristor works in one direction only, so if it works on one half of a sine wave, it cannot work for the other half, with its reversed current.

SCRs are particularly useful because they can handle large amounts of current, especially as compared to other solid-state devices. Commonly available SCRs can handle continuous currents of hundreds of amps.

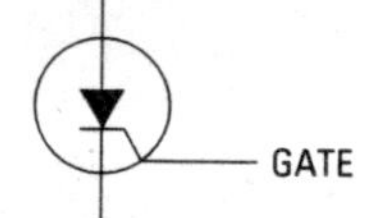

Figure 7-23 Thyristor symbol.

SCRs are also used to produce variable DC voltages from an AC source by drawing power from one half of a sine wave.

Figure 7-23 shows the symbol for thyristor (SCR). Figure 7-24 shows two common types of SCRs.

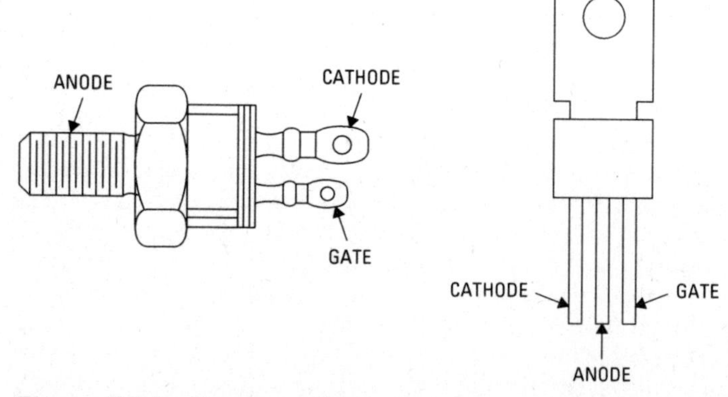

Figure 7-24 Common SCRs.

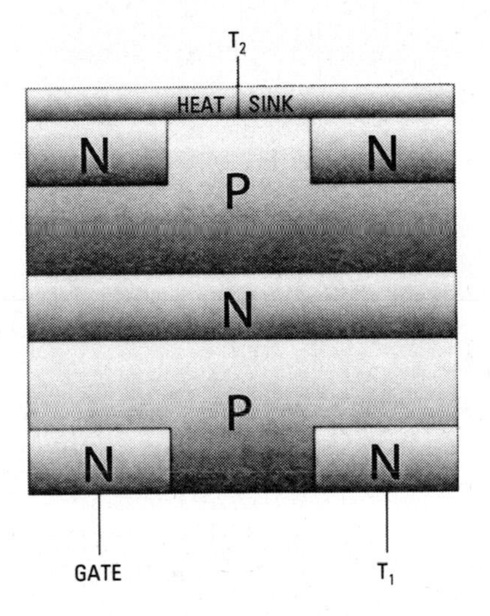

Figure 7-25 A triac.

Triacs

Triacs are modified SCRs, as shown in Figure 7-25. The triac blocks current flow in either direction until a current is sent in or out of its gate. Once one of these currents begins, current will be allowed to flow in either direction through the triac.

Since the triac can clip an AC sine wave in both directions, triacs are the functional components inside most dimmer switches and similar devices.

Summary

Diodes are the simplest semiconductor devices, consisting of a single PN junction. Normal diodes can be either forward-biased to allow current flow through them or reverse-biased, disallowing current flow.

There are also many specialized types of diodes. Zener diodes allow current to flow backward in a controlled manner and are useful in surge protection and for other applications. Avalanche diodes operate very similarly to Zeners, but use a different method of operation. Gold-doped diodes are used for high-frequency applications, photodiodes are activated by light, and light-emitting diodes (LEDs) give off light as electrons pass through them.

Transistors are three-part devices, with two PN junctions. (There are exceptions.) Tremendously useful, transistors can function as automatic switches or amplifiers. Standard transistors can provide 100-to-1 signal amplification, and specialized types can go much higher. Because of this, transistors are used in almost every conceivable form of audio or video device, and in innumerable other applications.

Perhaps more importantly, transistors can be made very small and used in the integrated circuit (IC) chips that make microprocessors possible.

Field-effect transistors (FETs, especially the MOSFET) are among the most common transistors. Not only are they very efficient, but they can operate at very high speeds as well. Another type of FET is the junction FET (JFET). While it does not have all the attractive characteristics of the MOSFET, the JFET is very effective and is used in a great many applications.

Thyristors (also called silicon-controlled rectifiers, SCRs) are four-layer semiconductor devices, and are very effectively applied for using a portion of an AC signal. Similar to the thyristor, triacs can do what the thyristor does, but in two directions, not just one.

In the next chapter we will be examining semiconductors that deal not only with electricity, but with light as well.

Review Questions

1. What is leakage current?
2. Why are voltage ratings important?
3. Name two applications of diodes.
4. What does the term point contact refer to?
5. How are Zener diodes connected to circuits?
6. What is a breakdown voltage?
7. Which diodes produce photons?
8. What is a photon?
9. What is an inversion channel?
10. What is a depletion zone?
11. What is meant by "clipping a sine wave"?
12. How do triacs operate differently than thyristors?

Exercises

1. Draw a MOSFET in a circuit, operating some device. Specifics not yet covered in this text are not important. Show a MOSFET operating and controlling something. Give as much detail as you can.
2. Draw the operation of an LED. Show electron movement and photon production. Also, show the LED coupled with a photodiode and their operation together as an opto-isolator.

Chapter 8

Optoelectronics

Optoelectronics is the study and use of electronic devices that interact with light. Bear in mind, however, that light includes not only visible light but other wavelengths of light (such as infrared and ultraviolet) that are not visible to the human eye.

This chapter discusses some basic concepts about light, and then describes several optoelectronic devices.

The Origins of Light

Before you can understand the applications of light to optoelectronics, it is helpful to understand exactly where light comes from.

Like electricity, light begins in the outer electron shells of atoms. The essential particle of light is the *photon*. Photons, which share many characteristics with electrons, are expelled from the outer shells of electrons when an electron moves from one energy level to another, as if the excess energy of the electron is thrown off in the form of a photon.

The scientific study of the production of both electricity and light is called *quantum electrodynamics* (*QED*). The final theory of QED was developed by Richard Feynman (pronounced FINE-man), a much-renowned American physicist. This theory, which won Feynman the Nobel Prize (along with Julian Schwinger and Shin'ichiro Tomonaga) describes with great accuracy how energy and particles work in the electron shells of atoms.

Feynman's life also provides a great story, beginning in electronics. As an adolescent, Feynman was enthralled with radio, which was just being commercialized at that time. With obvious intellectual gifts, young Feynman threw himself into radio and electronics, and was offered one academic position after another. He made his way through electronics to mathematics to physics. Among other accomplishments, Feynman was one of the scientists who developed the atomic bomb, was given to safe-cracking and bongo playing, and ran the scientific investigation of the explosion of the Space Shuttle Challenger. (If you want to know more tales of Feynman's exploits, I recommend reading his *"Surely You're Joking, Mr. Feynman!": Adventures of a Curious Character* [New York: W.W. Norton, 1997].)

Quantum mechanics explains that, while electrons constantly vary in the amount of energy they possess, they do so in steps, not in smooth variations. It is this movement between steps that causes electrons to be expelled from atoms. Chapter 7 discussed photodiodes, and said that electrons give off photons as they pass through a PN junction. This is because the electron is reduced in energy at the junction and loses this energy in the form of a photon. Similar events cause the production of light in electric lamps—fluorescent, incandescent, and otherwise.

The Nature of Light

We have already stated that a photon is usually considered to be a particle of light, much as protons, neutrons, and electrons are atomic particles. But this is not entirely the case. Photons act as both particles and waves. If it seems strange to you that one thing can at the same time act as a particle and a wave, you are not alone. This has been a famously difficult concept ever since it was discovered. Nonetheless, it is true. We will not delve into quantum physics and attempt to explain this in a book on basic electronics, but you should keep this fact in mind.

At times, we will speak of light as having a certain wavelength or frequency—speaking of it as if it were a wave. Other times we will talk of it as being a particle, much like an electron. Both uses are equally true. We use one or the other because it is more illustrative for the particular setting. If this seems confusing, just remember that light is both particle and wave at the same time. Describing light as either particle or wave is correct, though not all-encompassing.

All light fits into the electromagnetic spectrum, as shown in Figure 8-1.

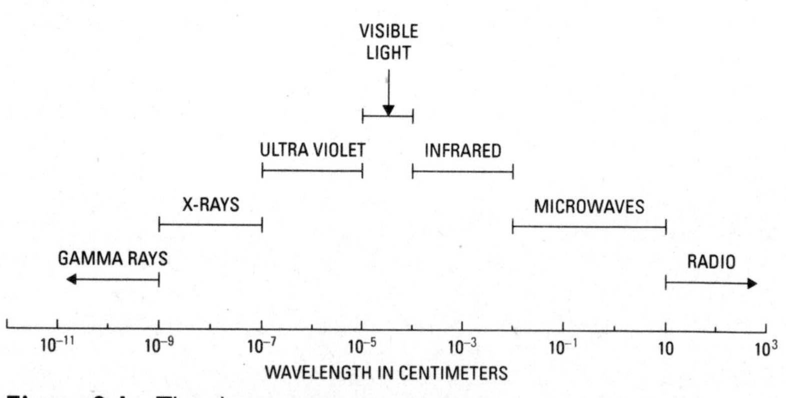

Figure 8-1 The electromagnetic spectrum.

Wavelength

The first characteristic of light that presents itself is color. Some light is red, some green, and so on. After years of research, it was discovered that each color of light has its own *wavelength*.

Figure 8-2 shows the concept of wavelength. Wavelength is the distance between two repeating units of a wave pattern. Notice in Figure 8-2 that the wavelength is the distance between two peaks of the sine wave that is shown. The basic formula for wavelength is as follows:

$$\lambda = \frac{c}{f}$$

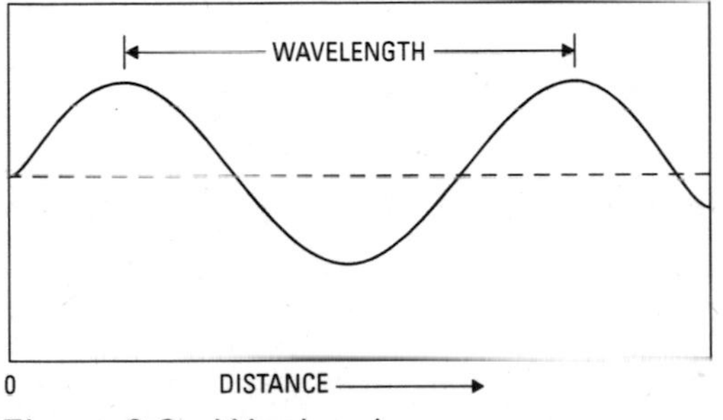

Figure 8-2 Wavelength.

In this formula we use the Greek letter lambda (λ) to represent wavelength, and c to represent the speed of the light (for radio waves and light) or the speed of sound (for sound waves). And while we won't delve into this subject here, be aware that the speed of light through glass, water, or some other medium is not the same as the speed of light in a vacuum. This difference of speeds (light is slower in these media) is called the *refractive index*.

Looking at Figure 8-2, you can see that wavelength is a function of frequency. If more waves occur in a certain period of time (one second is almost always used), then the waves have a higher frequency. Also, each wave will be shorter. So, frequency and wavelength have an inverse relationship to one another. Fewer waves per second makes for longer waves, as is shown in Figure 8-3A. More waves per second makes for shorter waves, as is shown in Figure 8-3B.

Table 8-1 shows the frequencies of various electromagnetic waves.

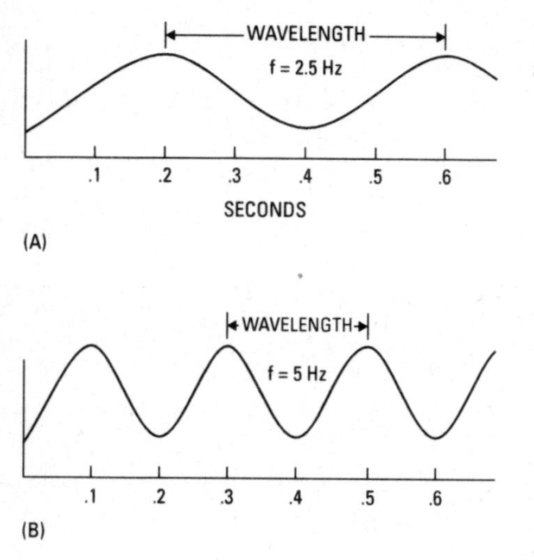

Figure 8-3 Wavelength and frequency.

Table 8-1 Frequencies of Electromagnetic Waves

Type	*Frequency*
Gamma Rays	More than 3×10^{19} Hz
X-Rays	3×10^{17} to 3×10^{19} Hz
Ultraviolet	7.5×10^{14} to 3×10^{17} Hz
Visible	4.3×10^{14} to 7.5×10^{14} Hz
Infrared	3×10^{12} to 4.3×10^{14} Hz
Microwave	3×10^{9} to 3×10^{12} Hz
Radio	Up to 3×10^{9} Hz (3000 MHz)

Every ray of light has its own frequency (measured in hertz). For instance, waves of red light have a lower frequency than blue light, and green is somewhere in between. Our eyes perceive different frequencies of light waves just like radios tune into different frequencies of radio waves. But rather than specifying light by its frequency, we signify the color by the wavelength of the light. So, a certain *wavelength* of light means a certain *color* of light.

We measure wavelengths in nanometers (nm), or billionths of a meter. For example, a deep red light would have a wavelength of about 600 nm.

Wavelength and frequency really signify the same thing, and can be interchanged by using a formula. Wavelength is specified rather than frequency, since it is generally easier to use.

Our eyes are tuned for the light of the sun. The sun puts out certain wavelengths of light, which our eyes pick up very well. There are, however, other wavelengths of light, and some of those are beyond our visual range. In other words, our eyes have limits, and beyond those limits we are blind. This is like a radio that can pick up the AM and FM bands but not the shortwave bands. The shortwave signals are out there waiting, but your receiver cannot tune in to them.

You have no doubt heard of *infrared light*. These light waves are useful, but we cannot see them. It is the type of light that is commonly used in optical signals, such as television remotes.

Ultraviolet light is much the same as infrared light. It is light just as much as blue and green are light, but our eyes do not recognize it. There are many sources of ultraviolet light, such as the sun; ultraviolet light causes sunburn.

Intensity

While the color of light is best understood by using the wave analogy, the intensity of light is best understood by using the particle analogy.

You are now familiar with the intensity of current flow, and you understand that the more electrons per second that pass through a circuit, the more intense is the current, and the more power is present in the circuit.

Light is much the same. The more photons that strike your eye per second, the more powerful (that is, brighter) is the light. A bright light pours huge numbers of photons into your eyes. A dim light pours very few photons into your eyes.

Just as electrical power is measured in watts, optical power is also measured in watts. However, the most levels of optical power are in the ranges of milliwatts (mW) and microwatts.

Optoelectronic Devices

Armed with some understanding of the nature of light, we can now proceed to optoelectronic devices. This will by no means be a comprehensive discussion, but this section addresses some of the important, basic optoelectronic devices.

Photodiodes

As explained in Chapter 7, photodiodes are PN junction devices that have transparent junctions. That is, the junctions are made with specialized, clear materials. When photons enter the device, they arrive and enough energy to push electrons across the junction. Thus, in these devices, light causes the flow of electricity.

Photodiodes are used to make solar cells, which are used to generate electricity from sunlight. They are also used in light meters, and more important, to receive light pulses from optical fibers and translate them into electrical pulses. Such photodiodes are fitted with ports for the connection of fiber-optic connectors.

Phototransistors

Phototransistors are normal bipolar-junction transistors that are enclosed in a clear material, so that light is permitted to reach the base-collector junction. Because of the amplification effect of transistors, a phototransistor works just as a photodiode does, except that the resultant output signal is much larger.

Light-Emitting Diodes (LEDs)

As electrons cross the PN junction of a diode, they emit photons. In standard diodes, these photons have infrared wavelengths (that is, not visible). They are also reabsorbed by the semiconductor material, much as a black material absorbs photons. (The fact that black items placed in sunlight heat up is an effect of absorbing photons and the energy of the photons. The dark-colored substance retains the light energy. This causes the molecules in this substance to vibrate and run against each other, which causes the heat. Actually, heat is nothing more than matter that is vibrating on a microscopic level.)

If diodes are manufactured in specific shapes and from specific materials, photons emitted from the junction are emitted as visible light.

By altering the operating voltages of LEDs, many specific colors are produced. Specialized LEDs are being developed continually, with combinations of LEDs being used to produce a wide variety of colors, and even white.

Laser Diodes

Special polishing and manufacturing techniques allow LEDs to function as *laser diodes*. This is a very useful device for fiber-optic technologies, because laser diodes are smaller than LEDs and produce a much narrower range of wavelengths than standard LEDs. As explained in Chapter 14, these are critically important characteristics for fiber-optic operations.

Optical Isolators

It can be very useful to pair an LED with a photodiode to create an *optical isolator* or *opto-isolator*.

An opto-isolator allows signals to be transferred from one electronic circuit to another without electrons passing through the junction. Photons are used to pass the signal at the junction instead. This is useful in many highly sensitive circuits.

Electricity is used for signal transmission of each side of the optical isolator, then to create light or receive light on either side of the gap. The electrical circuits never touch. Thus, one side of the device is isolated from the other.

Photoresistors

Photoresistors are, as their name implies, devices that vary in their resistance according to the amount of light that strikes them. Photoresistors are also called *photoconductors* or *light-dependent resistors*.

A photoresistor is built from a high-resistance semiconductor. When light strikes this material, the photons add their energy to the energy of the electrons already present and allow them to flow through the material more effectively. The effect is similar to that of adding extra widening to the conductive band in a field-effect transistor, as explained in the section "Field-Effect Transistors (FETs)" in Chapter 7.

Photoconductors are used in a wide variety of applications, including all sorts of light detectors, light-activated switches, and, especially, infrared detectors.

Photomultipliers

Photomultipliers are somewhat unique in that they are tubes, and no semiconductor has ever been developed to replace them.

When photons strike the cathode of the photomultiplier, the number of electrons transmitted through the device is multiplied enormously. Phototransistors can create a similar effect, but not on the scale of the photomultiplier. The photomultiplier tube can amplify the signal by a factor of 10^8, or 100 million times. Measurable signals can be obtained from a single photon.

Photomultipliers are used where extreme sensitivity is required, such as in scientific applications.

Charge-Coupled Devices (CCDs)

Charge-coupled devices (CCDs) are very common but complex devices, found in almost all digital photography equipment, still or video.

A CCD is usually approximately 2 centimeters square and flat. It usually contains several million elements, arranged in a grid. These elements are called *picture elements*, or *pixels*. Figure 8-4 shows a portion of such a grid, with an individual pixel identified.

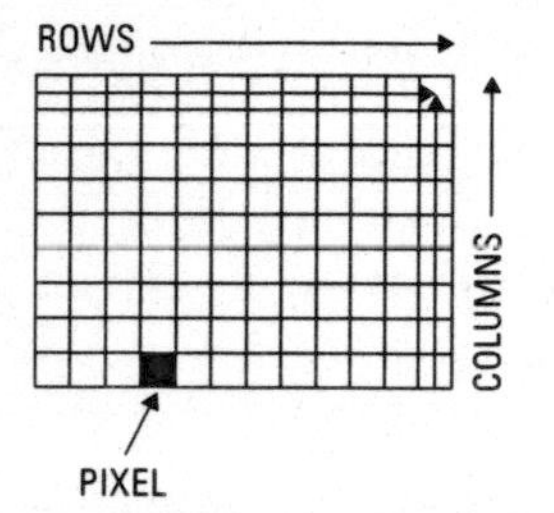

Figure 8-4 Portion of a CCD grid.

Each pixel is composed of a photodiode and a capacitor. If light strikes a pixel, it produces a small voltage, which causes electrons to flow into the linked capacitor. Used this way, the capacitors act as batteries. However, these batteries can be quickly and easily discharged. If a large amount of light strikes an element between discharges, it will accumulate a relatively high charge. If it receives little or no light between discharges, the electrical charge in the element will be very small.

The CCD generates a television signal by discharging the individual elements in the proper order (one row after another, from left to right). This is done very quickly, and each element is ready to detect light and collect electrons again after each discharge. One discharge of the entire CCD from top to bottom (every pixel in order) will allow one complete screen to be reassembled and shown on a display. By completing the process 30 times per second, standard television signals are obtained. The discharging process can be done either directly, or through *shift registers*, but the final result is the same—the conversion of photoelectric charges into a television signal.

This explanation pertains to black-and-white signals. Color photography is similar but more complex.

Summary

Optoelectronics is the study and use of electronic devices that interact with light.

Light begins in the outer electron shells of atoms. The essential particle of light is the *photon*. Photons are expelled from the outer shells of electrons when an electron moves from one energy level to another, as if the excess energy of the electron is thrown off in the form of a photon.

Photons act as both particles and waves. The action of light can be explained as either particle-like or wave-like. Both explanations are correct.

Visible light composes only a small part of the electromagnetic spectrum. Additionally, both ultraviolet and infrared are types of light that are not visible to the human eye. However, ultraviolet and infrared are very useful in optoelectronic devices.

The color of light is measured in wavelength. Although wavelength also translates into frequency, wavelength is the standard measurement of the color of light, and is usually expressed in nanometers.

The intensity of light (its brightness) is measure in milliwatts or microwatts.

Photodiodes are diodes with clear material surrounding the PN junction. When photons enter the device, they add enough energy to push electrons across the junction. The entering light causes the flow of electricity.

Phototransistors operate similarly to photodiodes, but they also amplify the resultant signal output.

LEDs are diodes that are arranged so that electrons crossing the PN junction of a diode emit photons in the visible light range.

Optical isolators are made by pairing an LED with a photodiode. This allows signals to be sent from one electrical circuit to another in the form of light. By using these compound devices, a signal is sent from end to end, but the two circuits are entirely isolated one from another.

Photoresistors are built from high-resistance semiconductors, and they vary in their resistance according to the amount of light that strikes them. When light strikes this material, the photons add their energy to the energy of the electrons already present and allow them to flow through the material more effectively.

The photomultiplier tube can amplify the signal by a factor of 10^8, or 100 million times. Measurable signals can be obtained from a single photon.

CCDs are integrated grids of picture elements. Each element consists of a photodetector with a capacitor coupled to it. The capacitors collect the electrons that are moved by light striking its photodiode, and special circuits discharge the capacitors one at a time in a rigid order. The resulting signal can be used to create either still or video images in a receiver.

In the next chapter we will shift gears a bit and cover all of the various materials that are used to construct electronic circuits.

Review Questions

1. Which scientific theory describes the production of light?
2. Where do photons originate?
3. What does "the dual nature of light" refer to?
4. How would you categorize light with a wavelength of 10^{-3} cm?

5. In which type(s) of diodes are photons created by an electron moving across a PN junction?
6. Which part of a phototransistor is covered with clear material to allow light to strike it?
7. What type of device is used to create solar cells?
8. What type of semiconductor device is designed to give off photons?
9. How is the operation of a photoresistor comparable to that of a MOSFET?
10. Describe the sensitivity level of a photomultiplier.
11. Why are the pixels in a CCD arranged in a grid?
12. Describe the operation of capacitors in CCDs.
13. How many photodiodes and capacitors might you expect to find in a CCD?

Exercises

1. Graph a wavelength of 3 centimeters.
2. Draw a photoresistor in a circuit. Identify the parts and show values.

Chapter 9

Circuit Components

Regardless of how familiar you may be with the intricacies of electronic components, their operations, and their applications, to actually build circuits and devices, you need skill with more mundane materials, such as wires, fuses, connectors, and switches.

This chapter explains the essential components of electronic circuits and applications. We will describe wiring materials and how they are used.

Conductors

By far, the most common type of conductor used in constructing anything electronic is insulated copper wire. Although there are many other types of conductors, copper wire is effective, cheap, readily available, standardized, and easy to use.

Copper Wire

Insulated copper wire should be familiar to all of us. These conductors have two parts: the conductors themselves and a layer of insulation surrounding and protecting the conductor.

The wire itself is measured in gauge. In almost all cases in the United States, the American Wire Gauge (AWG) is used. Table 9-1 shows some physical characteristics of copper wire. Note that under the AWG system, the larger the number, the smaller the wire. (Once the numbers reach zero, this reverses, and the larger the number, the larger the wire. However, this only applies to large building wires—the sizes used for carrying hundreds of amperes.)

Notice in the two far right-hand columns of Table 9-1 that the resistance of the wire is also shown. Two things are important to note:

- The larger the wire, the lower the resistance. The resistance of copper wire is very low. For example, #18 wire, a common size, has only 6.51 ohms per thousand feet, or 0.0065 ohms per foot. In most applications, the resistance of wire is negligible and is not included in design calculations.
- Temperature affects the resistance of copper wire. The hotter the wire, the more resistance it possesses. Again, this is negligible for most applications, but not in every case.

The insulation must also be considered carefully. Each type of insulation has its own characteristics. Some types protect quite well

Table 9-1 Characteristics of Copper Wire

AWG	*Area in Circular Mils (Thousandths of an Inch)*	*Diameter Millimeters*	*At 77°F Ohms per 1000 Feet*	*At 149°F Ohms per 1000 Feet*
40	9.9	0.080	1070	1230
38	15.7	0.101	673	776
36	25	0.127	423	488
34	39.8	0.160	266	307
32	63.2	0.202	167	193
30	101	0.255	105	121
28	160	0.321	66.2	76.4
26	254	0.405	41.6	48.0
24	404	0.511	26.2	30.2
22	642	0.644	16.5	19.0
20	1020	0.812	10.4	11.9
18	1620	1.024	6.51	7.51
16	2580	1.291	4.09	4.73
14	4110	1.628	2.58	2.97
12	6530	2.053	1.62	1.87
10	10400	2.588	1.02	1.18
8	16500	3.264	0.641	0.739
6	26300	4.115	0.403	0.465
4	41700	5.189	0.253	0.292
3	52600	5.827	0.201	0.232
2	66400	6.544	0.159	0.184
1	83700	7.348	0.126	0.146
0	106000	8.255	0.100	0.116
00	133000	9.271	0.079	0.092
000	168000	10.414	0.063	0.073
0000	212000	11.684	0.050	0.057

against some chemicals or some types of damage, but not others. In most electronic applications, any common type of insulation will do, but this is not always the case. Table 9-2 shows the characteristics of various types of conductor insulation.

Note also that conductor insulation is heat-rated. This rating is usually in degrees centigrade for the *ambient* temperature (the temperature in the immediate area of the conductor). Using a conductor in an area with higher ambient temperatures than its insulation is rated for leads to insulation failures, and, potentially, to shocking hazards and/or to fire.

Table 9-2 Conductor Insulation Characteristics

Insulation Material	Temp.°C		Resistance To:						
	Min.	Max.	Ozone	Flame	Moisture	Oil	Alcohols	Gasoline	Sunlight
PVC	–40	105	G	G	G	F	G	P	G
Polyethylene	–55	80	G	P	E	G	E	F	E
Crosslinked Polyethylene	–55	125	G	G	E	G	E	G	E
Foamed Polyethylene	–55	80	G	P	G	G	E	F	E
Rubber	–40	75	P	P	G	P	G	P	F
Nylon	–55	115	G	G	F	E	P	F	E
Duralon	–24	105				E		G	
PTFE Teflon	–70	260	E	E	E	E	E	E	E
FEP Teflon	–70	200	E	E	E	E	E	E	E
Foamed FEP	–70	200	E	E	G	E	E	G	E
Polypropylene	–40	105	G	P	E	F	E	F	E
Silicone Rubber	–80	150	E	F	G	F	F	F	E
Kapton	–70	20()	E	E	G	E	E	G	E
Tefzel	–70	150	E	E	E	E	E	E	E
XML-125	–55	125	G	E	G	G	G	G	G
Irravin	–40	105	G	G	E	E	G	F	G
Exane	–55	110	G	E	E	E	E	G	E

Note: E = Excellent; G = Good; F = Fair; P = Poor

Table 9-3 Conductor Ampacities

AWG	*Ampacity*
27	500 mA
20	5 A
18	7 A
16	10 A
14	15 A
12	12 A

Table 9-3 shows the ampacities for insulated copper conductors. If conductors are overloaded with current, they will overheat, possibly leading to shorts, arcing, and circuit failures.

Other Conductors

While aluminum conducts electricity well enough to be used in building wiring, it is very seldom used for electronics work. Because large conductors are so seldom required, the cost differential between copper and aluminum is negligible. Copper is usually chosen because it has better conductivity and heat expansion qualities. Aluminum's primary attractiveness for large conductors is that it is light (easy to transport) and less expensive.

A *bus* can also be used, though such use is rare. A bus is simply a solid piece of conductor, most typically a copper or aluminum bar. A copper bus bar typically is rated at 1000 amps per square inch, and an aluminum bus bar at 700 amps per square inch.

There are, however, other uses of the term bus. A bus is considered any common conductor. In most electronic work, the term bus is applied to the connecting means between components, regardless of the type of conductor being used to make these connections.

Cables

In addition to individual conductors, a wide variety of specialized cables are used in electronic applications. Frequently these cables are composed of conductors that are carefully paired together. Cable designs can be very intricate.

We will be covering specific cable types with the systems they pertain to. For example, we will cover fiber-optic cables in Chapter 14.

Connections

It is critical in any type of electrical or electronic work that connections be made both mechanically and electrically secure. Two common means for securing connections are soldering and the use of crimp connectors.

Soldering

The primary method of connection for electronic work is, of course, soldering. Soldered connections are quite sound electrically, and are strong enough mechanically, if they are in an enclosed location. However, soldered connections are not strong enough if they are exposed to any type of physical damage. They require some sort of case or enclosure and protection from the outside environment.

Soldering is not a difficult skill to master, though it can take time and practice. If you are not already experienced in soldering, spend time practicing. Get a good grade of soldering iron (properly called a *soldering pencil*), some rosin-core solder, and an old circuit board to practice on. Make connections, then test them both mechanically and electrically. That is, first pull on them and then connect an ohmmeter across them.

A soldering pencil should be rated between 25 and 40 watts for electronics work. Too much wattage results in too much heat, which can damage some items. We won't take the space here to go through all the details of soldering. A good soldering iron should come with soldering instructions. Sorry, there are no shortcuts. You simply must practice until you have a good feel for what you are doing.

You may also want to get a desoldering tool. Desoldering tools are often necessary for removing components from circuit boards. As with the soldering iron, practice until you get it right.

Crimp Connectors

While soldering is by far the most common means of connecting conductors and devices in electronic work, conductors are sometimes connected with solderless connectors as well. Crimp-type lugs and splices are used. *Crimp connectors*, as shown in Figure 9-1, are first stripped of about half an inch of insulation at their end, then placed inside the connector. Then, a special crimping tool (about the size and shape of a large pair of pliers) is used to crush the barrel of the connector. The type of metal used in the connector will crimp under pressure, then hold that position. The result is a strong, permanent connection.

Another type of connection that is used to connect two or more wires together is the standard *electrical wire nut*, also called a *solderless connector*.

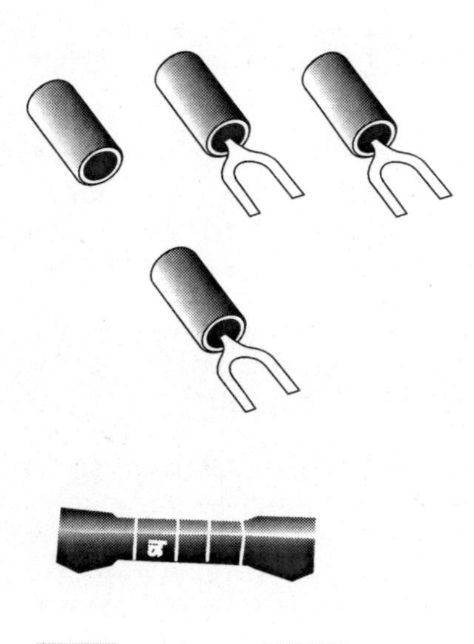

Figure 9-1 Crimp connectors and splices.

Switches

Electronic switches do the same job as switches in your home—they allow current to flow through a circuit, or they prevent current from flowing through a circuit. In other words, they make contact or break contact.

Figure 9-2 shows the basic configurations for switches. Figure 9-2A shows the basic *single-pole switch*, which makes or breaks contact for a single switch. Figure 9-2B shows a single-pole switch that can connect one end of the switch to either of two other circuits. This is called a *single-pole double-throw switch*. In other words, it can be thrown in either of two directions.

Figure 9-2C shows a *two-pole switch* (also called *double-throw switch*). It is, essentially, two single-pole switches that are tied together for simultaneous operation. Figure 9-2D shows a *double-pole, double-throw*. This is essentially two single-pole, double-throw switches tied together. Figure 9-2E shows a *selector switch*, which is used to connect one side of the switch to several different circuits.

Figure 9-3 shows another type of switching characteristic. This is the concept of *normal position*. The symbols shown here are those of push-button switches. You are no doubt familiar with these in the

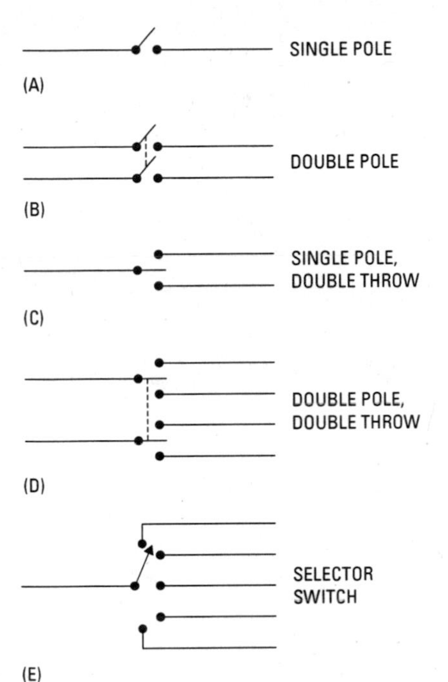

Figure 9-2 Switch configurations.

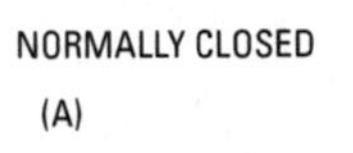

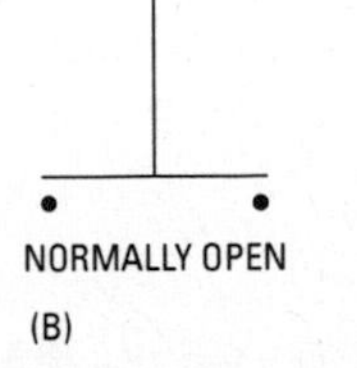

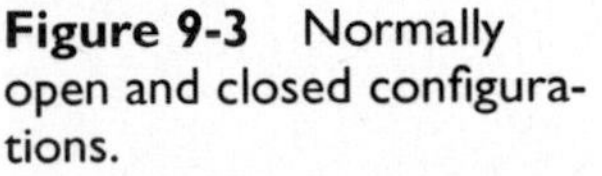

Figure 9-3 Normally open and closed configurations.

form of start and stop switches. These switches are shown in their *normal* position (that is, the position they remain in before you push them, or before they are somehow activated). Note that pushing the normally closed switch will break current flow in a series circuit. We use these as stop switches (and for many other uses as well). The normally open switch allows current to flow through its circuit only when activated. This is the classic start switch.

There are many types of switches: small, large, single-circuit, multicircuit, high-temperature switches, low-temperature switches, and almost any other type you might imagine. But they

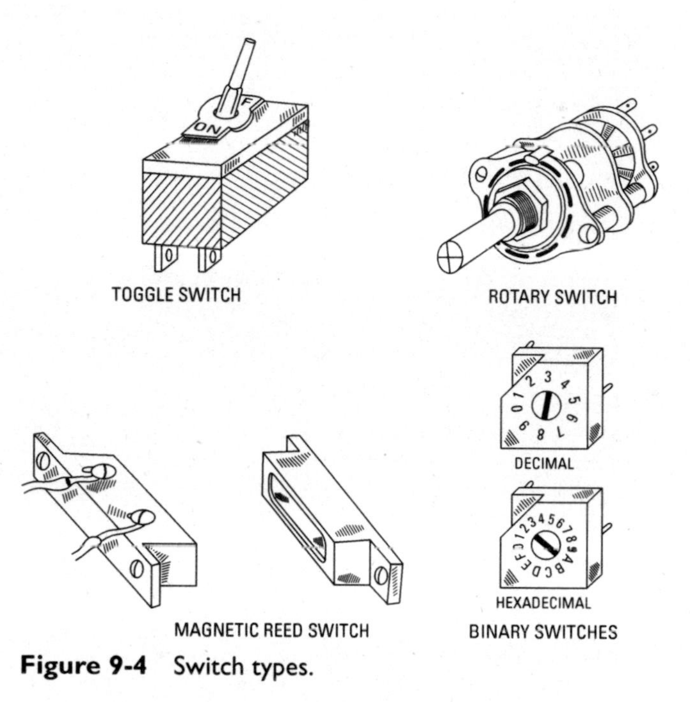

Figure 9-4 Switch types.

all make and break contact inside of circuits. Some switches are designed to maintain contact after they are thrown (such as a standard light switch). These are called *maintained contact* switches. Others only maintain contact for the moment they are activated. These are called *momentary contact* switches.

Figure 9-4 shows several common types of switches.

Remember that switches must be applied according to the voltage and current ratings. You cannot use a 5-volt switch on a 48-volt circuit. It will not work properly if you do. The same goes for current. You cannot use a switch rated 100 milliamps to break a 2-ampere current. They must be applied properly.

Fuses

Fuses are circuit elements that prevent too much current from flowing through a circuit. This excess current is called *overcurrent*, and fuses fall into the classification of *overcurrent protective devices*. Too much current flowing through a circuit tends to overheat and damage a variety of components and can lead to fire (among other unpleasant effects).

Electronic fuses are typically hollow glass cylinders with metal connection rings on each end and a metal (typically lead) filament going from end to end through the glass ferrule. This lead filament is designed to melt at a specific level of current. If an overcurrent condition occurs, the fuse will melt, breaking the flow of current.

Figure 9-5 shows several types of fuses.

Fuses are always installed in series with the circuit they are to protect, never in parallel (as should be obvious).

Fuses are either soldered in place or are clipped into special holders that may be bolted or screwed into place.

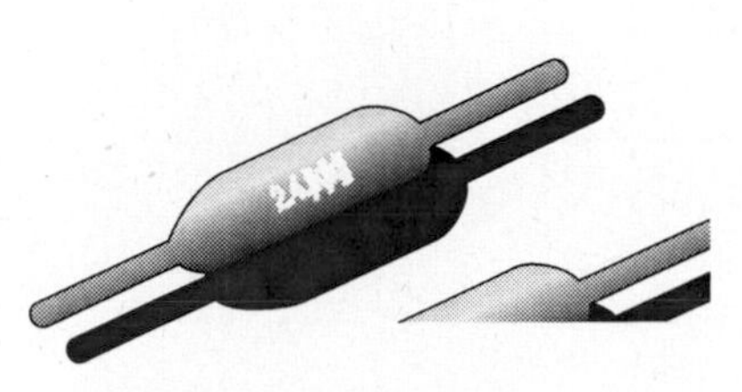

Figure 9-5 Types of fuses.

Resistors

Resistors are used frequently in electronics work. They are installed whenever resistance is required for circuit operations.

Resistors come in a huge variety of shapes and sizes. Almost all of them are designed to be soldered into place.

Resistors are made of materials that are known to have a certain and verifiable value of resistance. As with other electronic devices, it is critical to ensure that resistors are used within their specified limits. Resistors are obviously rated in ohms, but they are also rated in watts. Running too many watts of electricity through a resistor will cause it to burn out, opening the circuit and stopping its operation.

Because resistors are frequently too small to print on legibly, they are marked with a color code that expresses their value in ohms. Figure 9-6 shows the color codes. Notice that the first two rings express numerical values, the third ring expresses a multiplier, and the fourth ring expresses the resistor's tolerance.

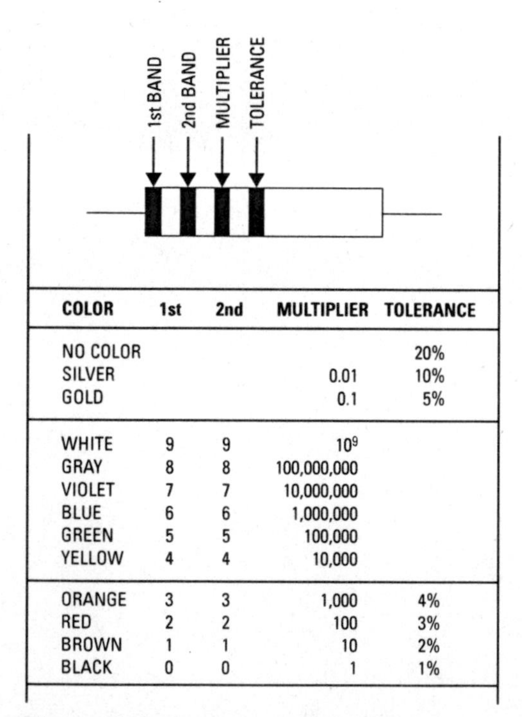

COLOR	1st	2nd	MULTIPLIER	TOLERANCE
NO COLOR				20%
SILVER			0.01	10%
GOLD			0.1	5%
WHITE	9	9	10^9	
GRAY	8	8	100,000,000	
VIOLET	7	7	10,000,000	
BLUE	6	6	1,000,000	
GREEN	5	5	100,000	
YELLOW	4	4	10,000	
ORANGE	3	3	1,000	4%
RED	2	2	100	3%
BROWN	1	1	10	2%
BLACK	0	0	1	1%

Figure 9-6 Resistor color codes.

For example, if the first two rings of a resistor were violet and green, that would give a numerical value of 75. If the third ring were orange, we would multiply 75 by 1000. This would tell us that the resistor possessed 75,000 ohms of resistance. If the final ring were silver, we would know that the tolerance for this resistor was 10 percent. This means that the actual resistance of the component is 75,000 ohms, give or take 10 percent. So, measured with a high-quality ohmmeter, the resistance would be measured in the range of 67,500 to 82,500 ohms.

You will frequently encounter variable resistors, also called *potentiometers*, because they can be used to control the potential (voltage) made available to a circuit or partial circuit.

Figure 9-7 shows several types of resistors.

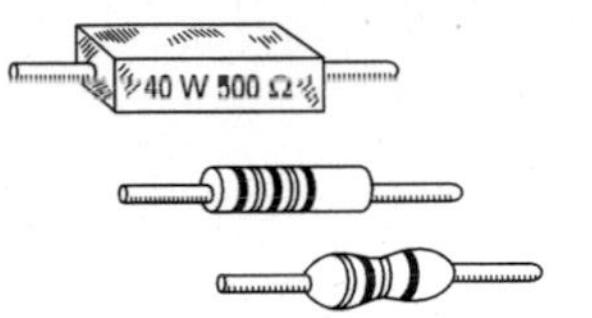

Figure 9-7 Types of resistors.

Capacitors

We have covered *capacitors* at some length in other discussions in this book, but it is important to note here that capacitors come in many different types, and in a wide variety of ratings.

Capacitors are almost always manufactured in some sort of small cylindrical package, with the electrolytic type being the favored type for most electronic work.

The first rating you would expect of a capacitor would be the amount of capacitance it adds to a circuit. This is expressed in *farads*. Because 1 farad is an enourmous amount of capacitance, you will far more commonly see capacitors rated in *microfarads* or *picofarads*. (For some modern audio systems, capacitors rated above 1 farad are used. They resemble large soda cans.)

The second rating for a capacitor is *voltage*. Because capacitors are composed as large surfaces in close proximity to one another, an excess of voltage can cause them to arc over and short out internally. This ruins the effect of the capacitor and, in electrolytic capacitors, generally causes them to explode (though not usually with great violence).

Figure 9-8 shows a listing of several common types of capacitors and their common ratings.

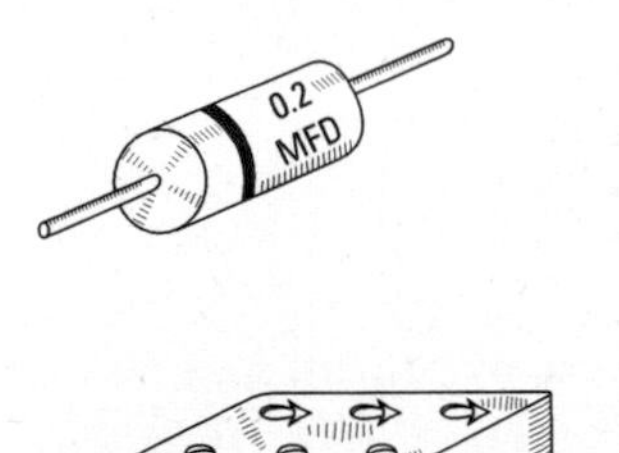

Figure 9-8 Types of capacitors.

One very important type of capacitor is the *variable capacitor*, as shown in Figure 9-9. You may recognize this as the old type of radio tuner mechanism. If you recall our coverage of resonance in Chapter 5, you will remember that the resonance of a circuit can be changed by varying the capacitance in the circuit. This is precisely how the variable capacitor was (and still is) used to make the radio circuitry resonant at specific frequencies.

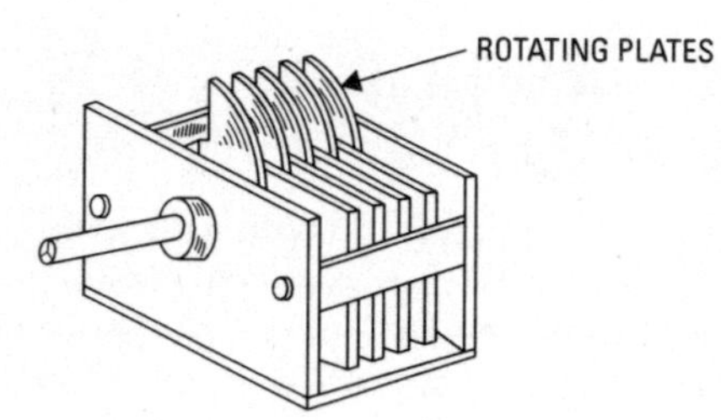

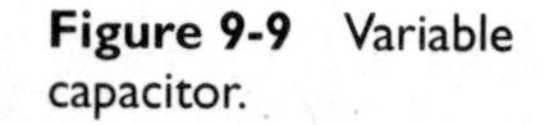

Figure 9-9 Variable capacitor.

Capacitors are nearly always soldered in place.

Inductors

The *inductors* used in electronic work are usually small coils of wire that add specific levels of inductance to a circuit.

Inductors with iron cores are often used for low-frequency applications, while air-core inductors are used for high-frequency applications. This is because of the tendency of the iron core of an inductor to overheat at high frequencies because of a magnetic effect called *hysteresis*.

Bear in mind that rather than saying, "there is an inductor in this circuit," many people will say, "there is a *coil* in this circuit." In other words, the two terms can be used interchangeably. You may also hear people refer to an inductor as a *choke*. This term is generally used to refer to an inductor that is being used in a circuit to suppress fluctuations in a direct current. Because inductors oppose a change in current level, they are very effective in this application.

Relays

Relays use a control current from one source to control a separate circuit by using electrical, magnetic, and mechanical principles. Figure 9-10 shows a very basic relay. The circuit that is controlling the relay is connected to terminals 1 and 2, and the circuit being controlled is connected to terminals 3 and 4. When the controlling circuit is energized, a current will flow through the magnetic coil, which is attached to terminals 1 and 2. This will cause the *solenoid* (defined later) to pull closed and so close the switch, which will complete the circuit that is being controlled.

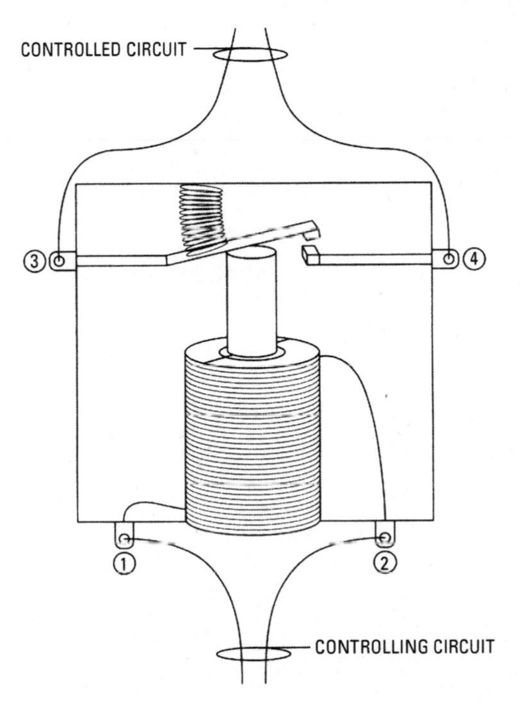

Figure 9-10 Basic relay.

The *solenoid* is nothing more than a coil of small copper wire with an opening in its middle into which the iron bar slides in and out. There is a small spring on the bottom of the iron bar to keep it pushed up (keeping the switch open) at all times, except when the solenoid is energized. When the solenoid is energized, it becomes an electromagnet and pulls the iron into its middle.

Figure 9-11 shows the most common type of control relay. This type of relay is also called an *eight-pin relay*, because it has eight terminals. The terminals are in the form of metal pins that fit into the relay's base, where wire can be connected. This type of relay is very small in size, standing only about 3 inches tall including the case, and is common and inexpensive. It can be found at electric supply houses, commercial supply houses, electronic supply houses, and even at some hardware stores. Figure 9-12 shows the circuitry for this type of relay. The power to the relay coil (the controlling

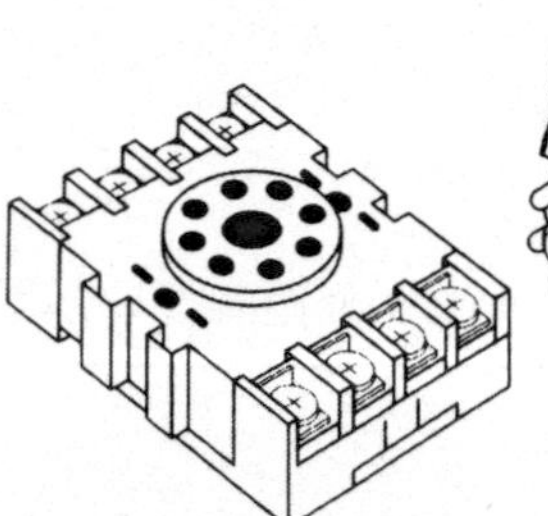
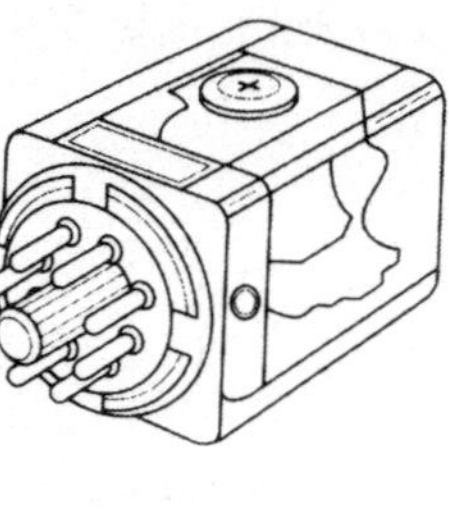

Figure 9-11 Control relay.

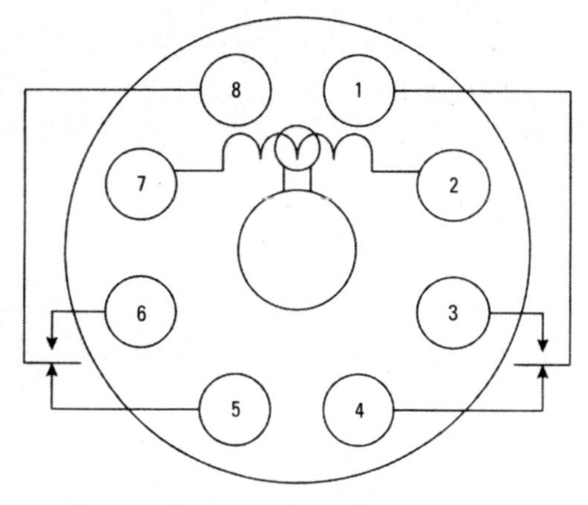

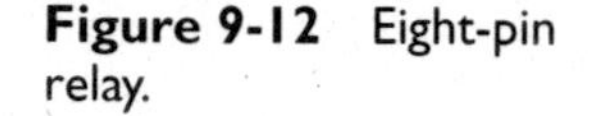

Figure 9-12 Eight-pin relay.

circuit) connects to terminals 2 and 7. Terminals 1 and 8 are what we can call *common* terminals (because they are common to two separate sets of contacts), 6 and 3 are *normally open* contacts, and 4 and 5 are *normally closed* contacts.

The operation of the relay, shown by using contacts 1, 3, and 4, is as follows. When power is sent to the relay coil (terminals 2 and 7), the coil energizes, and the various contacts in the relay either open or close. Contact 3 is not touching contact 1 before the power comes to the coil (which is why it is called a *normally open contact*). When the coil energizes, the contacts close, completing a circuit with terminals 1 and 3. Before the coil is energized, contacts 1 and 4 are touching (which is why they are called *normally closed contacts*). However, when the coil is energized, these contacts come apart (called *opening*), breaking the flow of any current flowing through the connection.

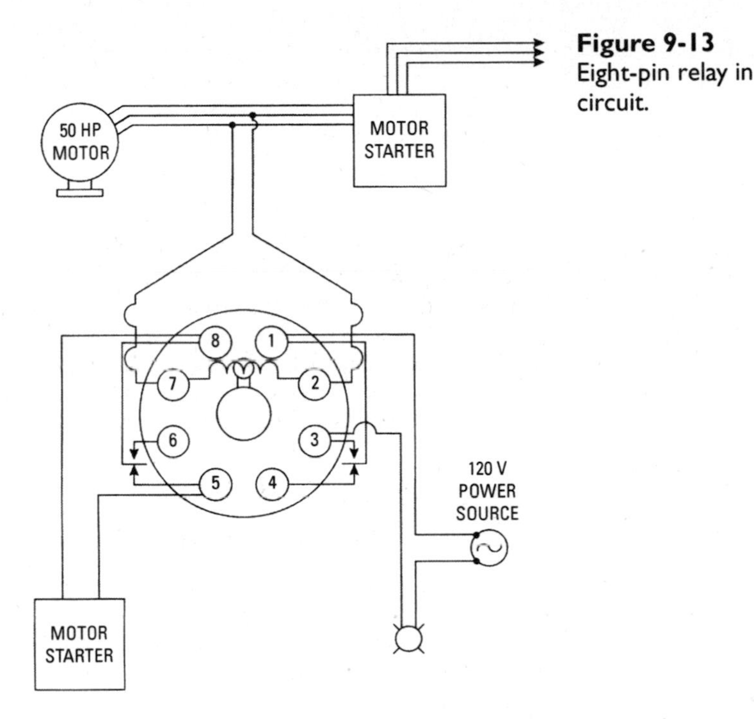

Figure 9-13
Eight-pin relay in circuit.

Figure 9-13 shows this relay in a traditional use. The relay is controlled according to a certain motor. When the motor is turned off, the relay coil stays deenergized; it becomes energized when the motor is turned on by its starter. When the motor turns on, the coil will become energized, and contacts 1 and 3 will close. This action causes a pilot light to turn on, which shows the factory's operations manager (in a remote location in the factory) that this particular motor has been turned on and is running. Contacts 5 and 8 are touching before power comes to this relay's coil, but when the relay is energized the contacts open, breaking the circuit going through these contacts. In this instance, when the contacts open, they break a control circuit that controls another motor in the factory. For safety reasons, this motor must be turned off when the first motor is operating. When contacts 5 and 8 break, the control circuit for the second motor is broken, shutting off the motor.

The relays just described are usually called *control* relays. But, beyond these, there are many other types of relays. Some of the more useful and common relays are as follows:

- *Miniature relays*—These operate in the same way as eight-pin relays, except that they are smaller and deal with lower levels of current. Their coils typically operate at between 5 and 24 volts DC. Figure 9-14 shows an example.
- *Reed relays*—Reed relays are light-duty relays that switch low levels of current. But they are very fast, with switching times of 2 milliseconds or less. Operating voltages are usually between 5 and 24 volts. Figure 9-15 shows an example.
- *Solid-state relays*—Solid-state relays are actually transistors and thyristors that are packaged as relays. They have superb response times but higher resistances than traditional relays. This sometimes requires design consideration. In addition, they have a low tolerance to overloads, burning out much more easily than mechanical relays. Figure 9-16 shows an example.

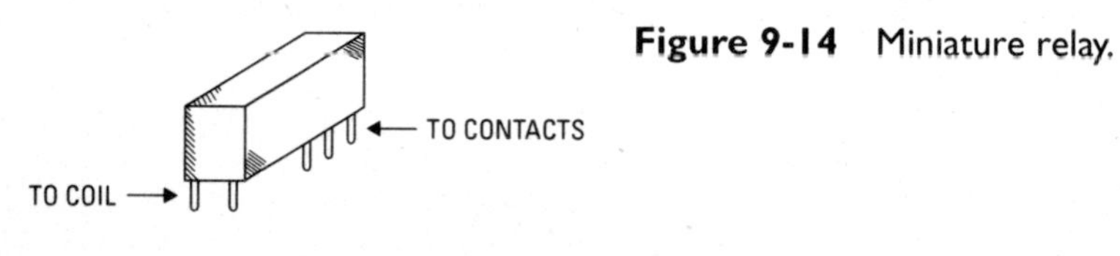

Figure 9-14 Miniature relay.

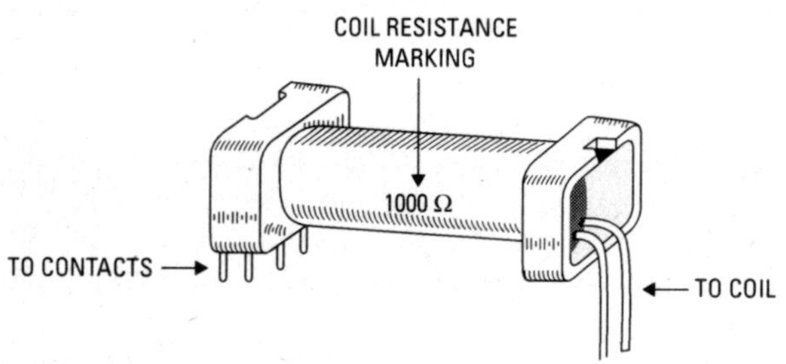

Figure 9-15 Reed relay.

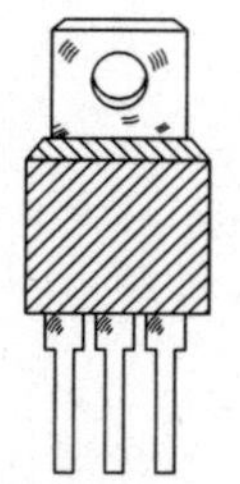

Figure 9-16 Solid-state relay.

- *Current relays*—These are specially made to trip when a certain level of current is present in the control circuit.
- *Time-delay relays*—These provide a delay of anywhere from a couple of seconds to a couple of minutes between the time when the solenoid actually pulls in and the time when the contacts open or close.
- *Momentary-contact relays*—The relays covered so far have been *maintained* contact relays. That is, once the relay is tripped, the contacts make or break contact and maintain that condition. *Momentary* contact relays are different in that, once they trip, they make or break contact for a moment and then return to the previous condition. They are especially effective when used with other control mechanisms and with control systems that require a momentary burst of energy instead of a constant one.
- *Mechanically held relays*—The standard type of relay is *electrically* held. That is, once the solenoid is pulled into position, it is held there by a continued electrical current. If the current were to stop, the solenoid would return to its original position. Mechanically held relays have a built-in mechanism that locks the solenoid into place once the relay is tripped. The control current can then be discontinued without affecting the position of the relay. This is often a desirable characteristic because it eliminates noise and waste of energy. Note that this type of relay needs a burst of power to turn on and a burst of power to turn off.

Transformers

Transformers were covered in depth in Chapter 4, but there are other installation concerns that we will cover here. We will expend more coverage regarding the installation of transformers simply because more and more specialized requirements apply.

The primary concern for the installation of transformers is the type of power supply they are exposed to. There are no special requirements for transformers that are on the load side of a Class 2 or Class 3 power supply. The class ratings are typically marked on power supplies, and refer to specific levels of votage and current that can be supplied by the devices. The concern over voltage and current levels are the threats of fire and other damage. So, because Class 2 and 3 circuits operate at low levels, special fire-prevention requirements are dropped.

But for transformers that connect to either building electrical systems or Class 1 power supplies, there are safety requirements that must be followed.

Overcurrent Protection

Transformers feed in from either power systems or Class 1 power supplies. If they are rated 600 volts or less (more restrictive requirements apply to transformers operating at over 600 volts), they can be protected by an overcurrent protective device on the primary side only (not in both primary and secondary sides). This must be rated at least 125 percent of the transformer's rated primary current. If the specified fuse or circuit breaker rating for transformers with a rated primary current of 9 amperes or more does not correspond to a standard rating, the next larger size can be used. For transformers with rated primary currents of less than 9 amperes, the overcurrent device can be rated up to 167 percent of the primary rated current (see Figure 9-17).

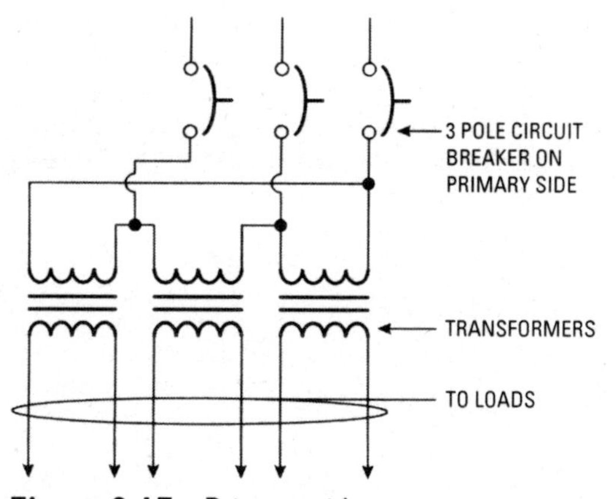

Figure 9-17 Primary side overcurrent protection.

Transformers operating at *less than* 600 volts are also allowed to be installed with overcurrent protection in the secondary only. This must be sized at 125 percent of the rated secondary current if the feeder overcurrent device is rated at no more than 250 percent of the transformer's rated primary current.

Transformers with thermal overload devices in the primary side don't require additional protection in the primary side unless the feeder overcurrent device is more than six times the primary's rated current (for transformers with 6 percent impedance or less) or four times primary current. If the specified fuse or circuit breaker rating for transformers with a rated primary current of 9 amperes or more does not correspond to a standard rating, the next larger size can be used. For transformers with rated primary currents of less than 9 amperes, the overcurrent device can be rated up to 167 percent of the primary rated current.

Autotransformers rated 600 volts or less must be protected by an overcurrent protective device in each ungrounded input conductor. This must be rated at least 125 percent of the rated input current. If the specified fuse or circuit breaker rating for transformers with a rated input current of 9 amperes or more does not correspond to a standard rating, the next larger size can be used. For transformers with rated input currents of less than 9 amperes, the overcurrent device can be rated up to 167 percent of the rated input current (see Figure 9-18 and Figure 9-19).

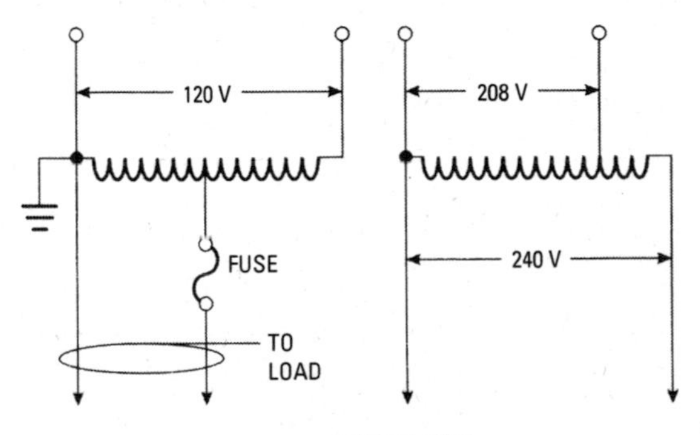

Figure 9-18 Autotransformers.

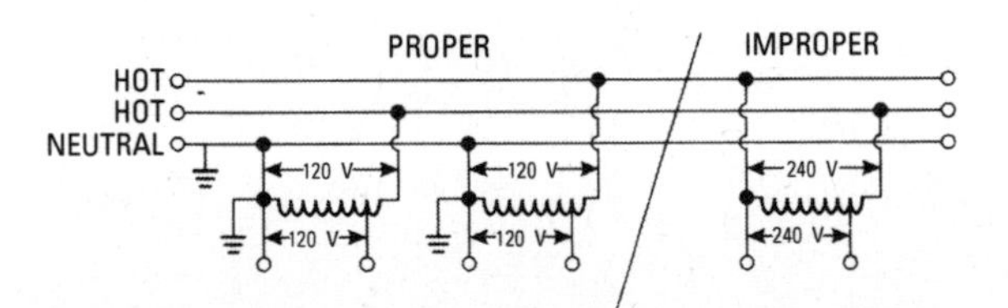

Figure 9-19 Proper and improper autotransformer connections.

Installation

Transformers connected to Class 1 or power circuits must be installed in places that have enough ventilation to avoid excessive heat build-up.

All *exposed noncurrent-carrying parts* of transformers must be grounded.

Transformers must be located in *accessible locations*, except as follows:

1. Dry-type transformers operating at less than 600 volts that are located in the open on walls, columns, and structures don't have to be in accessible locations.
2. Dry-type transformers operating at less than 600 volts and less than 50 volt-amperes are allowed in fire-resistant hollow spaces of buildings as long as they have enough ventilation to avoid excessive heating.

Indoor dry-type transformers must be separated by at least 12 inches from combustible materials. Fire-resistant, heat-resistant barriers can be substituted for this requirement. Transformers that are completely enclosed are exempt from this requirement.

The balance of the components we'll cover in this chapter are much simpler to apply than transformers.

Power Supplies

Figure 9-20 shows several types of *power supplies*. Figure 9-21 shows the interior of a standard power supply.

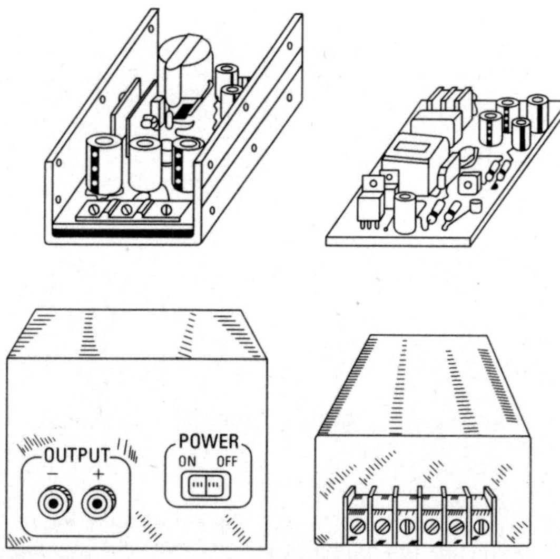

Figure 9-20 Power supplies.

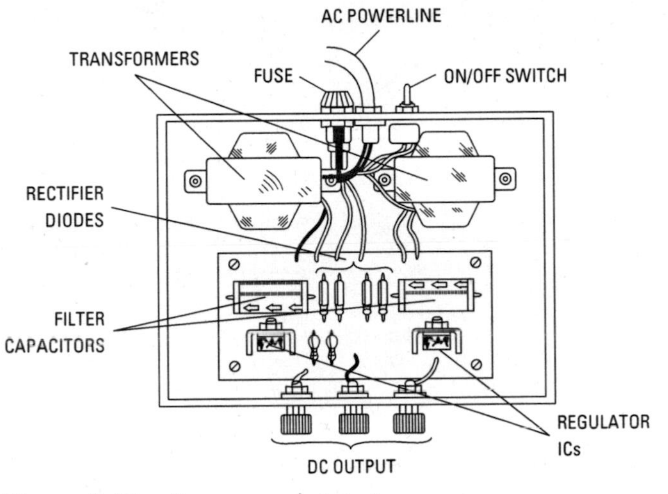

Figure 9-21 Power supply interior.

Figure 9-21 shows a standard type of DC power supply. This is done in stages. First, line power is reduced in voltage with one or more transformers. Second, transformers are used where a variety of output voltages are desired.

Next, the low AC voltages are rectified (turned into DC) with diodes. Then the rectified DC is filtered (that is, a capacitor is used to smooth out the rectified DC). This provides DC power at specified voltages.

A final stage is frequently added to this assembly—*voltage regulation*. This serves to raise or lower the voltage in the case of unusually low or high voltages being supplied to the power supply. This is done by using microprocessors that sample the voltage at fixed intervals, compare it to preset values, and then raise or lower the voltage as required. Figure 9-22 shows the basic circuitry for such a device.

Figure 9-23 shows the internal circuitry for a power supply.

Working with Electronic Components

Electronic devices must be handled with care, and used at or below their rated voltage and wattage.

Most electronic parts are very durable and, thus, not often damaged by normal treatment. Nevertheless, you may want to pay a little extra attention to the temperatures at which they are stored or operated. High temperatures can have a deteriorating effect on

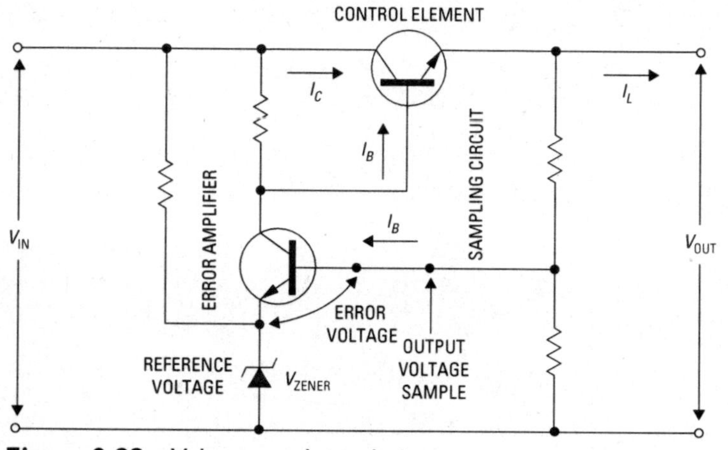

Figure 9-22 Voltage regulator design.

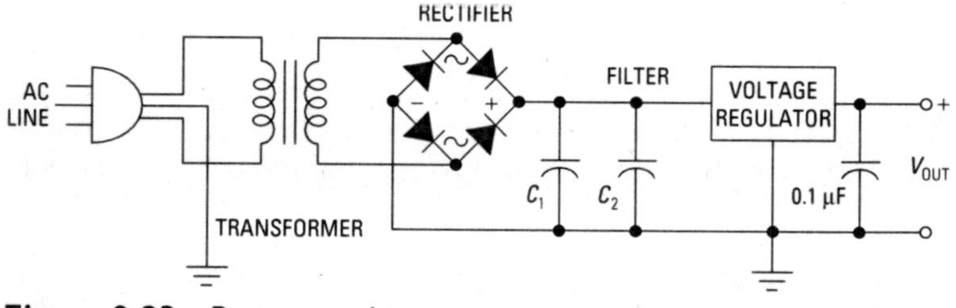

Figure 9-23 Power supply circuitry.

certain electronic items. Also beware of installing parts with pins. Take care not to bend the pins; insert them straight into their places and don't twist or turn them. They simply can't take the stress.

Voltage and wattage ratings are critical. You must keep all items within their limits. Failure to do so will usually result in an instant problem. Although electronic parts can be extremely effective, they are not at all forgiving. They will promptly blow out if you apply them incorrectly.

Printed Circuit Boards

There are two main concerns when working with printed circuit boards. The first is that you install and remove them properly. They should always be inserted and/or removed with an end-to-end motion rather than with a side-to-side motion (see Figure 9-24).

The second concern with circuit boards is handling repairs or replacements. Because of their complexity, many of the components

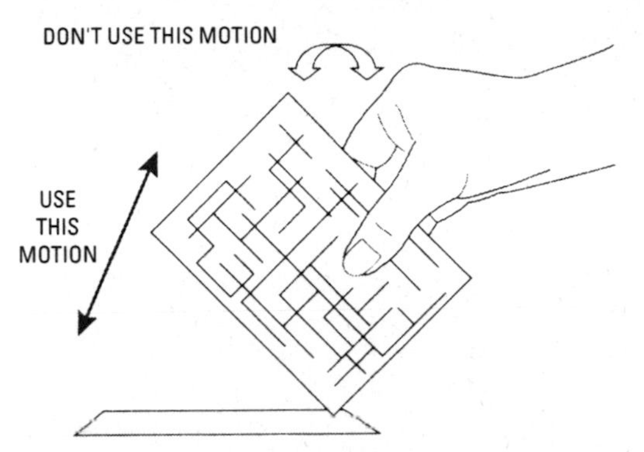

Figure 9-24 Proper method of removing circuit boards.

on these boards are nearly impossible to troubleshoot. In addition, manufacturers generally replace the entire board if you return it to them. But once a board has been worked on, the manufacturer has no way of knowing if the board was damaged because of a manufacturing error or because of your work on it. In these cases, manufacturers don't replace the board without payment.

Grounding

Anytime you interface with a power system, grounding is a concern. *Grounding* refers to a connection that eventually leads to a direct connection to the earth.

Grounding is critical for power system safety. Essentially, there are two purposes for grounding:

- To provide a reliable return path for errant currents.
- To provide protection from lightning.

At first, it would seem better *not* to provide a good return path for errant currents, thus making it harder for them to flow. Although this method would certainly be effective in reducing the size of *fault currents* (currents that flow where they are not intended), it would also allow them to flow more or less continually when they do occur.

Because fault currents pose the greatest danger to people, our primary concern is to eliminate these currents entirely. You do this by providing a clear path (one with virtually zero resistance—a "dead" short circuit) back to the power source so that these currents will be large enough to activate the fuse or circuit breaker. The overcurrent protective device will then cut off all current to the affected circuit, eliminating any danger. This also ensures that the circuit cannot be operated while the fault is present, which makes speedy repairs unavoidable.

You could say that you use grounding to make sure that when circuits fail, they fail all the way. Partially failed circuits are the dangerous ones because they can go unnoticed; therefore, grounding is essential.

Grounding Requirements

The requirements for grounding are found in the *National Electrical Code* (*NEC*). This document does not apply to most electronics work, but if you interface with an electrical power system, it can come into play. Our purpose here is not to explain the NEC, only to make you aware of the fact that it can apply to some of the work you may do. If you do need to meet NEC requirements, specialized knowledge will be required. Power systems are inherently dangerous, and the proper precautions must be taken.

The following systems are exempt from NEC requirements:

- Two-wire DC systems operating at 50 volts or less between conductors.
- Two-wire DC systems taken from a rectifier, with the AC supplying the rectifier being from a properly grounded system.
- Two-wire DC fire-protective signaling circuit that has a maximum current of 0.03 amperes.
- AC circuits operating at less than 50 volts, if the circuit is supplied by a transformer, and the transformer supply circuit is grounded.

Summary

By far, the most common type of conductor used in constructing anything electronic is insulated copper wire. It is critical that copper wire be of sufficient size to carry the circuit's current without overheating. It is also necessary that the wire is rated for the heat level of the environment where it is installed, and that the insulation is suitable for the area where it is installed.

The term bus may be used to refer to a solid bar of conductor or to a common current path between two points. In elecronic work the latter use is more common.

A wide variety of specialized cables are used in electronic applications. Frequently these cables are composed of conductors that are carefully paired together.

Electronic connections between conductors and other conductors or to devices must be mechanically and electrically secure. The primary method of connection for electronic work is soldering. However, soldered connections are not strong enough to be exposed to physical damage. They require some sort of case or enclosure and protection from the outside environment.

Electronic switches either allow current to flow through a circuit or prevent current from flowing through a circuit. Said another way, they make contact or break contact. Switches are designated by their number of poles and throws, by their normal position (open or closed), by momentary or maintained contacts, or by other specialized characteristics.

Fuses are circuit elements that prevent too much current from flowing through a circuit. This excess current is called overcurrent, and fuses fall into the classification of overcurrent protective devices. The lead filament inside a fuse will melt in the event of an overcurrent, thus interrupting the circuit it protects.

Resistors are rated in both ohms and in watts. They must be applied correctly. If the wattage of their circuit is too high, they will burn out. The value of a resistor is marked with a color code rather than in print. Without a knowledge of the color code, there is no way (except removing the resistor from the circuit and testing it) to determine its value.

Capacitors are rated in farads and in designed voltage. Subjecting a capacitor to an overvoltage will frequently cause it to explode.

Inductors used in electronic work are usually small coils of wire that add specific levels of inductance to a circuit. Inductors with iron cores are often used for low-frequency applications, while air-core inductors are used for high-frequency applications. An inductor is sometimes referred to as a coil or, in some applications, a choke.

Relays use a control current from one source to control a separate circuit by using electrical, magnetic, and mechanical principles. They generally have a variety of normally open and normally closed circuit contacts. Solid-state relays are actually transistors and thyristors that are packaged as relays. They have superb response times, but higher resistances than traditional relays. They also have a low tolerance to overloads, burning out much more easily than mechanical relays. There are many specialized types of relays available.

The primary concern for the installation of transformers is the type of power supply they are exposed to. There are no special requirements for transformers that are on the load side of a Class 2 or Class 3 power supply. The concern over transformer installations is voltage and current levels that could lead to fire. Because Class 2 and 3 circuits operate at low levels, special fire-prevention requirements are dropped. There are specific overcurrent protection and location requirements for transformers that are connected to standard power circuits.

The typical electronic power supply reduces voltage with one or more transformers, rectifies AC current to DC current, and filters (smooths) the current waveform. It may also have a voltage-regulating function.

Any time that a circuit or installation connects to a building power system, grounding is a concern. Grounding refers to a connection that eventually leads to a direct connection to the earth, and it is critical for power system safety.

Chapter 10 will introduce you to filters.

Review Questions

1. How is wire measured?
2. What is ambient temperature?
3. What are the two critical factors regarding electronic connections?
4. What type of tool is used for the most common electronic connections?
5. Where might you use a momentary contact, normally open switch?
6. How are switches rated?
7. What classification of device is a fuse?
8. What is signified by the third ring on a resistor?
9. What is the most common type of resistor used in electronic work?
10. Describe the shape of the most common type of capacitor.
11. What type of inductor is used for high-frequency applications?
12. What is a solenoid?
13. What is the most common type of relay?
14. When protection is required, how are transformers protected?
15. Why are circuits grounded?

Exercises

1. Imagine you were building a simple power supply on a table or bench. Draw this installation. Label the parts, including conductors and connections.

Chapter 10

Filters

As explained briefly in Chapter 5, filter circuits operate based on resonance, which is a characteristic of alternating-current circuits that allows us to isolate specific frequencies. This is accomplished by balancing inductive and capacitive reactance.

Since both capacitive and inductive reactance are determined by frequency, we can modify circuits to maximize or minimize both forms of reactance to achieve specific results. The formula for resonant frequency is as follows:

$$F_R = \frac{1}{2\pi\sqrt{LC}}$$

For this formula, F_R is the frequency of resonance, L is inductance measured in henrys, and C is capacitance measured in farads.

Parallel circuits and series circuits reach resonance at the same values of inductive and capacitive reactance, but they behave differently. Series circuits present a high impedance to all frequencies *except* the resonant frequency. Parallel circuits present a high impedance *only* to its resonant frequency.

Figure 10-1 illustrates this. The parallel resonant circuit shown in Figure 10-1A will present a high resistance to one frequency (or small range of frequencies) only. So, to all other frequencies but the F_R, the circuit looks like Figure 10-1B (a direct, nearly zero-impedance path to ground). In this case, almost no current will flow through the higher-impedance radio circuitry. However, to the resonant frequency, the circuit looks like Figure 10-1C (the path to ground having a very high impedance). In this case, the available voltage will push most of the current through the radio circuitry, rather than sending current through a much higher impedance to ground.

You can see from this example how we can use a parallel resonant frequency to separate out one frequency for a specific use. Actually, this is exactly how radio tuners work. If you take apart an older style of radio tuner, you will find a variable capacitor in it, which changes the resonant frequency of the parallel tuning circuit to separate out the frequency of each specific radio station.

This is an example of a basic filter circuit. You can see from this discussion how filters operate. Note that it is not just the basic resonant circuit that can be useful but the relation of other circuit elements to the basic characteristics of resonance. In the previous example, we acknowledged the fact that resonance shows a high

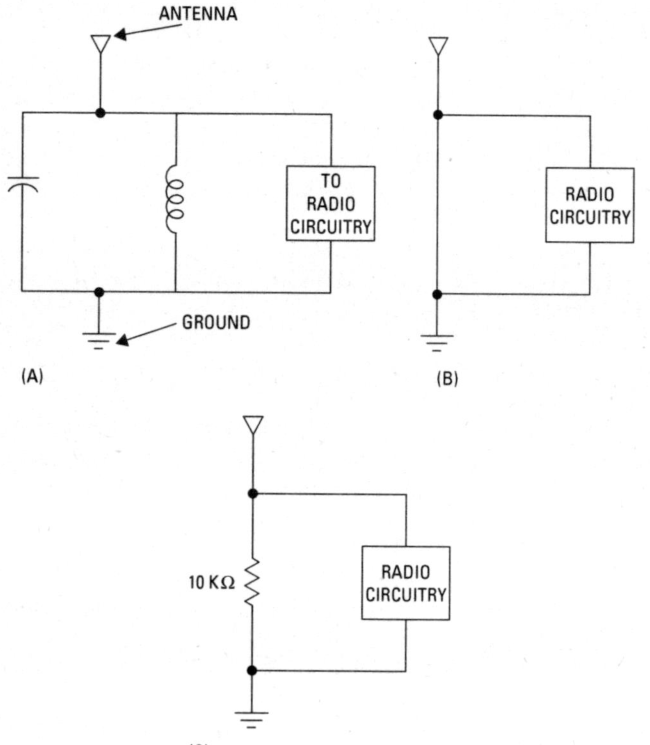

Figure 10-1 Parallel resonant circuit.

voltage to F_R, and used that phenomenon to shunt off circuit voltage in a specific way.

Series resonant circuits present a low resistance to the resonant frequency and allow for higher voltages to each circuit element. Parallel resonant circuits present a high resistance to the resonant frequency and allow for higher currents to flow through each circuit element.

Filters are circuits that allow a specific range of frequencies to pass, while blocking all others. Note the word range. Figure 10-2 displays the reactance of a specific circuit over a *range* of frequencies. Notice that the reactance is a curve, not a step pattern.

Some filters can accept or exclude a narrow range of frequencies and some can accept or exclude only a wider range. The term of

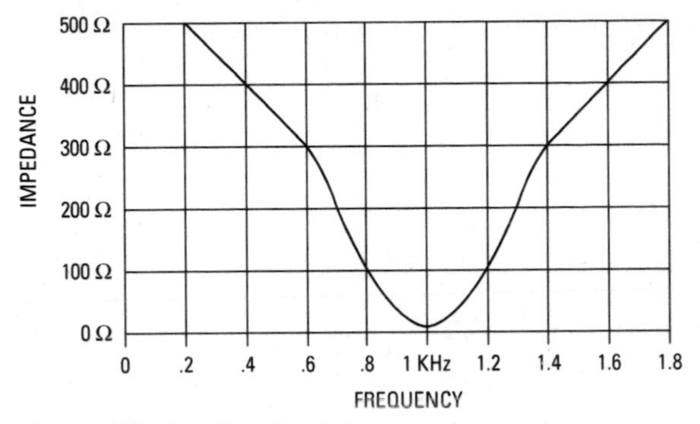

Figure 10-2 Graph of circuit resonance.

measurement of this frequency range is *Q*. Figure 10-3 shows graphs for two circuits. Figure 10-3A shows a resonant circuit with a high *Q*. Notice that the range of frequencies accepted by the circuit is small. Figure 10-3B is for a circuit with a low *Q*; notice that the range of frequencies is much broader. The lower the resistance in a circuit, the higher will be the *Q* value. So, resonant circuits with a very low resistance are useful for isolating a very narrow range of frequencies for either acceptance or exclusion.

Types of Filters

The four primary types of filters are as follows:

- Low-pass filters
- High-pass filters
- Band-pass filters
- Notch filters

All of these filters are used to accept (or *pass*) one narrow range of frequencies and to exclude (filter out) all others.

As we go through these types of filters, be aware that we use a trick that we have shown only once in previous chapters—using one circuit characteristic to gain the opposite overall effect. We do this by draining current away from the circuit in certain circumstances. By draining current away, we can use the current that is left behind. Keep this in mind as we go through these filters. You will find this technique used quite a bit in electronics work.

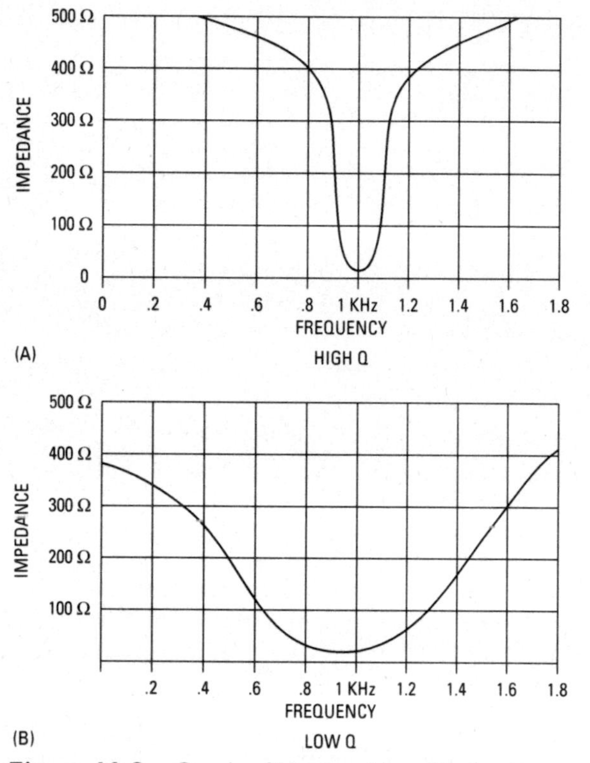

Figure 10-3 Graph of high- and low-Q circuits.

Low-Pass Filters

Low-pass filters are relatively simple filter circuits, designed only to filter out higher frequencies, while passing lower frequencies.

For this use, we can use either inductive reactance or capacitive reactance; we do not need both. Notice in Figure 10-4 that the low-pass filter does half of the work of the circuit graphed in Figure 10-2.

Figure 10-5 shows a low-pass filter. This one is an RC circuit (one with resistance and capacitiance). What you are looking at in this figure is a circuit element, not a complete circuit (that is, the filter portion of a larger circuit). Let's take a look at the operation of this filter.

Remember that the capacitor in this circuit will have a low resistance to high frequencies and a high resistance to low frequencies.

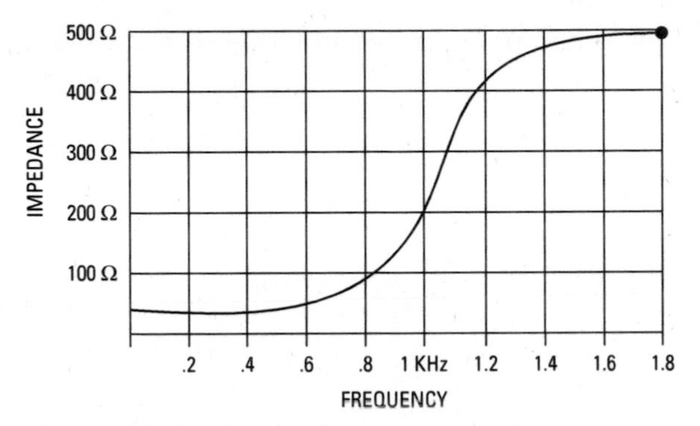

Figure 10-4 Graph of low-pass circuit.

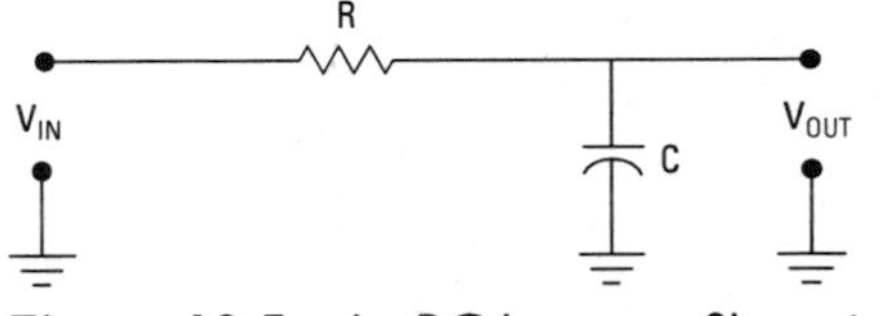

Figure 10-5 An RC low-pass filter circuit.

This circuit is designed to take advantage of that fact. At low frequencies, the capacitor's impedance is high. In this situation, only a small amount of current will pass through the capacitor. Thus, the capacitor presents very little voltage drop to the circuit and has a very limited effect on the circuit.

At high frequencies, however, the capacitor has a low resistance. In this case, quite a bit of current will flow through the capacitor. This adds a significant voltage drop to the circuit, and drains away voltage to ground. And if the voltage is drained away to ground, little of it will reach the far end of the circuit.

So, this RC low-pass filter does not block high frequencies; rather, it drains them away from the circuit. Nonetheless, the effect is the same—it does not pass high-frequency signals.

The critical factor for a low-pass filter is called the *cutoff frequency*. This is the frequency below which frequencies are passed, and above which they are cut off. The cutoff frequency in Figure 10-4 would be at 1 kilohertz (kHz). The formula for the cutoff frequency of an RC circuit is as follows:

$$f_c = \frac{1}{2\pi RC}$$

Note also from our graph in Figure 10-4 that this is not a sharp cutoff point. Rather, it is the center of the rise. We wouldn't want to use the circuit of Figure 10-4 to filter frequencies at precisely 1 kilohertz, though it would be a very good filter to separate frequencies of 0.4 kilohertz from frequencies of 1.4 kilohertz. Circuits can be made with a sharper cutoff point than Figure 10-4 illustrates, but they are seldom a step-like pattern.

An RL low-pass filter is shown in Figure 10-6. (An RL circuit is one with resistance and inductance.) Inductors have a low impedance at low frequencies and a high impedance at high frequencies. Notice that the inductor is connected in series with this circuit. So, at low frequencies, the inductor adds very little impedance, and most of the current will pass through to the far end of the circuit. At high frequencies, the inductor presents a lot of impedance to the circuit, and, as a result, most of the current flows to ground rather than through the inductor. In this way, the circuit passes low frequencies and filters out high frequencies.

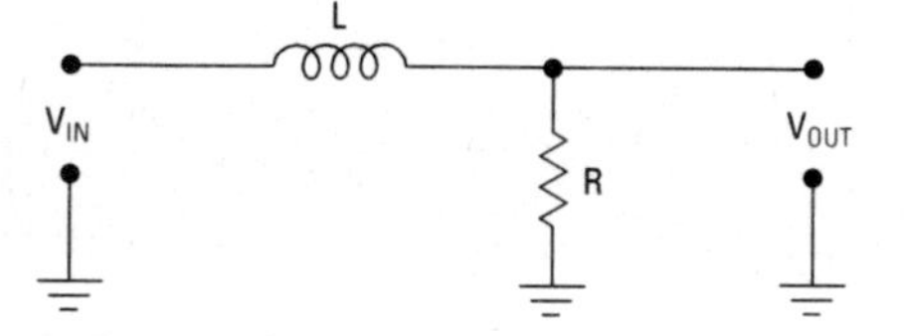

Figure 10-6 An RL low-pass filter circuit.

The formula for the cutoff frequency for RL circuits is as follows:

$$f_c = \frac{R}{2\pi L}$$

High-Pass Filters

High-pass filters allow high-frequency signals to pass through and filter out low-frequency signals. Obviously, this is the reverse of the low-pass filter's operations. You can see this illustrated in Figure 10-7.

A high-pass filter can be built with either RC or RL components, just as we could construct the low-pass filter from the same components. However, they are obviously connected differently—in the reverse.

Figure 10-8 shows a high-pass filter. Again, this is a circuit element, shown removed from a larger circuit. Let's take a look at the operation of this filter.

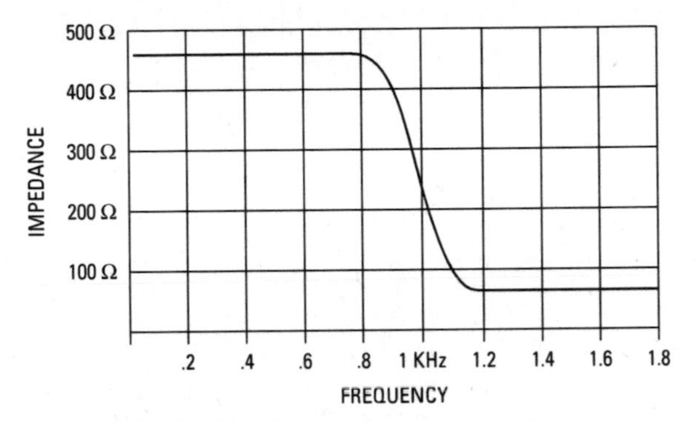

Figure 10-7 Graph of high-pass circuit.

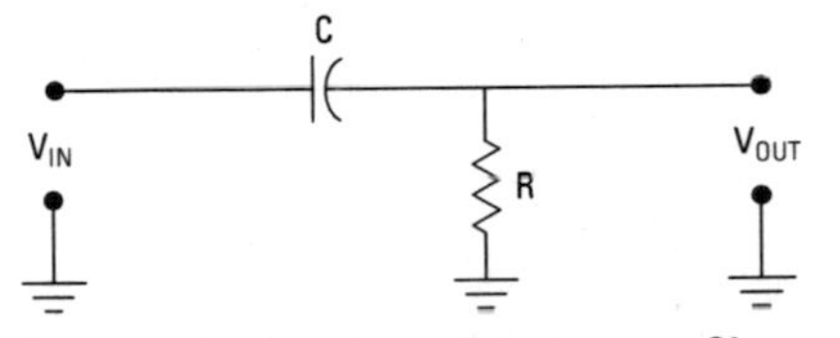

Figure 10-8 An RC high-pass filter circuit.

The imedance of a capacitor is low for high frequencies and high for low frequencies. So, this circuit places the capacitor in a series connection. In this position the capacitor will present a low impedance to high frequencies, allowing them to pass from end to end with minimal opposition. Low frequencies, however, will face a high impedance from the capacitor. As a result, low frequencies will be drained to ground, which is a much easier path than through the high-impedance capacitor. You may also hear this called shunted away to ground, which is just a different way or saying the same thing.

As with low-pass filters, the critical point for this circuit is called the cutoff frequency. For RC high-pass filters, the cutoff frequency is determined by the following formula:

$$f_c = \frac{1}{2\pi RC}$$

An RL high-pass filter is shown in Figure 10-9. The operation of this filter is as follows. An inductor (coil) will present a low impedance to a low-frequency signal, and a high impedance to a high-frequency signal. Notice that this circuit is arranged to take

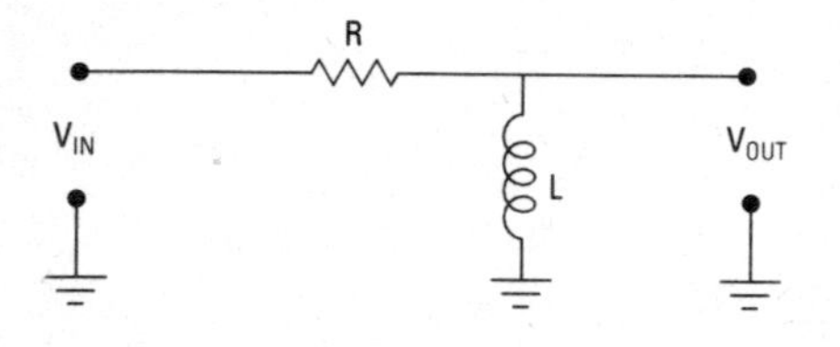

Figure 10-9 An RL high-pass filter.

advantage of that fact. A low frequency presented to this circuit will pass easily through the inductor and drain away to ground. Thus, low frequencies are drained away from this circuit. A high-frequency signal, however, will not pass through the inductor very easily. With very little current drained off to ground at the inductor, most of the circuit's high-frequency current will flow from end to end.

The formula for cutoff voltage for an RL high-pass filter is as follows:

$$f_c = \frac{R}{2\pi L}$$

Band-Pass Filters

You will remember that at the beginning of this chapter we said that series RCL circuits present a high impedance to all frequencies *except* the resonant frequency, and that parallel RCL circuits present high impedance *only* to its resonant frequency.

Figure 10-10 shows a band-pass filter. The band-pass filter uses a parallel CL element to present a high impedance to its resonant frequency to ground, thus forcing that signal to pass through the entire circuit. And since the parallel circuit element will present a lower impedance to higher or lower frequencies, those frequencies are shunted to ground.

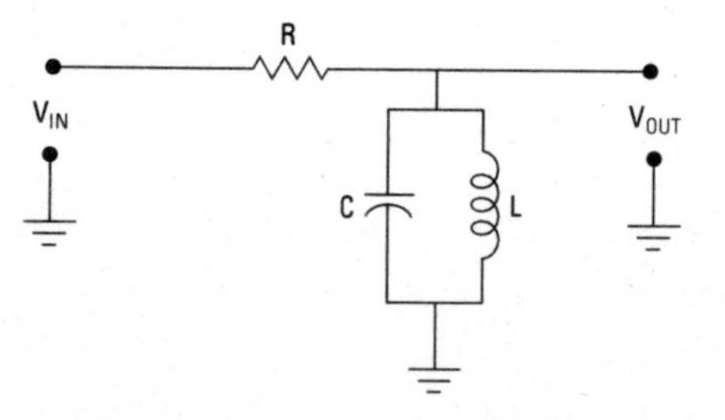

Figure 10-10 A band-pass filter.

Figure 10-11 shows a graph of the impedance of the band-pass filter's CL element (an element with capacitance and inductance).

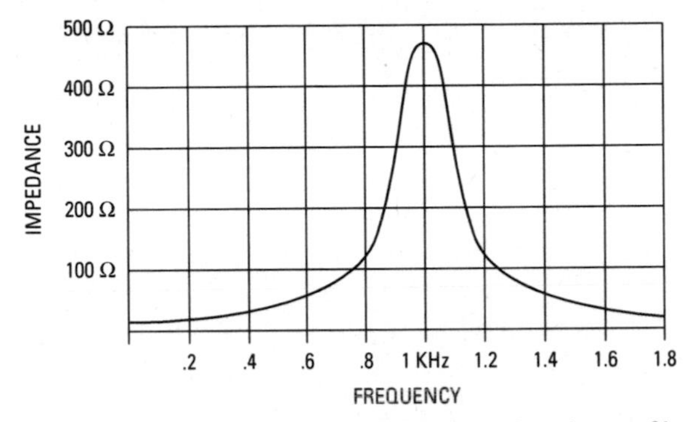

Figure 10-11 The impedance of the band-pass filter's CL element.

Note that this is showing the frequency of signals that are passed through the circuit (which is the opposite of the similar graphs we have been showing). The frequencies on either side of 1 kilohertz will be drained away from the circuit to ground.

As we have mentioned, inductors have a low impedance at low frequencies. In a band-pass filter, low frequencies travel into the circuit and through the inductor to ground. Capacitors have a low impedance at high frequencies. In a band-pass filter, high frequencies travel into the circuit and through the capacitor to ground. Only the resonant frequency encounters a high overall impedance. Not being able to travel to ground, it passes through the circuit from end to end.

Notch Filters

Notch filters take advantage of the fact that series resonant circuits present a high impedance to all frequencies *except* the resonant frequency. In other words, the series element in this circuit presents a low impedance to its resonant frequency. Because of the connections to ground combined with a low impedance at resonance, that frequency of signal is drained away from the circuit. Other frequencies will be presented with a high frequency and will be passed through the circuit from end to end.

Figure 10-12 shows a notch filter circuit element.

Figure 10-13 shows the impedance of the series element only. Again, this is the reverse of the early graphs in this chapter; we are showing the impedance of the CL circuit element. The low-impedance frequencies are the ones we are removing from the full circuit.

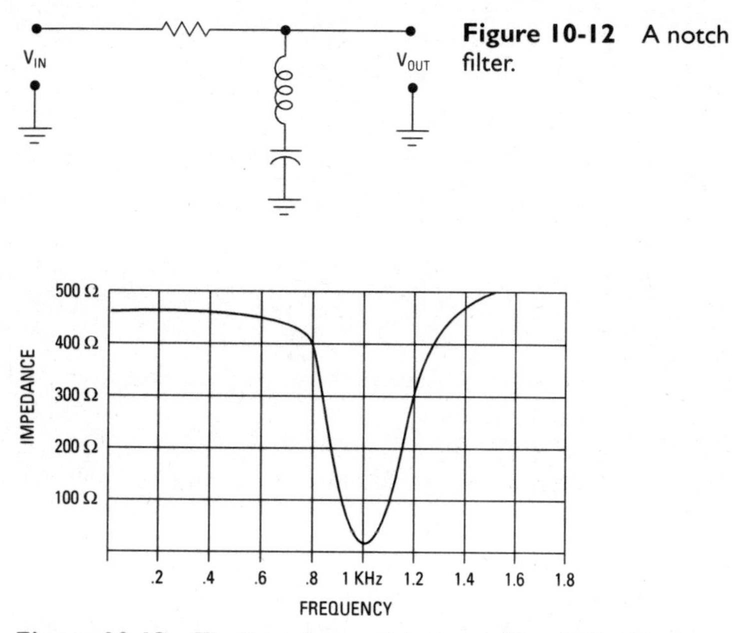

Figure 10-12 A notch filter.

Figure 10-13 The impedance of the notch filter's CL element.

Summary

By intelligently manipulating the effects of inductance and capacitance, filters can be constructed to allow (pass) or reject (filter) specific frequencies from an electronic circuit. This is done by constructing circuits that not only affect the circuit with inductance, capacitance, and resistance but also provide the option of drawing current away from the circuit to ground.

Low-pass filters use only resistance and inductance (no capacitance or resistance) or capacitance (no inductance) to reject high frequencies and allow low frequencies to pass through them.

High-pass filters use only resistance and inductance (no capacitance or resistance) or capacitance (no inductance) to reject low frequencies and allow high frequencies to pass through them.

Band-pass filters use resistive, capacitive, and inductive circuit elements to pass only a specific frequency (or band of frequencies) through them.

Notch filters allow all frequencies *except* a narrow range of frequencies to pass through them.

In the next chapter, we will examine amplifiers.

Review Questions

1. Why are ground connections important to filter circuits?
2. What is a cutoff frequency?
3. Which low-pass filter uses an inductor or a capacitor to drain high-frequency signals from the circuit?
4. Which high-pass filter uses an inductor or a capacitor to drain high-frequency signals from the circuit?
5. Which filter types are RCL circuits?
6. Which filter types use serial RC elements?
7. Which filter types use parallel RC elements?

Exercises

1. Draw a low-pass filter. Show values. Calculate the cutoff frequency. Show your calculations.
2. Draw a notch filter. Show values. Calculate the frequency that will be filtered. Show your calculations.

Chapter 11

Amplifiers

As you will remember from Chapter 7, transistors act as amplifiers. Base-emitter current is frequently 100 or more times greater than collector-base current. While a simple transistor works quite well for basic amplification, amplifiers that are used for specific purposes (such as for audio circuits) are designed to maximize signal quality.

In other words, for many uses, amplifying a signal is not enough. The signal must also be kept free of distortion and other impediments to maximum signal quality. The signal must not only be loud, it must also be clear.

In all the examples we will go through in this chapter, note the existence of batteries. We need to add DC voltage to these circuits to make them operate correctly. This is because transistors are essentially DC devices. If transistors are to be used with AC signals, the AC input must be combined with a DC voltage to keep the transistor in its active mode during the entire signal cycle. This is called *biasing*. The proper DC voltage required to bias a transistor varies by transistor type and circuit type.

There are two types of biasing:

- *Fixed-biasing*, which would be a battery power source.
- *Self-biasing*, which is achieved by placing a resistor in the base-emitter circuit.

Signal Distortion

Distortion of a signal occurs primarily because the materials used for amplifying the signal are imperfect. All transistors and semiconductor devices are limited. At certain levels or frequencies they operate differently. We simply do not have perfect semiconductor devices. Distortion occurs when the signal coming in to the device is altered when it leaves the device.

This section discusses various types of signal distortion, including the following:

- Amplitutde distortion
- Frequency distortion
- Clipping
- Crossover distortion

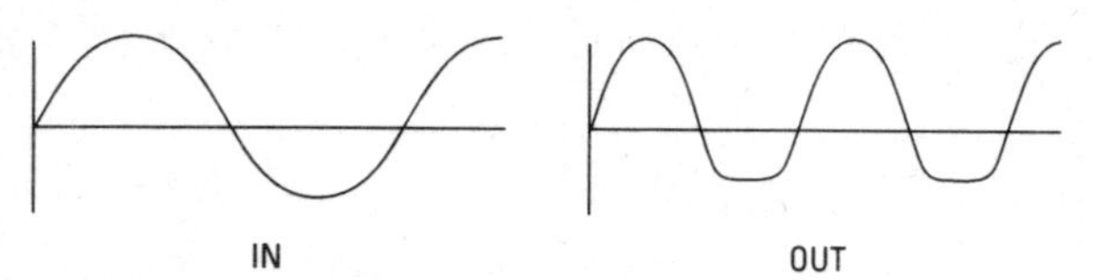

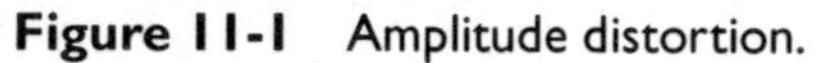

Figure 11-1 Amplitude distortion.

- Phase distortion
- Heterodyning

Amplitude Distortion

Figure 11-1 shows a common type of signal distortion—*amplitude distortion*. Notice that the signal in and the signal out are of the same frequency and polarity. So far, so good. But notice the signal out: The negative portion of the wave is significantly reduced in amplitude. The amplitude is distorted. The input waveform is smooth and even. Half of the output waveform is smooth and even, but the lower portion of the output waveform is not even.

Frequency Distortion

Figure 11-2 shows *frequency distortion*. Notice that the amplitude (or the strength) of the signal is not distorted. Input and output amplitude match. Also notice that the output waveform has two peaks for each alternation, not one as in the input waveform. In effect, it has double the frequency. So we say that the frequency of the signal is distorted. If this signal were a musical signal, for example, the output sound would be an octave higher than the input signal.

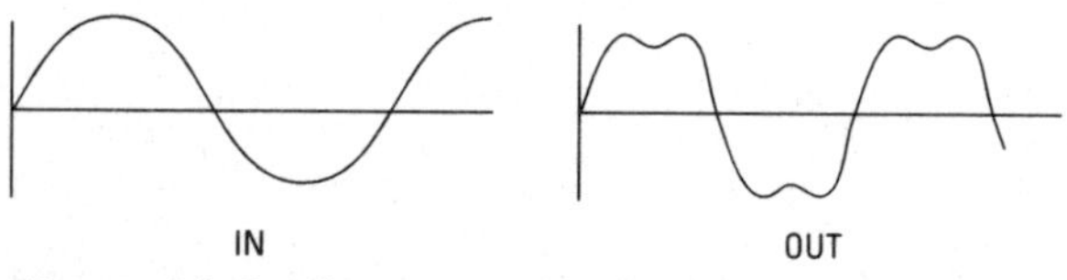

Figure 11-2 Frequency distortion.

Clipping

Figure 11-3 shows a *clipped* signal. Notice that the input waveform is smooth but that the output waveform has the top of each alternation clipped off. This occurs when the output circuitry is unable to produce a voltage or current above a certain level. All other

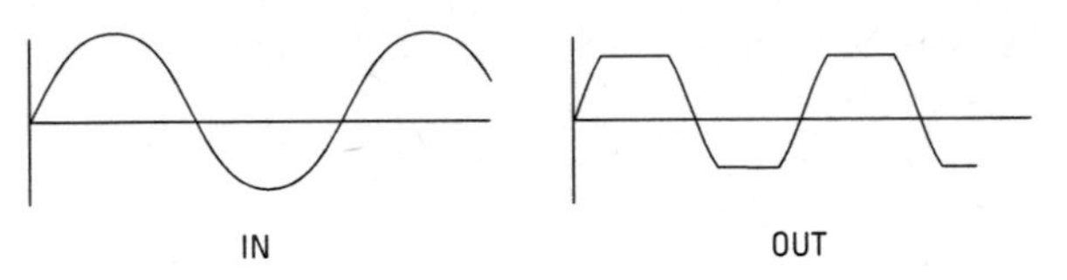

Figure 11-3 A clipped signal.

portions of the signal reproduction may be perfect, but the peaks are cut off.

Crossover Distortion

Figure 11-4 shows what is called *crossover distortion*. This is a distortion of a signal as it crosses over the zero-voltage line. That is, as it crosses from positive to negative voltage and current flow, the signal is not reproduced accurately. Again, this is a result of less-than-ideal materials.

Figure 11-4 Crossover distortion.

Phase Distortion

Figure 11-5 shows *phase distortion*. Notice that the input and output waves are of about the same amplitude but that the peak has shifted to the right in the output wave. We say that the signal has *phase-shifted*.

Figure 11-5 Phase distortion.

Heterodyning

Heterodyning (also called *intermodulation distortion*), occurs when an amplifier handles two signals at the same time, and the signals more or less merge.

Amplifier Classes

Audio amplifiers are nearly always referred to as *Class A*, *Class B*, *Class AB*, or *Class C*. These classifications are based on the conductance of the transistors in the amplifier as follows:

- If the transistors are conducting in less than 50 percent of the signal cycle, the amplifier is classified as Class C.
- If the transistors are conducting in 50 percent of the signal cycle, the amplifier is classified as Class B.
- If the transistors are conducting in slightly more than 50 percent of the signal cycle, the amplifier is classified as Class AB.
- If the transistors are conducting in 100 percent of the signal cycle, the amplifier is classified as Class A.

Class A Amplifiers

Figure 11-6 shows a standard type of Class A amplifier. Notice that two transistors are used. During the positive half of the input signal (more or less a sine wave), one of the transistors will put out more signal than the other because of its biasing, as shown in Figure 11-7A. During the negative portion of the wave, the other transistor will output more signal, as shown in Figure 11-7B. But the combined signal, as shown in Figure 11-7C, will be a faithful reproduction of the original—only greatly amplified.

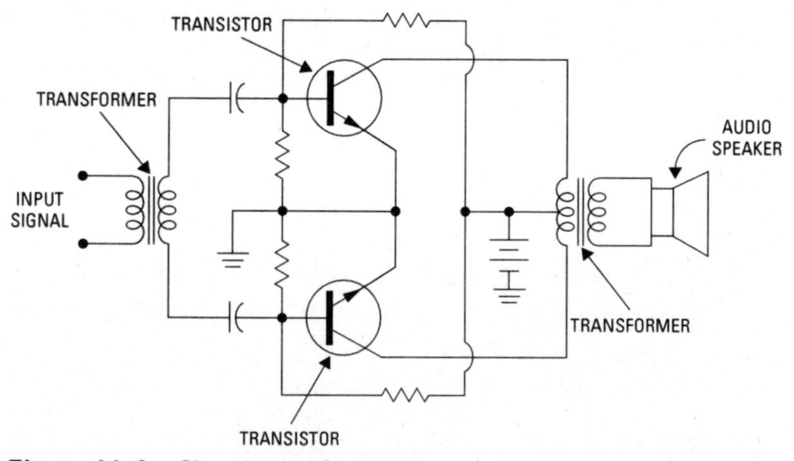

Figure 11-6 Class A amplifier circuit.

Amplifiers such as these are commonly called *push-pull* amplifiers.

Class A amplifiers are generally superior in the quality of signal output.

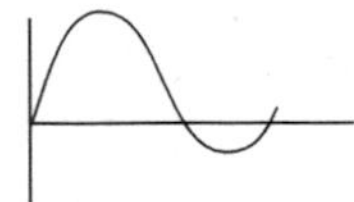

TRANSISTOR 1 SIGNAL

(A)

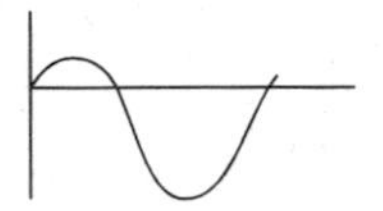

TRANSISTOR 2 SIGNAL

(B)

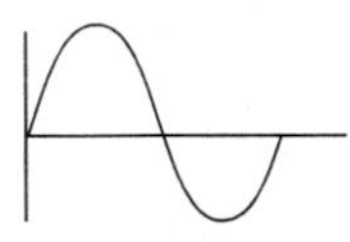

COMPOSITE SIGNAL

(C)

Figure 11-7 Outputs of a Class A amplifier.

Class B Amplifiers

Figure 11-8 shows a standard Class B amplifier circuit. The transistors in this circuit are conductive 50 percent of the time, so current flows during one half of the signal cycle. The quality of a Class B amplifier's output signal is not as good as a Class A device. As a result, Class B devices are found more in power amplification than they are in audio amplification.

Class C Amplifiers

Figure 11-9 shows a Class C amplifier circuit. The single transistor in this circuit is conductive less than 50 percent of the time. While Class C amplifiers offer a simple design, they are used less commonly than other types.

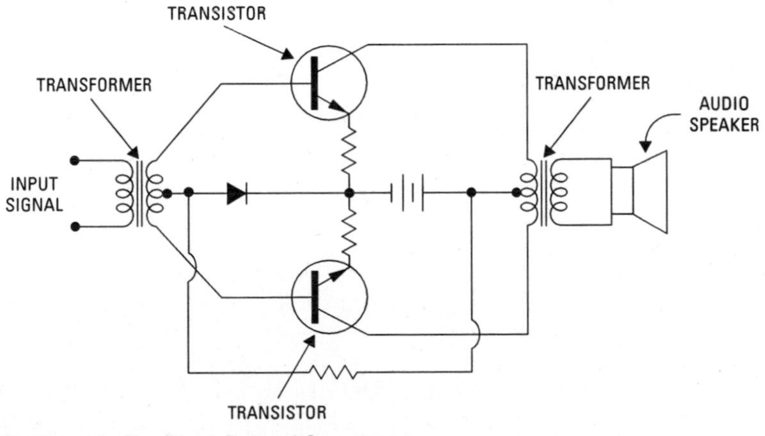

Figure 11-8 Class B amplifier circuit.

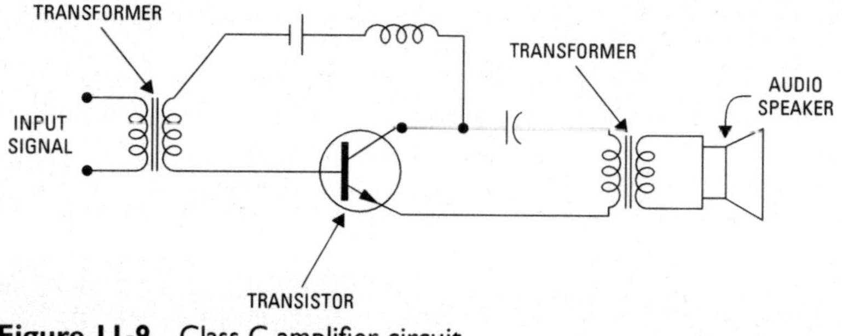

Figure 11-9 Class C amplifier circuit.

Transistor Coupling

Many amplifiers use two transistors together to provide high levels of signal amplification. In effect, this amplifies the signal two or more times. The tricky part, however, is coupling the transistors in such a way to get the best quality of output.

When two transistors are coupled together this way, we call the first amplification (at the first transistor) the *first stage* and the second amplification the *second stage*. There are several types of connections used to couple transistors, including the following:

- Direct coupling
- RC coupling
- Transformer coupling
- Impedance coupling

Direct Coupling

Figure 11-10 shows a circuit for directly coupling two transistors. Notice that this arrangement uses both NPN and PNP transistors. The resistor in the center of the schematic is not necessary if the collector current of the first transistor is less than the base current of the second transistor.

While directly coupled transistors are relatively cheap and easy to build, they are prone to problems with temperature instability, especially if a third stage (or more) is used.

RC Coupling

Figure 11-11 shows an *RC coupling* of two transistors. In this arrangement, the capacitor should have a high reactance and have very

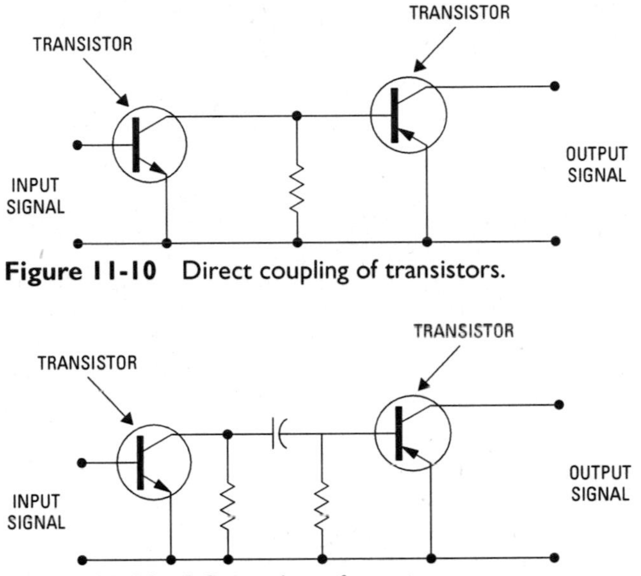

Figure 11-10 Direct coupling of transistors.

Figure 11-11 RC coupling of transistors.

little current flowing through, with a resulting low voltage drop. In this circuit, 90 percent of the voltage drop should be across the second-stage resistor. This provides sufficient voltage for the second stage's operation.

An RC coupling is very commonly used, primarily because it provides a high *gain* (gain is synonymous with amplification), takes relatively little space, and is economical. In addition, these couplings are functional from very small applications to very large applications. An RC coupling is less frequently used in devices that must be battery-powered, because of less economy of current usage than other types.

Transformer Coupling

Figure 11-12 shows a *transformer coupling circuit*. These circuits are frequently used for devices that must be battery-powered. Notice that there is no resitor in this circuit and, therefore, less excess heat loss.

But while transformer-coupled amplifiers are very efficient (approaching the theoretical maximum of 50 percent), the transformer they incorporate is invariably a fairly large, heavy component.

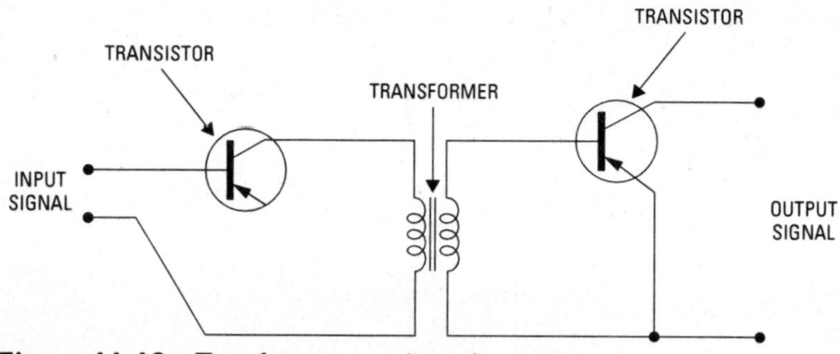

Figure 11-12 Transformer coupling of transistors.

Additionally, the frequency response of this type of coupling is not as good as the RC coupling.

Impedance Coupling

Figure 11-13 shows the *impedance coupling* of two transistors. Impedance-coupled amplifiers have frequency response characteristics that lie in between those of the RC-coupled amplifiers and the transformer-coupled amplifier. They are also quite efficient.

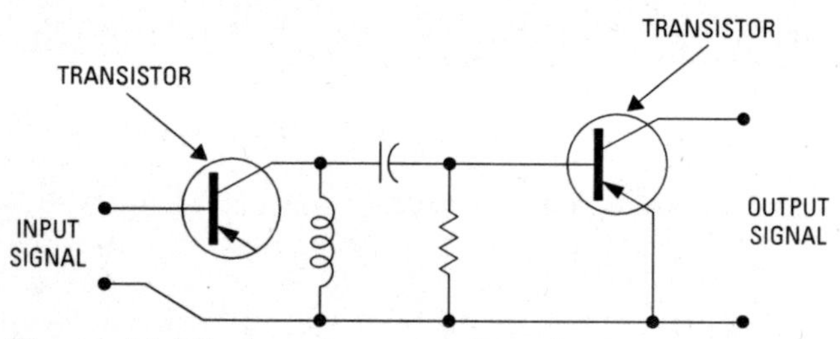

Figure 11-13 Impedance coupling of transistors.

Transistor Amplifiers

Transistor amplifiers are simple transistor circuits used for signal amplification. They are designed as *common-base circuits*, *common-emitter circuits*, or *common-collector circuits*. The designation *common* refers to that part of the transistor that connects directly to both input and output terminals.

Common-Base Circuits

Figure 11-14 shows a common-base transistor circuit. This is probably the least used of the common transistor types. Common-base

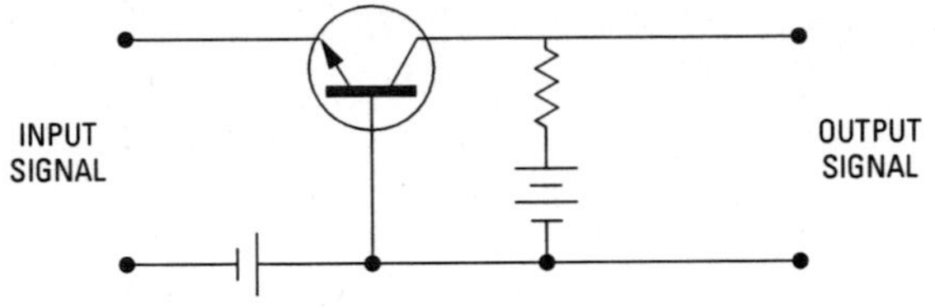

Figure 11-14 A common-base transistor circuit.

transistor circuits do not provide a current gain; they provide a voltage gain.

Common-Emitter Circuits

Figure 11-15 shows a common-emitter transistor circuit. Common-emitter transistors have the unusual characteristic that the output signal will be the opposite polarity as the input signal. For this reason, they are also called *inverting amplifiers*.

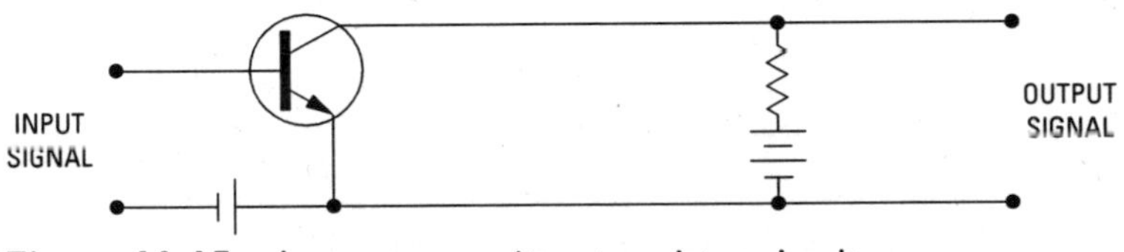

Figure 11-15 A common-emitter transistor circuit.

Common-Collector Circuits

Figure 11-16 shows a common-collector transistor circuit. Common-collector transistors have the best amplification (current gain) of the three types.

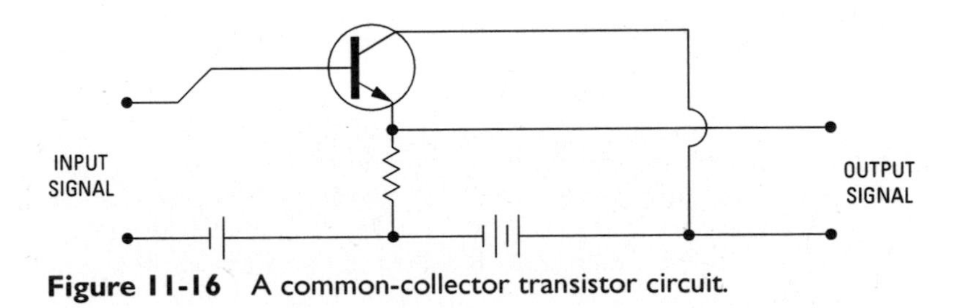

Figure 11-16 A common-collector transistor circuit.

Op-amps

Op-amps (an abbreviation for *operational amplifiers*) are probably the most useful devices used in nondigital (*analog*) circuitry. Op-amps are able (with very few ancillary devices) to perform a wide range of tasks. They are also affordable and durable. Tremendous gains in voltage (up to one million times) can be obtained with op-amps.

Op-amps use circuit *feedback* (signals sent from the circuit back to the op-amps, which then reacts) to gain a sort of automatic control.

Figure 11-17 shows a typical connection for an op-amp. This connection is called a *constant-gain amplifier*. Note the feedback resistor.

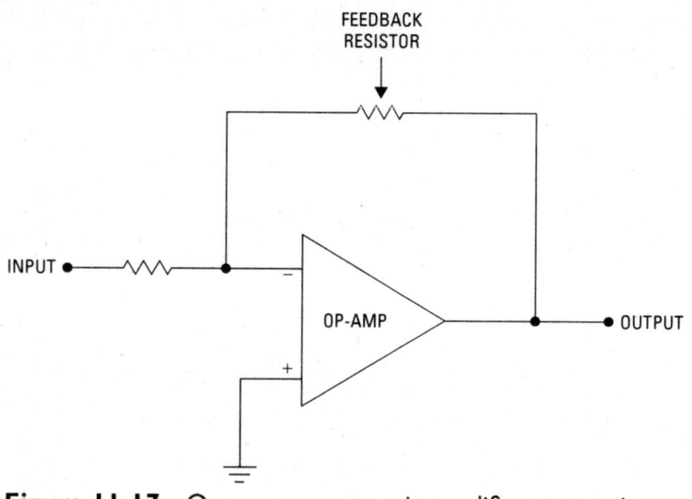

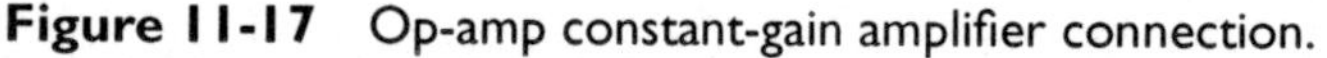

Figure 11-17 Op-amp constant-gain amplifier connection.

An op-amp is represented by the triangle shape shown in Figure 11-17. This symbol represents a prepackaged op-amp, as it would be purchased from an electronic supplier. The interior of such a unit, however, is quite complex. A typical op-amp may be composed of 20 transistors, perhaps a dozen resistors, and a capacitor. These are connected as one large combination of the circuits we have covered earlier in this chapter. The final op-amp is packaged as a single device with several leads. Many configurations are available. Figure 11-18 shows a schematic diagram of the interior of an op-amp.

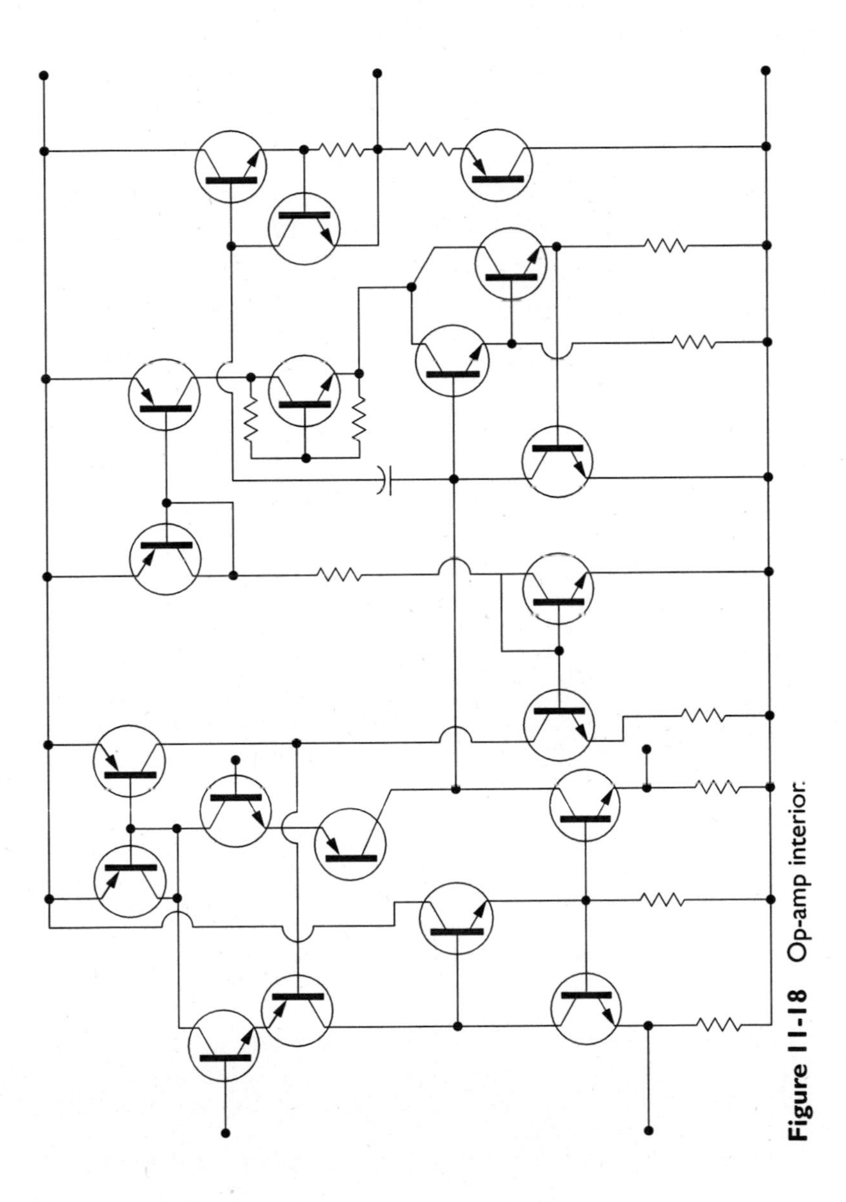

Figure 11-18 Op-amp interior.

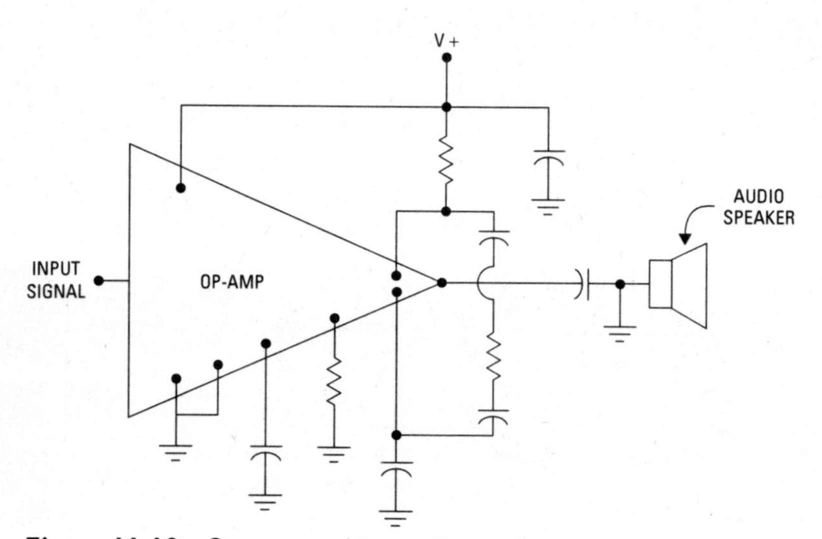

Figure 11-19 Op-amp used in a radio receiver.

Figure 11-19 shows an op-amp of a type normally used as the final amplifier in a radio (for output to the radio's speaker).

Summary

The basic device used in signal amplification is the transistor. By properly connecting transistors and other electronic components, very reliable and effective amplifiers can be built.

Many of these are multistage devices, having two or more transistors connected end to end. This provides large amounts of signal amplification but must be done properly to ensure effective operation.

The primary concern in achieving effective amplification is the avoidance of distortion. Amplification must be coupled with signal quality to be effective in most applications. There are several types of signal distortion (such as amplitude distortion, frequency distortion, clipping, crossover distortion, and others).

Audio amplifiers are classified according to the amount of time their transistors are in a conductive state. The common classifications are Classes A, B, C, and AB. Class A circuits have transistors that are conductive 100 percent of the time; Class B 50 percent of the time; Class C less than 50 percent; and Class AB just over 50 percent.

Transistors used with AC signals must be biased. That is, they must be fed with a DC voltage to keep them operating throughout the entire AC cycle. This is done either with a battery (fixed bias) or with a resistor placed in the transistor's base-emitter circuit (self-biased).

There are four methods of coupling transistors: direct coupling, RC coupling, impedance coupling, and transformer coupling. Each method has specific advantages and uses, but all are simply methods of connecting one transistor to another for increased signal gain.

The three basic transistor connections for amplifiers are common-base, common-emitter, and common-collector. The use of *common* signifies which part of the transistor is connected directly (or nearly directly) to both input and output (in other words, which part is common to both input and output).

Operational amplifiers (also called op-amps) are complex circuits that integrate numerous components into one grid-like circuit. Op-amps are among the most common and useful of electronic devices. One advantage of the op-amp is that is uses feedback to control the circuitry. This provides a level of automatic adjustment and control.

In the next chapter we will learn about oscillators.

Review Questions

1. Why are batteries used with transistors in audio amplifiers?
2. What is self-biasing?
3. What class of amplifier has the best signal quality characteristics?
4. What is a stage?
5. What is amplitude distortion?
6. What is frequency distortion?
7. What is clipping?
8. Why are transistors coupled?
9. What components are used in the direct coupling of two transistors?
10. Why are specific transistor parts called *common* in certain circuits?
11. What is unique about a common-emitter transistor amplifier?
12. What is the unique output characteristic of common-base transistors?

13. Approximately how many individual transistors might you expect to find in an op-amp?
14. What is feedback?
15. Why is feedback useful?

Exercises

1. Using what you learned in this chapter, create a circuit that would multiply its input voltage.
2. Draw a common-emitter transistor in a circuit and show input and output waveforms.

Chapter 12

Oscillators

We covered resonance and resonant circuits in Chapter 5. The uses of resonance, however, go beyond the basics that were covered in that chapter. One of the most useful characteristics of resonant circuits is *oscillation.*

To oscillate is to swing back and forth with a steady, uninterrupted rhythm; to vary between alternate extremes within a definable period of time. In electronic circuits, oscillation refers to current flowing in one direction, then in the other with a fixed rhythm.

We construct oscillating circuits for many reasons. One of the most obvious is to use as a timer. If a circuit oscillates at a fixed frequency, we can use that frequency to keep track of time. In fact, this is precisely how many types of electronic equipment (computers most prominently) control their operations.

In this chapter, we will begin with the basic operation of an oscillating circuit and then explain several common types of oscillators.

The Tank Circuit

The basic oscillating circuit is a parallel LC circuit. The old trade name for this circuit (and one that is still used) is a *tank circuit.*

Think of the tank circuit circuit as a closed piping system, with the coil as a pump and each side of the capacitor as a balloon that will expand to hold water. First, the water is pumped one direction to fill up the balloon, then it pumps the other way and fills up the other balloon.

Figure 12-1 shows this concept. Figure 12-1A shows the pump pushing water in one direction and filling up the upper balloon, which is connected to one plate of the capacitor. In Figure 12-1B, the pump is running in the other direction and is filling up the lower balloon connected to the other plate of the capacitor.

Figures 12-1C and 12-1D show how this water example has its parallel in an electrical circuit. Figure 12-1C matches Figure 12-1A. In this figure, the top plate of the capacitor is charged. Notice also that a magnetic field is built up around the coil. Once the capacitor approaches a full charge, current flow slows and the magnetic field around the coil begins to collapse. This action induces current flow but is limited also, and current flow soon ceases completely. Then, the fully charged capacitor begins to feed current back into the circuit, but in the other direction.

Figure 12-1D (which matches Figure 12-1B) shows what happens next. Current flows the opposite direction, charging the other

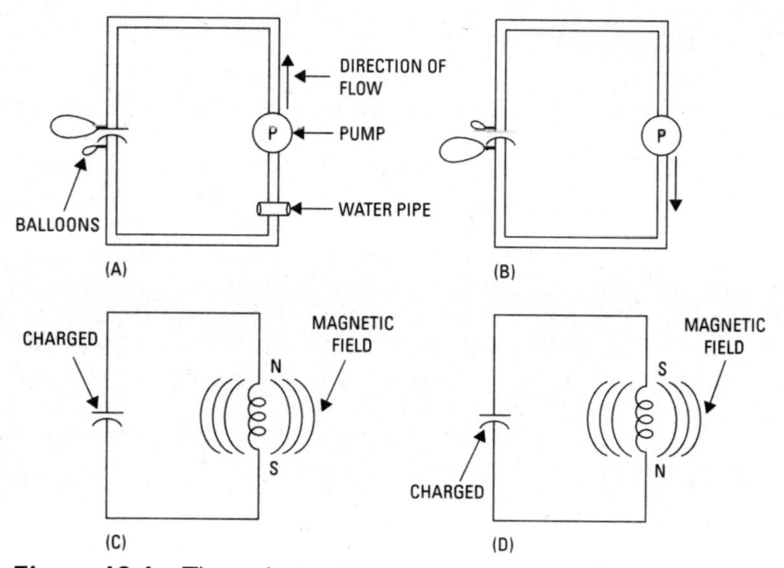

Figure 12-1 The tank circuit as a water system and as an electronic system.

plate of the capacitor and building a magnetic field (in the opposite polarity) around the coil.

As you may imagine, this process could continue at some length. The only thing that would impede this process would be a resistance that would bleed energy from the circuit in the form of heat. And, because every actual circuit does have some resistance, this is exactly what happens. This is the reason why any practical oscillating circuit must have an external source of power. Without such a power source, the oscillator would operate for a very short period of time.

Timing

You can see from this example that a tank circuit with a power source can oscillate indefinitely. Because the capacitor will charge and discharge at the same speed every time, it makes an ultra-reliable timer.

To set a specific oscillating frequency, we use a resistor, as is shown in Figure 12-2. Adding the resistance makes the capacitor charge and discharge more slowly. A smaller resistor will restrict the charge and discharge a small amount, and it will produce a higher frequency of oscillation. A larger resistor will restrict the charge and discharge a

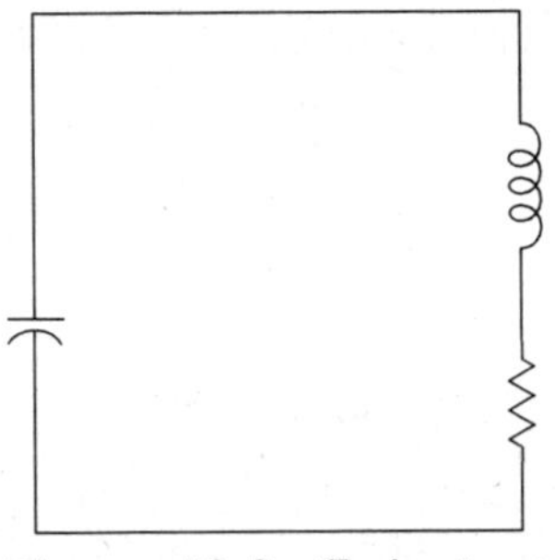

Figure 12-2 Tank circuit with resistor to control timing.

large amount, and it will produce a lower frequency of oscillation.

Positive Feedback

In our discussion of the tank circuit, we said that the tank circuit needs an external source of power to keep it going. Otherwise energy would be bled from the circuit in the form of heat caused by the circuit's natural resistance. But this external power must be supplied at the right time and in the right direction, or else it would fight against the tank circuit rather than keeping it going.

Figure 12-3 shows how this is done. Power being output is connected to the tank circuit through a resistor. The voltage drop at the resistor reduces the voltage reaching the tank circuit. However, this small amount of voltage added to the tank circuit with each alternation of output is enough to keep the circuit going.

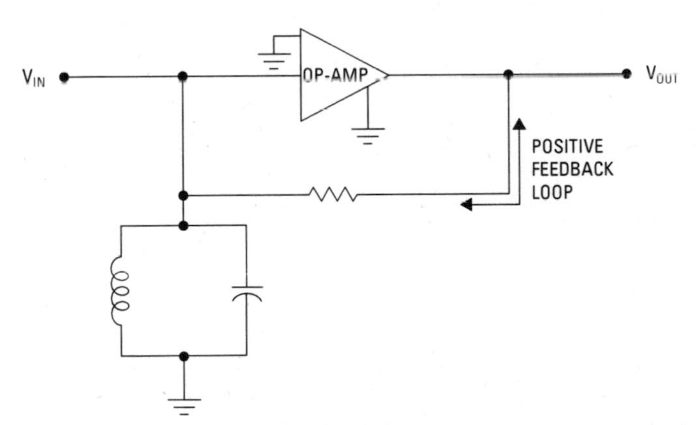

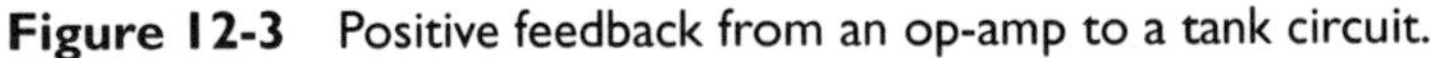

Figure 12-3 Positive feedback from an op-amp to a tank circuit.

There is some feedback element in every oscillator circuit.

RC Timers

While not an oscillator itself, the RC circuit is among the most useful timing circuits. It is built around a principle called the *RC time constant*. The formula for the RC time constant is as follows:

$$t = \text{RC}$$

Table 12-1 Time Constant Charge Levels

Time Constant	*Charge*
1	63 percent
2	87 percent
3	95 percent
4	98 percent
5	99 percent

So, the time for a capacitor to charge (measured in seconds) is equal to the resistance (in ohms) times the capacitance (measured in farads). Note that the RC time constant is also used to calculate the amount of time for the capacitor to reach 63 percent of full charge, not 100 percent. It may also be used to calculate the time to discharge the same capacitor from full charge to 37 percent. So, a 63 percent movement in either direction is calculated. Table 12-1 shows how much the capacitor will be charged over several time constant periods. From this you can see that to get a completely full charge would take about five time constants. Note, however, that capacitors do not have to be charged to 100 percent to be useful.

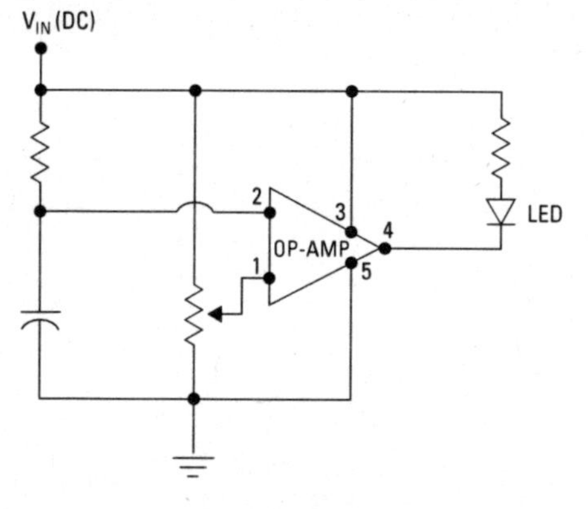

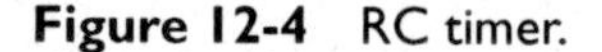
Figure 12-4 RC timer.

Figure 12-4 shows a simple RC timer for an LED. You will notice that the resistor and capacitor on the far left form an RC circuit. Also notice an external power source is fed to the circuit. As the capacitor approaches full charge, it reaches a level that is preset in the op-amp, and a transistor in the op-amp begins conducting. This allows current flow from terminals 4 and 5 on the op-amp. At this time, the capacitor can discharge through terminals 2 and 5 of the op-amp (which are now a low-resistance path), and the LED will glow.

Now, once the capacitor is discharged, terminal 2 at the op-amp is presented with a lower level of voltage, and the transistor ceases to conduct. Current flow stops, the LED does not produce light, and the capacitor begins to recharge. Once the capacitor reaches the op-amp's preset level, conduction will begin again, and so on, indefinitely.

Note that in this arrangement, the LED will blink at the circuit's speed of charge and discharge. However, this circuit could send cycles of power to any sort of device.

The precise cycle time of this circuit can be altered by changing the size of the RC components. Making the resistor of a greater value will slow the time, as will a larger value of capacitor (the reverse is true for smaller values).

Types of Oscillators

The standard oscillator circuit is composed of a tank circuit combined with an amplification stage and a path for feedback. The tank circuit provides a small oscillating current, and the amplifier increases the current and/or voltage. Because the output of a tank circuit is quite low, an amplifier is required to increase the signal level to a useable level.

Types of oscillators include the following:

- Tuned-base occillator
- RC relaxation oscillator
- Crystal oscillator

Tuned-Base Oscillator

Figure 12-5 shows a *tuned-base oscillator*. The left side of this circuit, along with the transformer and capacitor 1 (C_1), form the tank circuit. The right-hand side of the circuit is a common-emitter amplifier circuit.

R_1 is the biasing resistor to keep the transistor conducting properly. If capacitor C_1 were a variable capacitor, it could be used to easily modify the frequency of the tank circuit's oscillations. R_2 keeps the emitter properly biased. C_3 is the feedback capacitor, providing feedback between the tank circuit and the output.

RC Relaxation Oscillator

Figure 12-6 shows a fairly simple oscillator that uses an op-amp acting as both switch and amplifier. In this circuit, the capacitor will

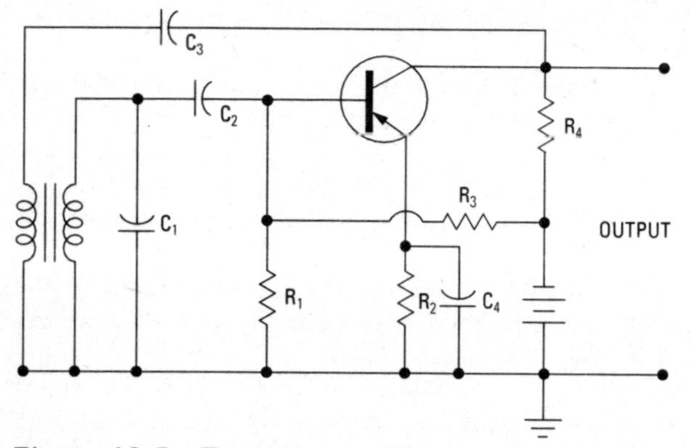

Figure 12-5 Tuned-base oscillator.

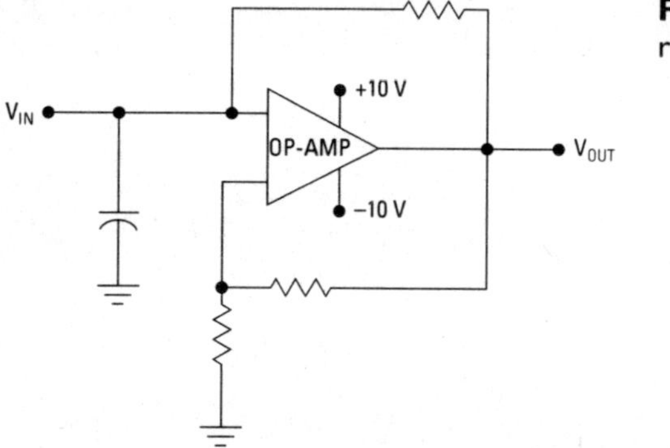

Figure 12-6 RC relaxation oscillator.

charge as quickly as the top resistor will allow. Once the capacitor reaches the op-amp's threshold voltage of +5 volts, the op-amp switches and the capacitor begins discharging (again, at a rate determined by the resistor) until it reaches –5 volts. At this time, the op-amp switches back again, and the process repeats.

The output of this oscillator, however, is amplified to 10 volts. Refer to Figure 12-7, where you will see the voltage across the capacitor in Figure 12-7A. Note the sawtooth pattern as the capacitor charges and discharges. In Figure 12-7B, however, you see the square-wave output of the op-amp. The cycling of the RC circuit triggers the op-amp, but does not supply output power. Output power is furnished by the 10-volt supply that is noted above and below the op-amp. The RC circuit turns current on and off at the correct times but does not supply output power.

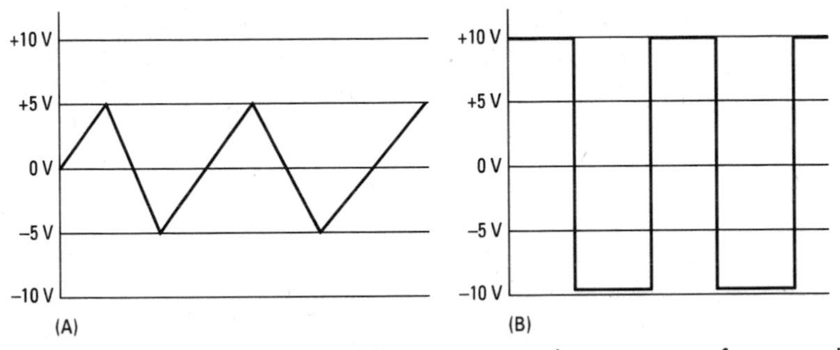

Figure 12-7 Relaxation oscillator input and output waveforms and amplitude.

Crystal Oscillators

Properly designed slices of crystals have a very interesting property: they are *piezoelectric* (that is, they can turn pressure applied to them into an electric charge). This is how record players used to operate. The needle would be moved up and down according to grooves in the record, which would put pressure on piezoelectric materials connected to the needle. This created a voltage that would (after processing) be turned into amplified sound at the audio speaker. Crystals are also used in many types of watches.

Crystal oscillators use a finely cut piece of quartz crystal with conductive plates on each side. Each crystal has a particular frequency of vibration. Once set in motion, small electric pulses will be produced at each side of the crystal's mechanical vibration.

Crystals are used to replace an LC tank circuit.

Figure 12-8 shows a basic crystal oscillator circuit. Notice that this is the same circuit as Figure 12-3, except that the tank circuit portion is replaced with the crystal.

Crystal oscillators can be very finely tuned to specific frequencies (much more so than oscillators using LC components). You will recall from Chapter 10 that this ability to affect only a very narrow range of frequency is measured as an amount called Q. The standard formula for Q with RL components is as follows:

$$\frac{X_L}{R}$$

With standard components, it is hard to obtain a high Q, since it would require an exceedingly low resistance. From the preceding formula, you can see that the lower R is, the higher Q will be.

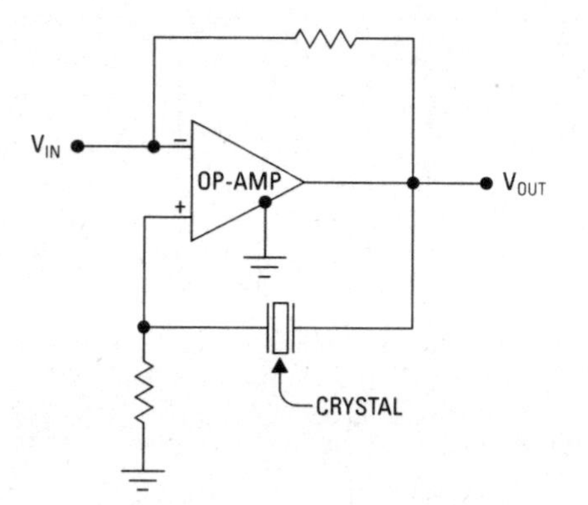

Figure 12-8 Crystal oscillator circuit.

For a crystal, however, the frequency is not generated electrically, but mechanically. The crystal oscillator's regularly reversing current is derived from a mechanical movement that is exceptionally regular.

Because crystals are small, relatively inexpensive, and ultra-precise, they are used frequently.

Additional Oscillators

The building of oscillators is a field of its own. Our purpose in this chapter was to explain basic oscillator operations. Be aware that there are many types of oscillators with many specialized applications.

Summary

In electronic circuits, oscillation refers to current flowing in one direction then in the other with a fixed rhythm. Oscillating circuits are used for many reasons. One of the most obvious uses is as a timer. If a circuit oscillates at a fixed frequency, we can use that frequency to keep track of time. This is precisely how computers control their operations.

The basic oscillating circuit is a parallel LC circuit, which is commonly called a tank circuit. Tank circuits, once charged, produce an oscillating current flow at the circuit's resonant frequency. However, the natural resistance of the circuit will lead ultimately to the extinguishing of the oscillating current. For this reason, feedback current is required to keep the tank circuit oscillating. This feedback must be fed to the tank circuit at the right time and at the right polarity.

Because the capacitor in the tank circuit will charge and discharge at the same speed every time, it makes an ultra-reliable timer.

Adding a resistance to the tank circuit makes the capacitor charge and discharge more slowly. A smaller resistor will restrict the charge and discharge a small amount, and will produce a higher frequency of oscillation. A larger resistor will restrict the charge and discharge a large amount, and will produce a lower frequency of oscillation. The same thing applies to the capacitor—a larger value of capacitor will take more time to reach a full charge and will reduce the number of oscillations per unit of time.

The RC circuit is among the most useful timing circuits. It is built around a principle called the RC time constant. The formula for the RC time constant is as follows:

$$t = \mathrm{RC}$$

The time for a capacitor to reach 63 percent of full charge (measured in seconds) is equal to the resistance (in ohms) times the capacitance (measured in farads). The RC time constant is also used to calculate the amount of time for the capacitor to discharge.

The standard oscillator circuit is composed of a tank circuit combined with an amplification stage and a path for feedback. The tank circuit provides a small oscillating current, and the amplifier increases the current and/or voltage. Because the output of a tank circuit is quite low, an amplifier is required to increase the signal level to a usable level. The transistors in amplification stages must be biased to keep them conductive while using alternating currents. In oscillator circuits this is usually done by using resistors to create self-biasing conditions.

Properly designed slices of crystals have piezoelectric qualities (that is, they can turn the pressure applied to them into an electric charge). Crystal oscillators use a finely cut piece of quartz crystal with conductive plates on each side. Each crystal has a particular frequency of vibration. Once set in motion, small electric pulses will be produced at each side of the crystal's mechanical vibration. Crystals are used to replace LC tank circuits, but they still require amplification and feedback.

Crystal oscillators can be very finely tuned to specific frequencies (much more so than oscillators using LC components). Since crystals are small, relatively inexpensive, and ultra-precise, they are used frequently.

Chapter 13 discusses digital electronic systems, including an examination of logic circuits and integrated circuits.

Review Questions

1. What important electronic effect is oscillation related to?
2. Why is oscillation useful?
3. What is a tank circuit?
4. What is feedback?
5. What are the three fundamental parts of an oscillator circuit?
6. What is the RC time constant?
7. What is the percentage of charge or discharge expected after three time constants?
8. How is the transistor in a tuned-base oscillator biased?
9. What are the critical elements in positive feedback?
10. How could you change the period (measured in the amount of time per pulse) of an RC timer?
11. What effect allows a crystal to be effective in an electronic circuit?
12. What are crystals in oscillators used to replace?
13. What level of *Q* is associated with a crystal?
14. How would you obtain the same level of *Q* in an LC circuit?
15. What is a square wave pattern?
16. What is a sawtooth wave pattern?

Exercises

1. Draw an RC circuit. Show values. Calculate the RC time constant for this circuit. Show your calculations.
2. Draw a tuned-base oscillator circuit. Show values for the tank circuit. Calculate the frequency of oscillations. Show your calculations.

Chapter 13

Digital Electronics

Digital electronics has been, without a doubt, of tremendous importance in the past few decades. This is largely because using digital pulses to represent digits allows us to transmit, modify, and codify data far better and easier than does *analog* electronics.

In this chapter we will first explain the difference between digital and analog electronics and will then describe digital devices and their applications.

To understand the importance of this field, consider that all computers operate with digital signals. So do all computer networking systems, and nearly all microprocessor-based devices. In addition to these, telephone, television, and other traditional electronic systems are changing over to digital.

Digital and Data

Digital electronic systems use voltages to represent digits (that is, numbers). The presence of a sufficient voltage represents the number 1, and voltage below a certain level represents the digit 0.

Analog systems feature continually changing levels of voltage and current, such as in traditional power systems, sound systems, and the like. A sine wave is perhaps the best example of an analog signal. It varies continually.

At first glance, limiting all data to zeroes and ones might seem a bit restrictive, using two digits only, while we normally use ten (0–9). But the real value of this is certainty. "Yes or no," "zero or one," and "all or nothing" are much easier than differentiating between 10 levels of voltage.

Binary Code

Binary is a way of counting using a *base* 2 method, rather than our usual *base 10*. Under our normal system, we move one space to the left, adding a new digit, when we hit 10. Under a base 2 system, we move to the left and add a digit when we hit 2. By doing this, the numbers 1 through 5 are translated into binary numbers as follows:

$$0 = 0$$
$$1 = 1$$
$$2 = 10$$
$$3 = 11$$
$$4 = 100$$
$$5 = 101$$

Table 13-1 Binary Coding

Decimal	*Binary*	*Binary-Coded Decimal (BCD)*
0	0	0000 0000
1	1	0000 0001
2	10	0000 0010
3	11	0000 0011
4	100	0000 0100
5	101	0000 0101
6	110	0000 0110
7	111	0000 0111
8	1000	0000 1000
9	1001	0000 1001
10	1010	0001 0000
11	1011	0001 0001
12	1100	0001 0010
13	1101	0001 0011
14	1110	0001 0100
15	1111	0001 0101

So, by using a system that contains only zeroes and ones, we can still deal with any sort of number. Table 13-1 shows how we do this. The first column on the left shows standard numerals. The middle column is simple binary numbering like we showed previously. The third column is called a *binary-coded decimal* (*BCD*). If you look at the BCD column carefully, you will see that it is a method of counting to 10 in the right-hand grouping of four digits and using the left-hand grouping of four to count tens. There are other systems of binary counting, but these are the most common.

Sending Binary Numbers

There are two primary methods of sending binary code from point to point.

One, as displayed in Figure 13-1, is *series* transmission. This is simply sending a stream (that is, a series) of voltage pulses, with low voltage representing 0 and higher voltages representing 1.

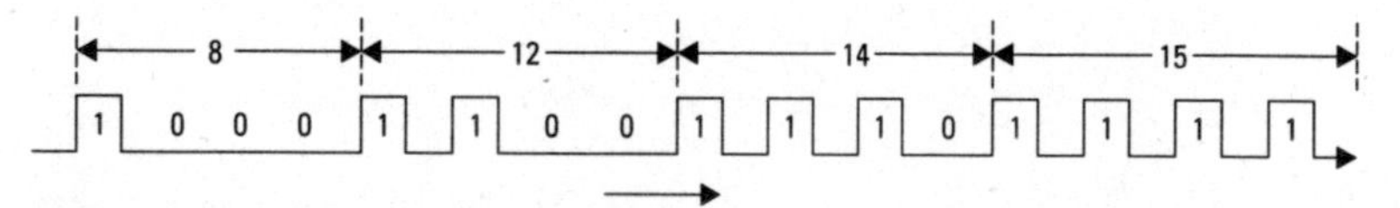

Figure 13-1 Serial transmission.

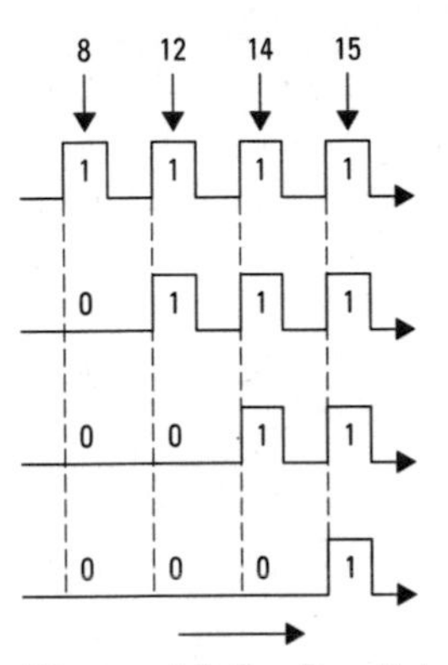

Figure 13-2 Parallel transmission.

The other method, as shown in Figure 13-2, is called *parallel* transmission. In this figure you see four streams of binary numbers being sent at the same time.

It is rather obvious that parallel transmission would allow you to send more data per unit of time. And, in fact, this is how a great deal of data processing is done (though serial transmission is used a great deal also).

Figure 13-3 illustrates how binary digits are used. Eight parallel data paths are shown. Each path is transmitting one digit at a time, all at the same rate. For this example, let's say the eight paths are each spitting out one digit every millisecond. So, every millisecond, a new batch of eight digits come out of this array of conductors. You have heard the term *byte*. The circled group of eight digits coming out of the conductors is a byte. Keep the following in mind:

- A *bit* is a single binary digit (a zero or a one).
- A *byte* is a group of eight bits.

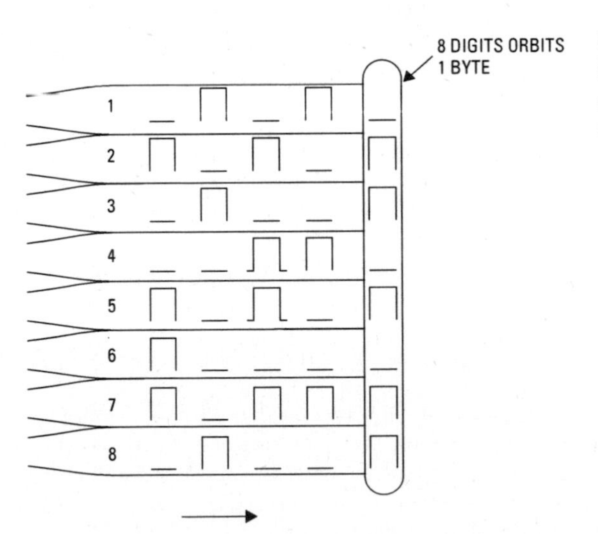

Figure 13-3 Sending groups of eight bits, making one byte.

Computers and many microprocessors use bytes as a standard binary group.

Basic Logic Circuits

Logic circuits are able to send out signals in certain conditions. The basic conditions are AND, OR, and NOT. In other words, we build circuits called *gates* that will put out a high voltage (that is, a voltage representing the number 1) this way:

- If both A and B are ones, this is called an AND gate.
- If either A or B are ones, this is called an OR gate.
- If A is a one, no voltage is output, but if A is a zero, a one will be put out. This is called a NOT gate.

AND GATE

A	B	OUT
0	0	0
0	1	0
1	0	0
1	1	1

OR GATE

A	B	OUT
0	0	0
0	1	1
1	0	1
1	1	1

NOT GATE

IN	OUT
0	1
1	0

Figure 13-4 Truth tables.

Figure 13-4 shows listings of how these simple gates operate. These are called *truth tables*.

The circuits that make gates are not nearly as complex as you might think. Figure 13-5 shows how AND, OR, and NOT gates could be built with simple normally open and normally closed switches.

Figure 13-5A is an AND gate, with a truth table shown next to it. The circuit is built with two normally open switches, and an LED is used as the output device. The truth table next to the circuit charts the operation of the LED in all of the conditions of A and B. Take a moment to analyze this. The LED will not light when the circuit is incomplete, as it would be in the first three conditions of the truth table. Only when both A *and* B are activated will the LED light.

Figure 13-5B is an OR gate, with a truth table shown next to it. The circuit is built with two normally open switches, and an LED is used as the output device. This time, however, the switches are connected in parallel, not in series. In this configuration, the LED will light when either A *or* B are activated.

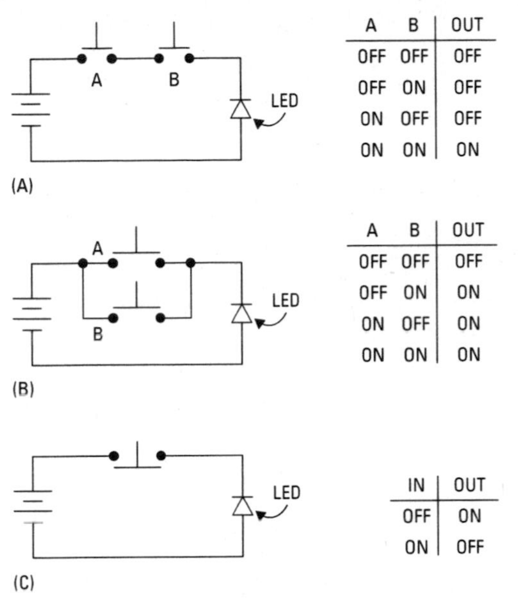

Figure 13-5 Building gates with switches.

Figure 13-5C is a NOT gate, with a truth table shown next to it. The circuit is built with one normally closed switch, and an LED is used as the output device. Notice that this circuit sends power to the LED when A is *not* activated, and doesn't send power to the LED when A is activated. In other words, an action upon A causes the opposite action on the LED. For this reason, the NOT gate is also called an *inverter.* It performs the opposite action.

You can see from these examples that building a logic gate is not as frightening as it sounds. These are just switches in certain configurations. And, in real applications, they are not much more complex than in these examples. Instead of switches, we use diodes or transistors; we may use larger groupings of devices, but the operation is essentially the same.

Figure 13-6A shows an OR gate built with forward-biased diodes. When current (in this case, at 6 volts) passes through either the A or B diode, voltage will show up at the gate's output. (In these examples we are using input voltages of 6 volts and output voltages of 5.5 volts. The reduced output voltage is caused by voltage drops at the devices.) Note that the operation of this circuit is shown on the truth table set next to it.

Figure 13-6B shows an AND gate built with reverse-biased diodes. Notice also that a battery is added to the circuit. When both

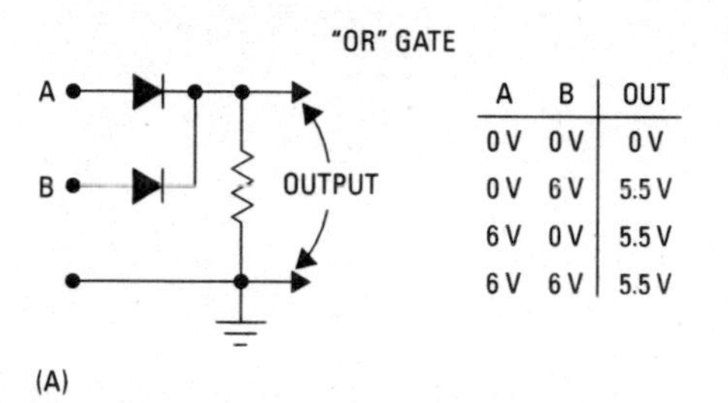

A	B	OUT
0 V	0 V	0 V
0 V	6 V	5.5 V
6 V	0 V	5.5 V
6 V	6 V	5.5 V

"AND" GATE

A

B

\+ OUTPUT

A	B	OUT
0 V	0 V	0 V
0 V	6 V	.5 V
6 V	0 V	.5 V
6 V	6 V	5.5 V

(B)

Figure 13-6 AND and OR gates built with diodes.

Figure 13-6A and Figure 13-6B are more positive than ground, current will flow from the battery, through the resistor, and to the gate's output. If, on the other hand, the voltage at either Figure 13-6A or Figure 13-6B is at or near ground, one or both diodes conducts in its forward direction, and current flows away from the gate's output. Examine the circuit and its truth table until this becomes clear to you.

Transistor Gates

If you remember back to Chapter 6, where we introduced transistors, we said that transistors can be used as switches as well as for amplification. One of the places where we use transistors this way is logic gates. Both bipolar and field-effect transistors are used in these applications.

Figure 13-7 shows two transistors connected in series to create an AND gate. The letters "H" and "L" on the truth table represent high and low voltage. Notice how the truth table in this figure precisely matches the truth table in Figure 13-6B.

Figure 13-8 shows two transistors connected in a parallel configuration to make an OR gate. Notice how the truth table in this figure precisely matches the truth table in Figure 13-6A.

Figure 13-9 shows two transistors connected to make a NAND gate. The NAND gate is a combination of the AND and the NOT gate. Essentially, this is an AND gate with the NOT functions built in.

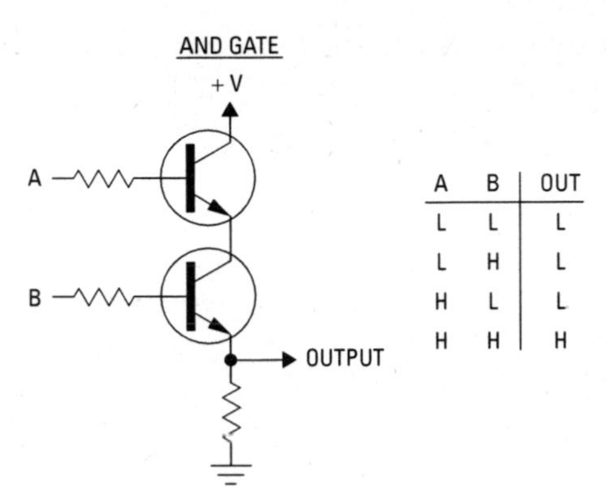

A	B	OUT
L	L	L
L	H	L
H	L	L
H	H	H

Figure 13-7 Transistor AND gate with truth table.

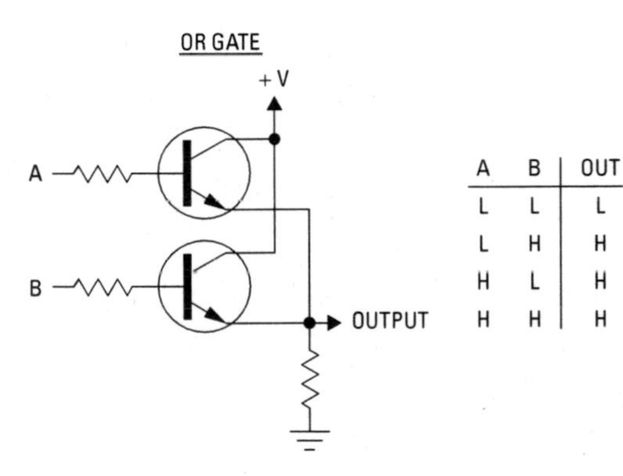

A	B	OUT
L	L	L
L	H	H
H	L	H
H	H	H

Figure 13-8 Transistor OR gate with truth table.

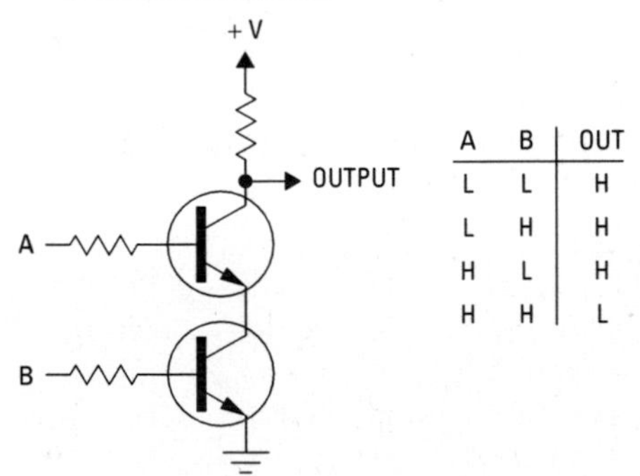

A	B	OUT
L	L	H
L	H	H
H	L	H
H	H	L

Figure 13-9 Transistor NAND gate with truth table.

Figure 13-10 shows two transistors connected to make a NOR gate. The NOR gate is a combination of the NOT and the OR gate. Essentially, this is an OR gate with the NOT functions built in.

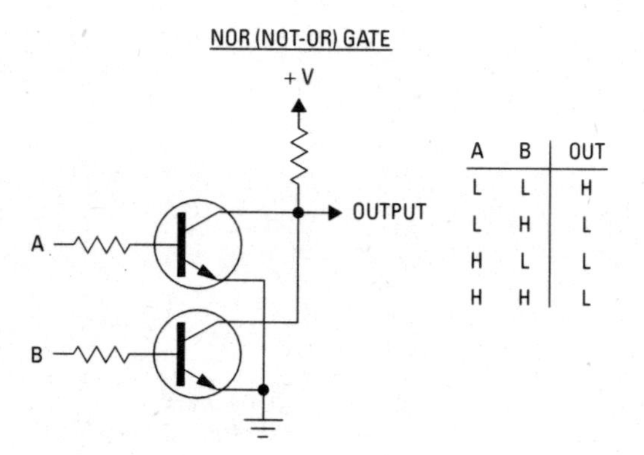

A	B	OUT
L	L	H
L	H	L
H	L	L
H	H	L

Figure 13-10
Transistor NOR gate with truth table.

Single-Input Gates

Earlier we presented the NOT gate, also called an inverter. Unlike the others, this gate (and others like it) is a single-input device. These devices are especially useful for combining other gates and for allowing other gates to operate beyond their normal ranges.

Figure 13-11 displays several types of single-input gates along with truth tables. In these truth tables, *X* represents not applicable (or doesn't matter), and *Hi-Z* represents high-output resistance.

Symbols

Figure 13-12 shows the symbols for the six most common two-input logic gates. Next to each symbol is a truth table.

Figure 13-13 shows symbols and truth tables for three-input logic gates. The three-lead logic gate has the obvious advantage of more descision-making possibilities. Note that all the logic gates we are showing in this chapter must also have connections to ground. In general, however, these connections are not shown in symbol drawings.

Combination Logic Circuits

Logic circuits are considered sequential or combinational. *Combinational circuits* do the same thing every time. In other words, after each operation, they return to their original state. *Sequential logic circuits* are different; they react based upon their previous state. They do not automatically return to their original state, and they use this characteristic to accomplish certain goals.

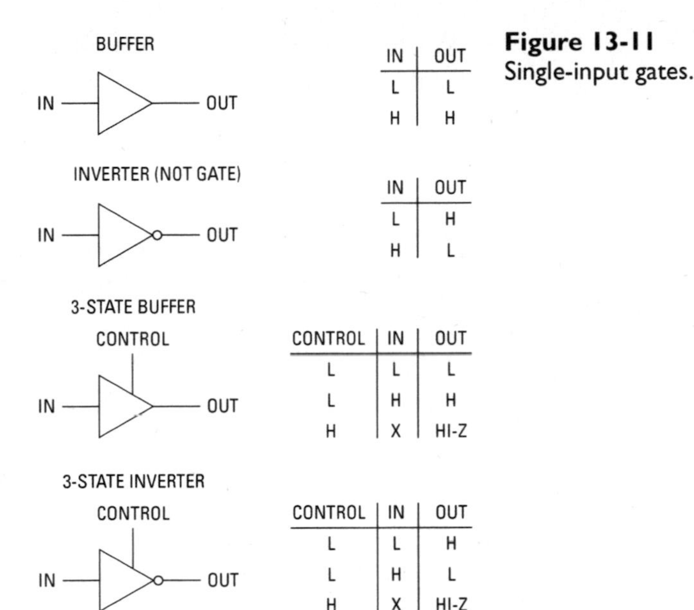

Figure 13-11
Single-input gates.

Combinational logic circuits are frequently very complex, and are usually built entirely of NAND and NOR gates. You will remember that the NAND gate incorporates the operations of both NOT and AND gates. Also remember that NOR gates incorporate the operations of both NOT and OR gates. Hence they are very uscful.

Figure 13-14 displays a variety of common combinational gate arrangements. Remember that many such small groupings can be put together to make a much larger group. From this you can see that very specific logical decisions can be made, even mimicking reasoned thought. (Actually, they simply act the ways reasonable people build them to react, but the effect is the same.)

You can also see that as the circuits get more complex, their truth tables become very important. As these digital circuits get more and more complex, a quick look at a schematic will not be sufficient to analyze their actions. Therefore, the truth table is a necessity.

Figure 13-15 shows a combinational circuit that is used to decode 2-bit binary numbers and turn them into decimal equivalents.

Figure 13-16 is a combinational logic circuit that chooses which data stream is allowed to pass and which is not.

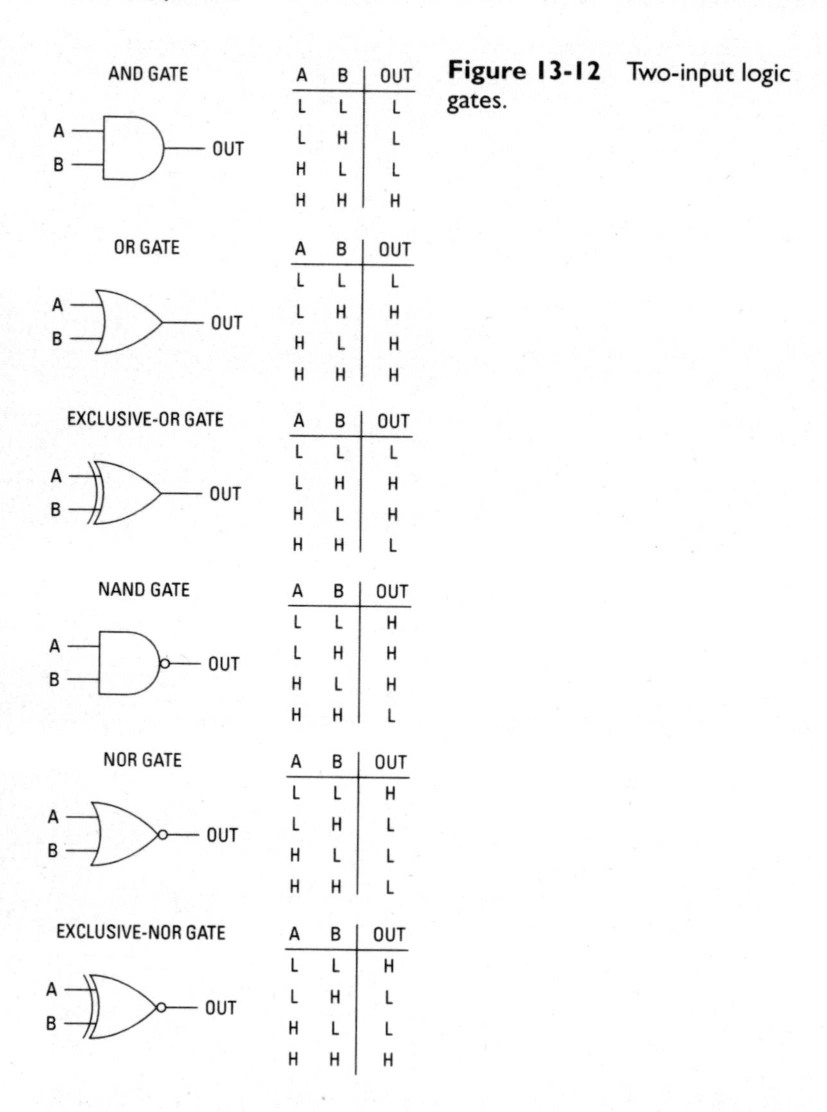

AND GATE

A	B	OUT
L	L	L
L	H	L
H	L	L
H	H	H

OR GATE

A	B	OUT
L	L	L
L	H	H
H	L	H
H	H	H

EXCLUSIVE-OR GATE

A	B	OUT
L	L	L
L	H	H
H	L	H
H	H	L

NAND GATE

A	B	OUT
L	L	H
L	H	H
H	L	H
H	H	L

NOR GATE

A	B	OUT
L	L	H
L	H	L
H	L	L
H	H	L

EXCLUSIVE-NOR GATE

A	B	OUT
L	L	H
L	H	L
H	L	L
H	H	H

Figure 13-12 Two-input logic gates.

In actual use, both of the circuits above may be expanded to a much larger size.

Sequential Logic Circuits

As we mentioned briefly earlier, sequential logic circuits act based upon their previous state. They do not automatically return to their

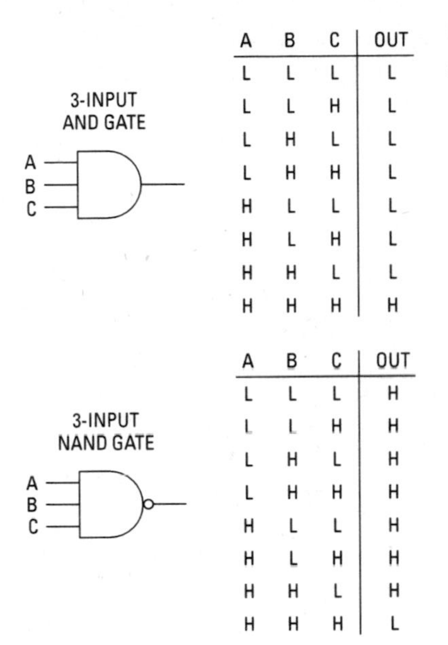

A	B	C	OUT
L	L	L	L
L	L	H	L
L	H	L	L
L	H	H	L
H	L	L	L
H	L	H	L
H	H	L	L
H	H	H	H

A	B	C	OUT
L	L	L	H
L	L	H	H
L	H	L	H
L	H	H	H
H	L	L	H
H	L	H	H
H	H	L	H
H	H	H	L

Figure 13-13 Three-input logic gates.

original state, and they use this characteristic to accomplish certain goals. So, if they were to begin in a normally open state, they would react one way. But if they began in a normally closed state, they would react in a different way.

Data (pulses) generally proceed through a sequential logic circuit step by step. In other words, one pulse sets a device (that is, puts it in a specific state of operation). The device then remains in that position until it is acted upon. In this way, we can store data in a sequential logic circuit. Then, when the next bit of data comes along (be it a thousanth of a second, or a number of minutes later), it will encounter the stored data, and a specific reaction to both pieces of information will occur. The final output will reflect this.

Figure 13-17 shows a basic alternating sequential logic gate. This one is usually called a *reset-set* or *RS* gate. These gates are sometimes generically called *flip-flops* or *latch circuits*. Their function is to alternate between conditions and to maintain those conditions until acted upon.

Figure 13-18 shows two more RS gates. Figure 13-18A shows a clocked RS gate, and Figure 13-18B shows a D (data or delay) RS gate.

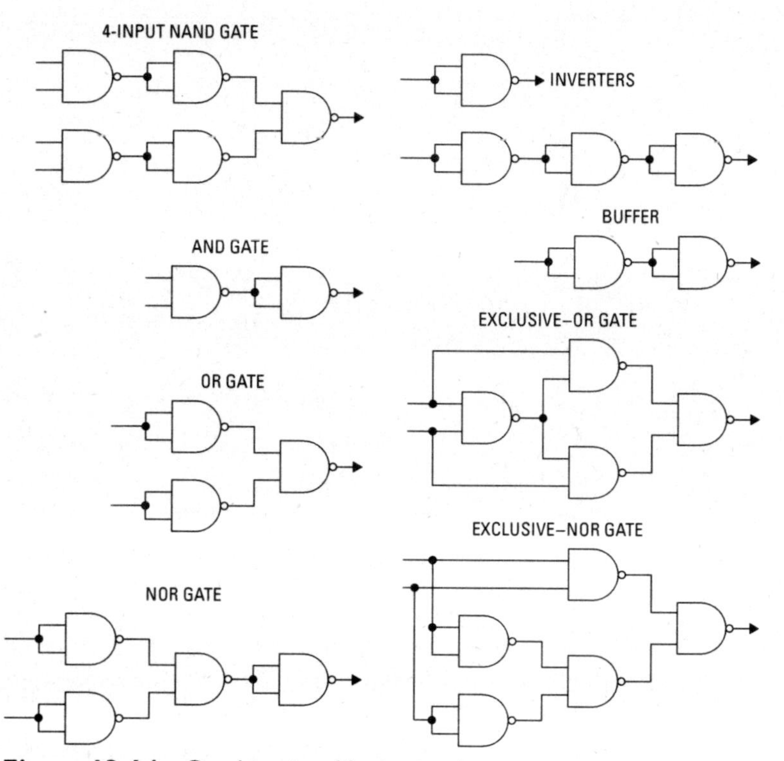

Figure 13-14 Combinational logic circuit arrangements.

Note the symbol Q and its counterpart. Q represents a specifc output, as shown in the following:

$$\overline{Q} = \text{Not } Q$$

Q and $\overline{Q}$ are always opposite states. So, the following is true:

if $Q = 1$,
$\overline{Q} = 0$

Figure 13-19 shows how a group of four D flip-flops are used to build a *storage register*, which is a digital memory device. Notice that 4 bits of binary data enter from the bottom. Timing pulses feed into a clock bus from the left. The data pulses affect the D flip-flops and are, in effect, stored at the D flip-flops until they are used by the circuits they are connected to.

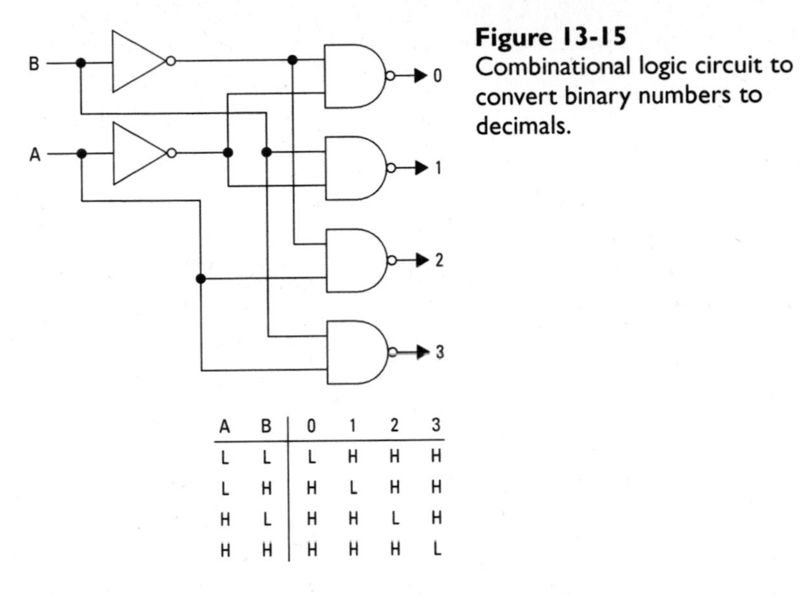

A	B	0	1	2	3
L	L	L	H	H	H
L	H	H	L	H	H
H	L	H	H	L	H
H	H	H	H	H	L

Figure 13-15 Combinational logic circuit to convert binary numbers to decimals.

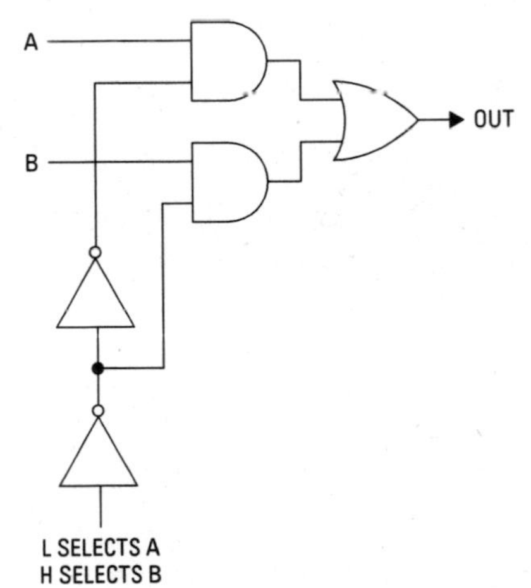

Figure 13-16 Combinational logic circuit to choose between data streams.

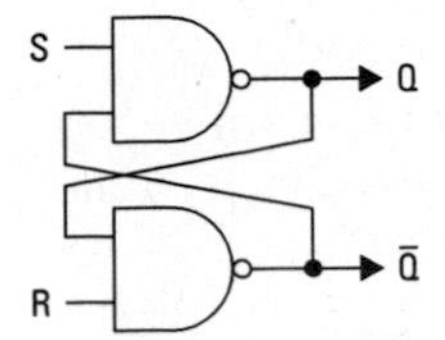

S	R	Q	$\bar{Q}$
L	L	(DISALLOWED)	
L	H	H	L
H	L	L	H
H	H	NO CHANGE	

Figure 13-17 Basic alternating sequential logic gate.

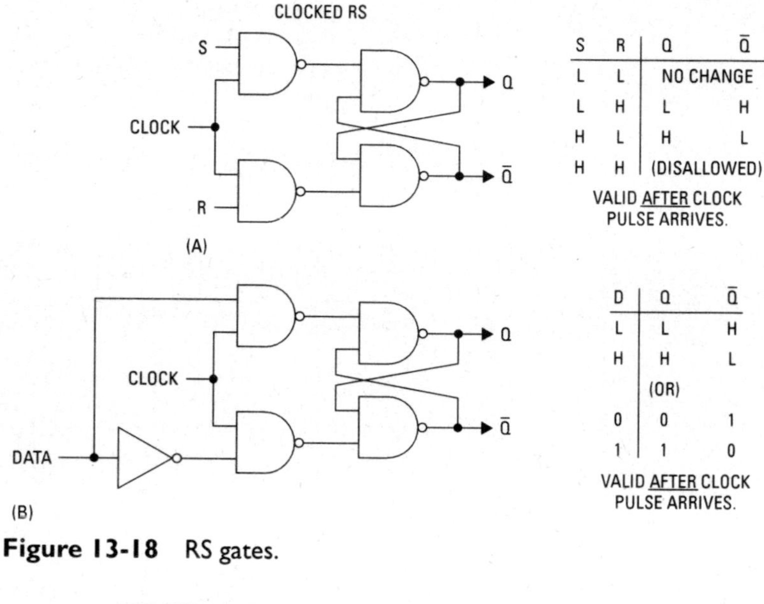

Figure 13-18 RS gates.

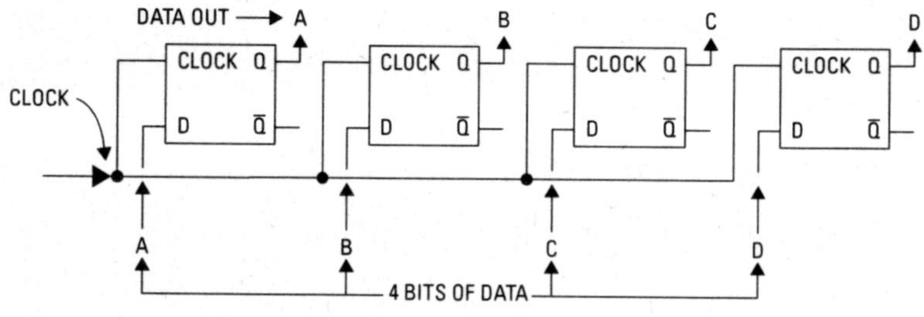

Figure 13-19 A storage register.

Figure 13-20 shows a similar set of four sequential logic gates being used as a 4-bit binary counter. In this case, a *T* (or *toggle*) flip-flop is used. This arrangement simply counts binary pulses. Figure 13-20A is the circuit, and Figure 13-20B is the associated truth table.

Combination Logic Circuits

It is probably no surprise that combinational and sequential logic devices can be used together. And, as you might expect, such circuits are made in all sizes and combinations.

Figure 13-21 shows one example. This is a counter and display. Digital pulses enter from the left in Figure 13-21A. They are fed into

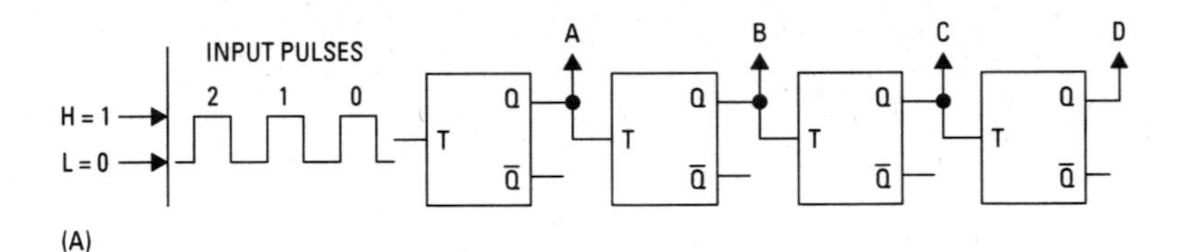

COUNT	D	C	B	A
0	0	0	0	0
1	0	0	0	1
2	0	0	1	0
3	0	0	1	1
4	0	1	0	0
5	0	1	0	1
6	0	1	1	0
7	0	1	1	1
8	1	0	0	0
9	1	0	0	1
10	1	0	1	0
11	1	0	1	1
12	1	1	0	0
13	1	1	0	1
14	1	1	1	0
15	1	1	1	1

(B)

Figure 13-20 A binary counter.

a coder, where they are counted. They then pass to a decoder, where they are translated into signals to drive an LED display. Figure 13-21B shows an LED display, which you are no doubt familiar with.

Circuits such as the one just discussed, and the ones we have covered throughout this chapter, are the precise circuits that make computers work. They count, they code signals, they store signals, and they modify signals according to prespecified stimuli. And, of course, they provide logical reactions. If you put enough of these operations together intelligently, you get a modern computer.

Given what you have been shown in this chapter, plus enough time and effort, you could build counting machines and, ultimately, computers. However, up to this point we've been working with normal-sized devices. If you were to build a modern type of computer with these devices, it would probably be as large as a truck, and

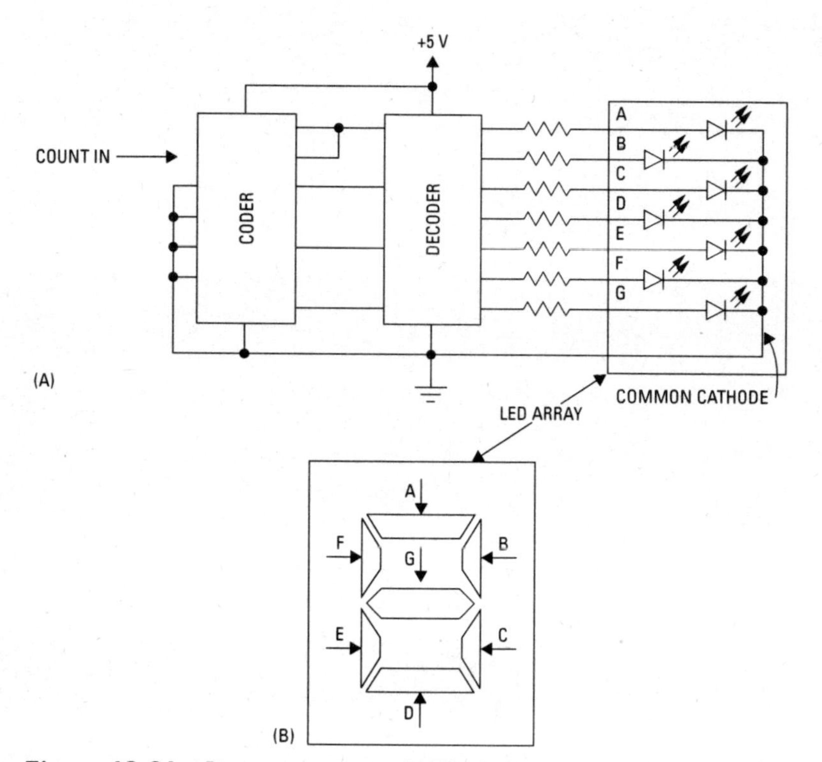

Figure 13-21 Binary counter and LED display.

perhaps as large as a house. The next section examines the process of shrinking these devices down to very small sizes.

Integrated Circuits

Integrated circuits (ICs) were first developed by Jack Kilby of the Texas Instruments company in 1958. Kilby was preceded by a British radar scientist, Geoffrey Drummer, who conceived the IC in 1952 but was not able to build a working model. Kilby's work was augmented by Robert Noyce of Fairchild Semiconductor, who patented an improved version in 1961.

A modern IC is a thin chip of silicon upon which are thousands or millions of semiconductor devices. These devices are primarily transistors, though there are usually a large number of resistors as well. Most chips are approximately 1 centimeter square, though many other sizes exist. The most important types of IC chips are microprocessors and memory chips.

Bear in mind that these chips are just large-scale combinations of the devices and circuits we have already covered in this text.

The process of manufacturing IC chips (that is, of putting millions of transistors on a chip the size of a thumbnail) is certainly fascinating, but we will not be covering it in depth here. It involves a process called *photolithography*, in which ultraviolet light is used to modify special materials in or on the surface of the chip.

Figure 13-22 provides a nice view of the construction of an IC chip. The bulk of the chip is made of pure silicon. The insulation layer on top is silicon dioxide (SiO_2), the same as for the field-effect transistors discussed in Chapter 7. Notice that layers of P and N semiconductor material are embedded in the silicon. Running on top of the silicon dioxide (and penetrating through to the P and N layers at times) is a layer of aluminum conductor.

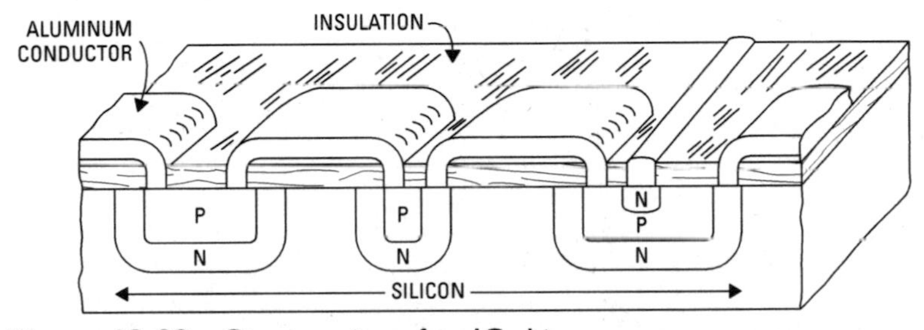

Figure 13-22 Construction of an IC chip.

From the left in Figure 13-22, the first device created by this arrangement is a resistor. Notice that the aluminum conductors attach to each side of the same P layer. This layer has a specific value of resistance, so placing these conductors at a precise distance will provide a precise amount of resistance.

The next device to the right is a diode. Notice that the aluminum conductor on the left attaches to the P layer, and the conductor on the right connects to the N layer. This creates a simple PN junction diode.

The third device shown in this figure is a transistor. The aluminum conductor from the left connects to the P layer. The conductor to the right connects to the N layer. In addition to these, a thin aluminum conductor connects to the top N layer and rides on the top of the silicon dioxide insulation.

All of this is done at a miniture size and is very carefully manufactured and tested. The really difficult parts of making IC chips are designing the chip in the first place (this requires teams of design engineers), building and using the specialized manufacturing processes, and testing millions of transistors to be sure that they all work properly. (You can imagine how many transistors could be ruined with just a very slight manufacturing inaccuracy.)

Figure 13-23 shows a relatively large and unsophisticated type of IC chip. Note that the chip is surrounded by numbered pins that are made to be inserted into special connectors.

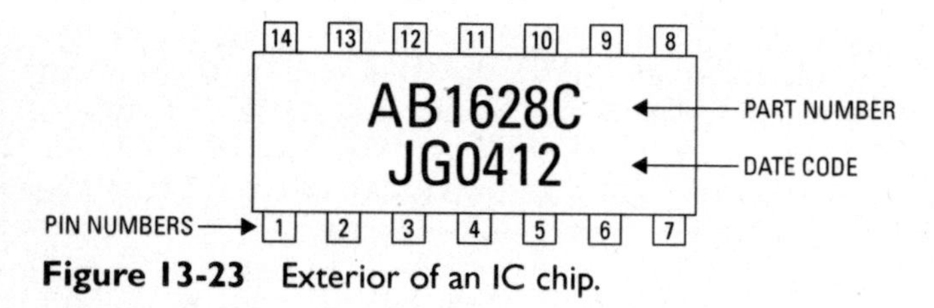

Figure 13-23 Exterior of an IC chip.

Linear and Digital

Although this chapter is about digital electronics, you should be aware that ICs can also be made to use nondigital circuitry. We call such circuits analog, or *linear*. A sine wave, for example, is an analog signal; it varies continuously. Digital signals, on the other hand vary only between zeros and ones; no other values matter.

Summary

Using digital pulses to represent digits allows us to transmit, modify, and codify data far better and easier than using analog electronics. Digital electronics systems use voltages to represent digits (that is, numbers). The presence of a sufficient voltage represents the number 1, and voltage below a certain level represents the number 0.

Binary is a way of counting using a base 2 method, rather than our usual base 10. Under our normal system, we move one space to the left, adding a new digit when we hit 10. Under a base 2 system, we move to the left and add a digit when we hit 2.

Binary mathematics use two digits only. Normally we use 10 digits (0–9). The value we gain from digital electronics and binary numbering is certainty. "Yes or no," "zero or one," and "all or nothing" are much easier than differentiating between 10 levels of voltage.

One type of digital signal transmission is series transmission. A series transmission is a stream of voltage pulses, with low voltage representing 0 and higher voltages representing 1. The other type of digital signal transmission is called parallel transmission, which consists of sending several streams of pulses at the same time.

A bit is a single binary digit (a zero or a one). A byte is a group of eight bits. Computers and many microprocessors use bytes as a standard binary group.

Logic circuits output signals in certain conditions. The basic conditions are AND, OR, and NOT. We build circuits (called gates) that will put out a high voltage:

- If both A and B are ones, this is called an AND gate.
- If either A or B are ones, this is called an OR gate.
- If A is a one, no voltage is output. But if A is a zero, a one will be put out. This is called a NOT gate.

Listings of exactly how gates operate are called truth tables.

Transistors can be used as switches as well as for amplification. Both bipolar and field-effect transistors are used in logic gates.

Logic circuits are considered sequential or combinational. Combinational circuits do the same thing every time. After each operation, they return to their original state. Sequential logic circuits are different. They react based upon their previous state; they do not automatically return to their original state.

Combinational logic circuits are usually built entirely of NAND and NOR gates. The NAND gate incorporates the operations of both NOT and AND gates. NOR gates incorporate the operations of both NOT and OR gates.

As logic circuits get more complex, truth tables become a necessity.

Sequential logic circuits act based upon their previous states. They do not automatically return to their original states, and they use this characteristic to accomplish certain goals. Data (pulses) generally proceed through a sequential logic circuit step by step. In other words, one pulse sets a device (puts it in a specific state of operation). The device then remains in that position until it is acted upon. In this way, we store data in a sequential logic circuit. Then, when the next bit of data arrives, it will encounter the stored data, and a specific reaction to both pieces of information will occur. The final output will reflect this. Sequential logic devices can be used together.

A modern integrated circuit (IC) is a thin chip of silicon upon which are thousands or millions of semiconductor devices. These devices are primarily transistors, though there are usually a large number of resistors as well. Most chips are approximately 1 centimeter square, though many other sizes exist. The most important types of IC chips are microprocessors and memory chips.

IC chips are just large-scale combinations of the devices and circuits that have been previously covered in this text.

The body of these chips is made of pure silicon. The insulation layer on top is silicon dioxide (SiO_2). Layers of P and N semiconductor material are embedded in the silicon. Running on top of the silicon dioxide (and penetrating through to the P and N layers at times) is a layer of aluminum conductor. From this combination of materials, resistors, diodes, and transistors are constructed and connected together.

While most ICs operate with digital pulses, they can also be made to use nondigital circuitry. We call such circuits analog, or linear.

Chapter 14 provides insights into the world of fiber optics.

Review Questions

1. What is binary numbering?
2. What advantage do digital signals have over analog?
3. What relationship do high and low have with zero and one?
4. How would you write the number 8 in binary numbering?
5. What is the base of our normal numbering system?
6. What is a bit?
7. What is a byte?
8. Describe parallel transmission.
9. Describe serial transmission.
10. What is a NAND gate?
11. What is a NOR gate?
12. What type of gate has a single input?
13. What is a linear circuit?
14. What is a truth table?
15. What type of diode gate would be built with forward-biased diodes?
16. What type of diode gate would be built with reverse-biased diodes?
17. How many layers of silicon would be necessary for an IC chip that contains transistors?
18. What type of insulation is used for IC chips?

19. What types of conductors are used for IC chips?
20. Describe the position of a conductor in an IC chip.

Exercises

1. Draw a diode logic gate. Show values and describe the gate's operation.
2. Draw a transistor logic gate. Show values and describe the gate's operation.

Chapter 14

Fiber Optics

Optical fiber is used to transmit all types of data and communications signals over short or long distances. Fiber can transmit huge quantities of data and does so at very low costs and very high reliability. This technology is becoming ever more important for the electronics industry. Although the signals used in these systems are light (and not electricity), electronic components are used on both ends of such communications links, as well as in the middle of many optical links.

Fiber-optic systems are especially important because of the general increase in communication signal usage worldwide. Not only are worldwide communications growing rapidly, but many systems that were formerly implemented using only copper wiring are now being implemented with fiber. Telephone systems in particular are rapidly changing over to fiber, as are cable television, local area networks, and many other means of communication.

Optical fiber is simply a better, more efficient system for sending information than metallic wire. At low transmission speeds, copper performs admirably, and it has a very low cost. But as signal speed (also called *throughput* in the optical fiber industry) increases, the cost differential changes, and fiber eventually becomes less expensive than copper. There are design factors that complement this end-cost advantage as well.

The effect of all this is that optical fiber has become the most important of all signal-transmission media. Aside from the advantages of speed, fiber is very durable, has complete immunity to electromagnetic interference, is incapable of sparking or excessive heat, and is immune to any type of electronic bugging.

Light Recap

Chapter 8 discussed the nature of light in some depth. We'll recap the main points here:

- Light begins in the outer electron shells of atoms. The essential particle of light is the *photon*. Photons are expelled from the outer shells of electrons when an electron moves from one energy level to another, as if the excess energy of the electron is thrown off in the form of a photon.
- Photons act as both particles and waves. The action of light can be explained as either particle-like or wave-like. Both explanations are correct.

- Visible light constitutes only a small part of the electromagnetic spectrum. Additionally, both ultraviolet and infrared are types of light that are not visible to the human eye. However, ultraviolet and infrared are very useful in optoelectronic devices.
- The color of light is measured in wavelength. Although wavelength also translates into frequency, wavelength is the standard measurement of the color of light, and is usually expressed in nanometers.
- The intensity of light, its brightness, is measured in milliwatts or microwatts.

Sending Light through Glass Fibers

Optical fiber acts as a conduit for light. As you can see in Figure 14-1, there are three concentric layers to an optical fiber. Light travels only through the glass *core* of the fiber. The *cladding* (which is a different type of glass) serves as a barrier to keep the light within the core, functioning much like a mirrored surface. The *coating* has nothing to do with light transmission; it is used only for mechanical strength and protection.

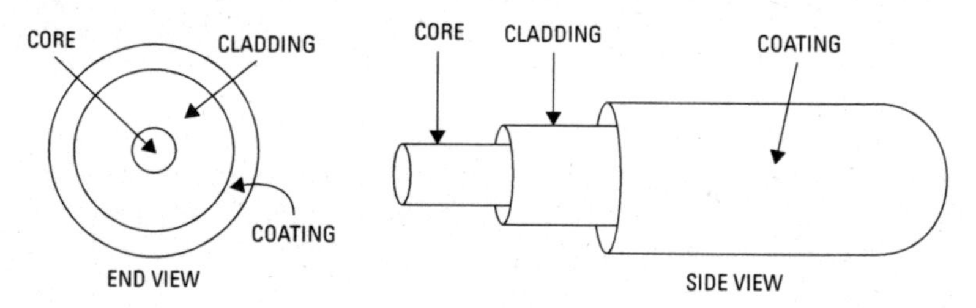

Figure 14-1 Layers of an optical fiber.

Light is kept to the core of the fiber and flows though the core as water would flow through a tube. We could even say that the fiber is a virtual tube. Light stays in the center of the fiber, not because there is a physical opening there but because the cladding glass reflects any escaping light back to the core.

Optical fibers are very thin strands of ultrapure glass. The core is composed of glass with a specific density. The cladding is a second grade (and density) of glass. The coating is made of plastic.

To suitably protect glass fibers, we package them in cabling. Actual fiber itself (not the *cable*, but only the thin *fiber*) is surprisingly flexible and will not break easily. (It is actually several times stronger than steel, when measured on a square-inch basis.) Fibers are, however, susceptible to internal cracking if twisted and/or pulled with too much tension. They are, after all, glass.

Fiber *cables*, however, are not at all fragile. In fact, they are often more durable than copper communication cables. Optical cables encase the glass fibers in several layers of protection, as is shown in Figure 14-2.

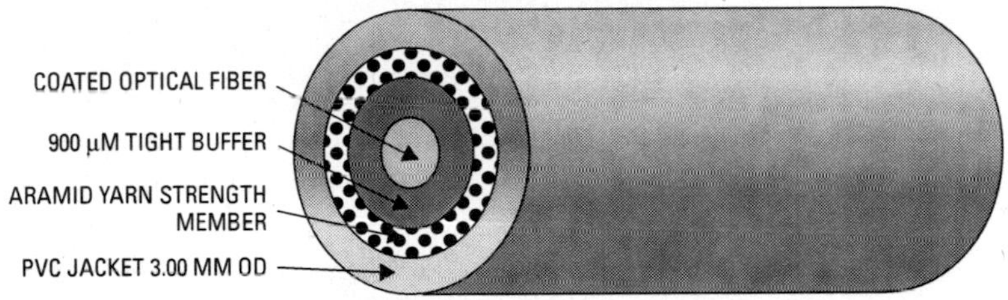

Figure 14-2 An optical cable.

Optical-Signal Transmission

Many factors affect the transmission of light through a long glass fiber. These are not terribly difficult concepts, but they are likely new to you. Go through them carefully and you should grasp them without too much trouble.

Attenuation and Dispersion

The two primary concerns when sending optical signals through a fiber are signal integrity and signal strength.

The simple weakening of an optical signal as it passes through a fiber is called *attenuation*. Such losses can be caused by the absorption of the light and its conversion to heat by molecules in the glass. Absorbers tend to be residual deposits of chemicals that are used in the manufacturing process to modify the characteristics of the glass. This absorption is determined by the elements in the glass and is most pronounced at the wavelengths around 1000 nanometers (nm), 1400 nm, and more than 1600 nm.

The larger cause of attenuation is *scattering*. Scattering occurs when light collides with individual atoms in the glass and is deflected from its original course, exits the core of the fiber, and is lost. However, loss is seldom a serious issue in fiber-optic systems, which usually have an abundance of power.

The more serious problem in optical-signal transmission is signal integrity. When a signal degrades, we say that it is *dispersed*.

This results in a phenomenon called *pulse spreading*. This is shown in Figure 14-3. Notice that a clean square wave signal enters the fiber. Then, because of the dispersion of the signal, the square pulses spread out. If this happens to a significant degree, the detectors on the far end of the link will not be able to distinguish between zeros and ones, and the signal will be unusable.

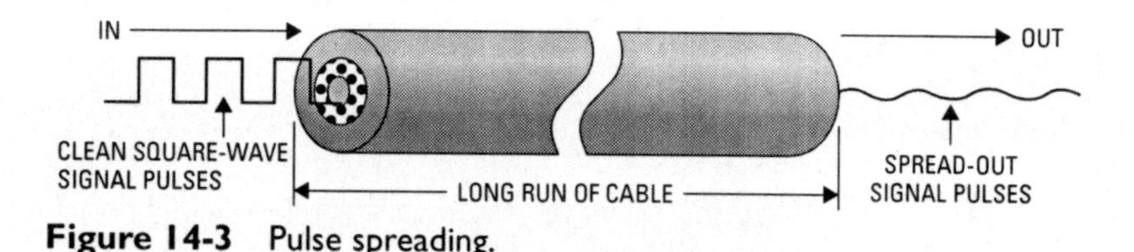

Figure 14-3 Pulse spreading.

There are two potentially confusing terms that you will come across in your readings: *chromatic dispersion* and *modal dispersion*. In both of these terms, *dispersion* refers to the spreading of light pulses until they overlap one another and the data signal is distorted and lost. *Chromatic* refers to color, and *modal* primarily refers to the light's path. Thus, we can state in simple terms that:

- Chromatic dispersion is signal distortion caused by color.
- Modal dispersion is signal distortion caused by path.

Note that dispersion is not a *loss* of light; it is a *distortion* of the signal. Thus, dispersion and attenuation are two different and unrelated problems. Attenuation is a loss of light; dispersion is a distortion of the light signals.

Chromatic dispersion occurs because various wavelengths (that is, colors) of light travel at different speeds through a glass fiber. (You may have heard that the speed of light is an absolute. This is true for light traveling in a vacuum but not for light traveling in some other medium, such as glass, water, or air.) So, over a long run, certain wavelengths will arrive sooner or later than others. This tends to spread the pulse, turning it from a square wave to a much flatter, more rounded wave.

Modal dispersion is caused by the path of various light rays in a fiber, as shown in Figure 14-4. This shows a side view of an optical fiber. Notice that the ray of light shown as a dashed line proceeds straight down the middle of the fiber. Then notice that the ray of light depicted as a solid line bounces from side to side as it travels the length of the fiber. This second ray, traveling a longer path, will arrive at the far end of the fiber later than the ray that goes straight. Again, this tends to spread the pulse, turning it from a square wave to a much flatter, more rounded wave.

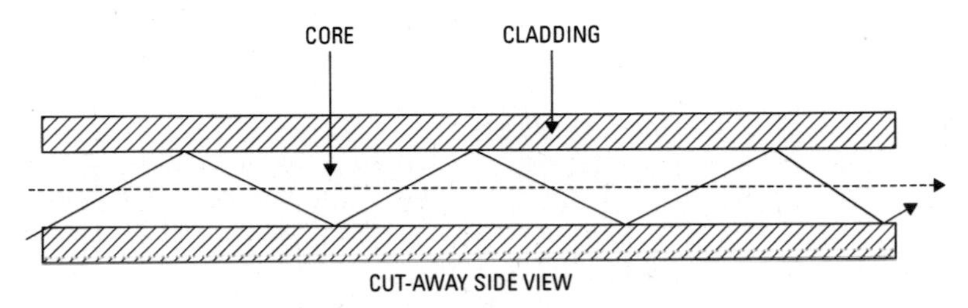

Figure 14-4 Modal dispersion.

Internal Reflection

Optical fiber functions well for signal transmission because of the principle of *total internal reflection.* When light goes from one material to another of a different density (in scientific terms, this difference in density is called *index of refraction*), the light's path will bend. No doubt you have seen the illusion of a stick bending when it is stuck into water. This is caused by air and water having different indexes of refraction. When the light bends at a certain angle (and this angle is different for different types and densities of materials), all of it is reflected, and none passes through the boundary between the two materials.

To illustrate this concept, think of standing next to a still mountain lake. As you look into the water near your feet, you will see rocks and perhaps small fish swimming in the water. But as you raise your eyes, you will reach a point where you are not seeing through the surface any more, but, instead, you see a reflection of the trees and mountains on the far side of the lake. There is a critical angle beyond which light is reflected from the lake's surface rather than passing through it. This phenomenon is used to bend the light at the core/cladding boundary of the fiber and trap the light in the core. This *critical angle* defines a primary fiber specification, the *numerical aperture* (*NA*) of a fiber.

The NA designates the angle called *the angle of acceptance*—the angle beyond which the light rays injected into an optical fiber are no longer guided and will pass through the core/clad boundary and be lost.

Types of Fibers

Following are the three main types of optical fibers that are commonly used today:

- *Single-mode fibers*—A single-mode fiber allows only one light wave ray to be transmitted down the core. The core is extremely small, usually between 8 and 9 microns. Because of quantum mechanical effects, the light traveling in the very narrow core stays together in packets, rather than bouncing around the core of the fiber. Thus, single-mode fiber has an advantage over all other types in that it can handle far more signal over far greater distances.
- *Multimode, graded-index fibers*—Graded-index fibers contain many layers of glass, each with a lower index of refraction as it moves outward from the center. Since light travels faster in the glass with lower indexes of refraction, the light waves refracted to the outside of the fiber are speeded up to match those traveling in the center. The result is that this type of fiber allows for high-speed data to be transmitted over a reasonably long distance. Multimode fibers are used with LED light sources, which are less expensive than the laser light sources used for single-mode. Graded index fibers come in core diameters of 50, 62.5, 85, and 100 microns.
- *Multimode, step-index fibers*—Step-index fibers are used far less than the other types, having a far lower capacity. They have a relatively wide core (like multimode, graded index fibers). However, since they are not graded, the light put through them bounces wildly through the fiber and exhibits high levels of modal dispersion (pulse spreading caused by path).

Figure 14-5 shows all three types of fiber.

The size of an optical fiber is referred to by the outer diameter of its core and cladding. For example, a size given as 62.5/125 indicates a fiber that has a core of 50 microns and a cladding of 125 microns. The coating is not typically mentioned in the size because it has no effect on the light-carrying characteristics of the fiber.

The diameters of typical fibers are as follows:

- *Core*—8 to 62.5 microns
- *Cladding*—125 microns
- *Coating*—250 microns

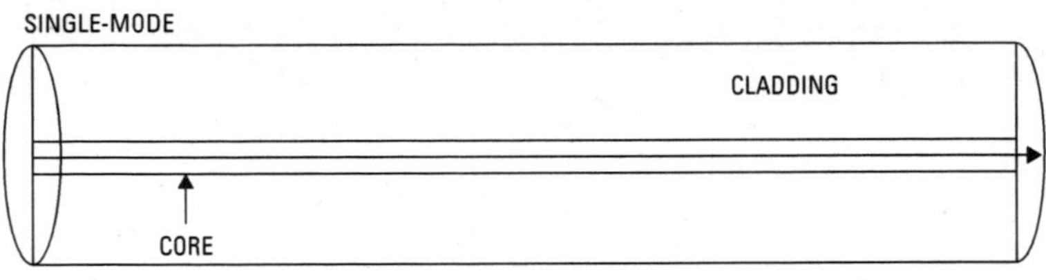

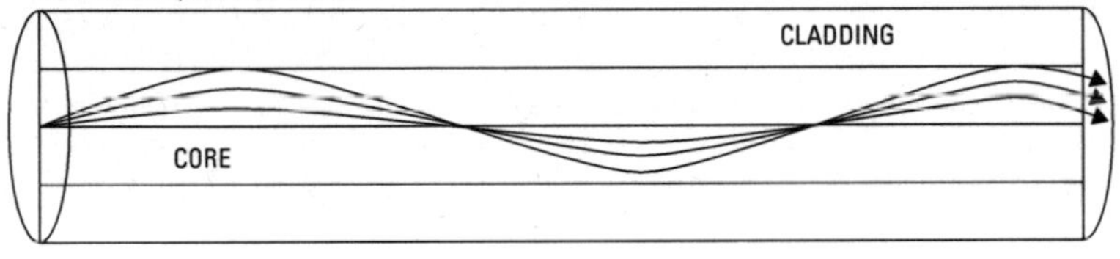

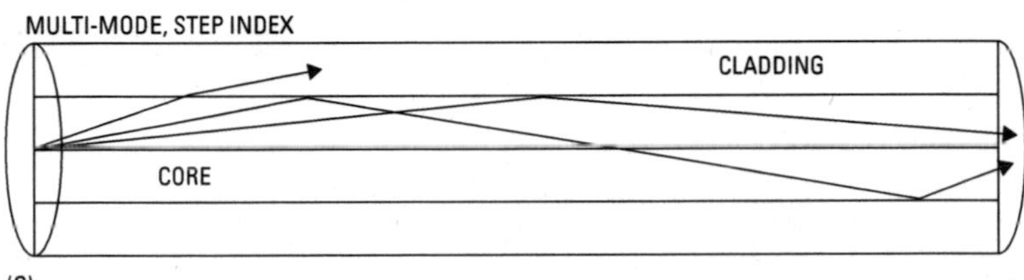

Figure 14-5 The three primary types of optical fibers.

Note that 1 micron is equal to one millionth of a meter. For comparison purposes, a sheet of paper has a thickness of approximately 25 microns.

The *core* is the part of the fiber that actually carries the light pulses that are used for transferring data. This core may be made of either plastic or glass. The size of the core is important, because core sizes of joined fibers must match. Larger cores have greater light-carrying capacity than smaller cores but generally cause greater modal dispersion.

The *cladding* sets a boundary around the fiber so that light running into this boundary is reflected back into the cable. This keeps the light from escaping. Claddings are made of glass and have a different density than that of the core. (If they did not have a different density index of refraction, they could not reflect escaping light back into the core.)

Coatings are typically multiple layers of acrylate plastic. This is necessary to add strength to the fiber, to protect it, and to absorb shock. Coatings can be stripped from the fiber (and must be for terminating) either mechanically or chemically, depending what type of plastic is used.

Transmission Devices and Methods

Figure 14-6 shows optical data transmission. The signal (of whatever type) is scanned by a digital encoder, which reduces the signal into binary code. The driver, which activates the LED or laser light source, transmits the appropriate pulses of light. The light travels through the optical fiber cable until it is received at its destination, amplified, and fed into a digital decoder. The decoder translated the digital signal back into the original electrical signal.

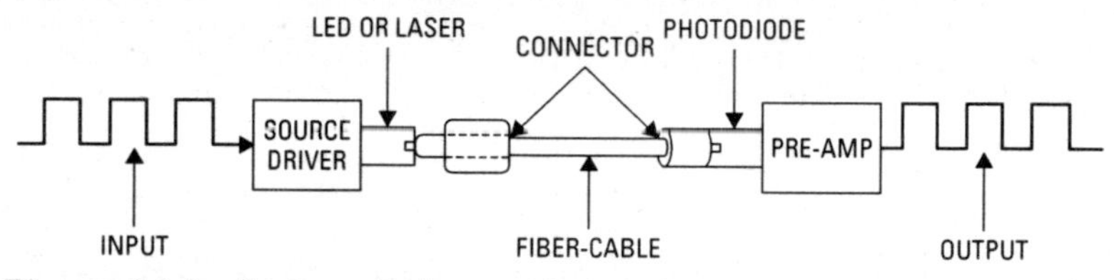

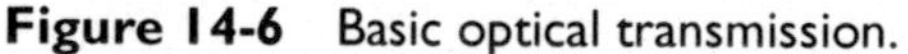

Figure 14-6 Basic optical transmission.

Transmitters and *decoders* are the devices that change electricity into light pulses and change light pulses back into electrical pulses.

Optical transmissions are classified as follows:

- *Simplex tranmission* is in one direction only.
- *Half-duplex systems* send signals in both directions, but not in both directions at the same time. Signal transmission is in one direction; then it stops and the signal is transmitted in the opposite direction, and so on.
- *Full-duplex transmission* uses two fibers to communicate. One fiber always transmits from A to B, while the other fiber is transmitting from B to A. Both ends of a full-duplex system have both transmitters and receivers.

Decibels

In optics, all energy and power levels, all losses or attenuations, are expressed in *decibels* (dB) rather than in watts. Transmission calculations and measurements on a fiber are almost always made as comparisons against a reference: received power compared to emitted power, energy in versus energy out, and so on. The decibel is a comparative measurement. The term means nothing unless you know what the reference point is for 0 decibels.

Keep the following in mind:

- Generally, energy levels (emission, reception, and so on) are expressed in *dBm*. This signifies that the reference level of 0 dBm corresponds to 1 milliwatt (mW) of power.
- Generally, power losses or gains (attenuation in a fiber, loss in a connector, and so on) are expressed in dB.
- The unit dB is used for very low levels.

Decibel measurement works as follows:

- A difference of 3 decibels equals a doubling or halving of power.
- A 3-decibel gain in power means that the optical power has been doubled. A 6-decibel gain means that the power has been doubled and doubled again, equaling four times the original power.
- A 3-decibel loss of power means that the power has been cut in half. A 6-decibel loss means that the power has been cut in half, then cut in half again, equaling one-fourth of the original power.
- A loss of 3 decibels in optical power is equivalent to a 50 percent loss (for example, 1 milliwatt of power in and 0.5 milliwatt of power out).
- A 6-decibel loss would equal a 75 percent loss (1 milliwatt in, 0.25 milliwatt out).

Optical System Materials

We have touched on fibers, and briefly on cabling. In addition to these, there are many other components used in optical-signal transmission. In general, these devices break down into two categories: active components and passive components.

Active components are those that alter, produce, or use the light signals in some way. Active components include light sources, receivers, amplifiers, and so on.

Passive components are those that do not change the light signal but merely pass it. Passive optical components would include couplers (T connections, which split a branch off of a continuous fiber), connectors, splices, and similar components.

We will cover the more important components here.

Light Sources

The first criterion of choice of an optical emitter (LED or laser) is its *wavelength*, which influences the attenuation in the link. The second criterion is its *spectral bandwidth*, which influences chromatic dispersion. No light emitter is entirely monochromatic. Light sources produce light energy centered around its *nominal wavelength* (the wavelength for which it is specified). The spectrum of emission of laser diodes is much narrower than that of LEDs. Chromatic dispersion is, therefore, much lower with a laser than with an LED.

The emission wavelength is dependent on the materials used to manufacture the light source. The spectral bandwidth is dependent on the structure of the emitter.

LEDs

LEDs are the most commonly used light emitters. You will recall that we covered LEDs in some depth in Chapter 8. Therefore, we will not cover them again here.

There are three types of LEDs that are commonly used for optical fiber work. The first two are the *surface emission* diodes and *transverse emission* diodes. The fundamental difference between these two types of diodes lies in the emission surface, which is much smaller in transverse emission diodes. The photons generated by current flowing between the 2 P and N regions are confined in the junction, which has a very small thickness. Therefore, light is emitted on one side of the diode.

The third LED type is the *Burrus diode* (named after Charles A. Burrus of AT&T Bell Laboratories), in which an optical fiber gathers the emitted light. The attractive characteristics of this diode are directivity and a high coupling rate.

The most common LEDs are:

- GaAs, GaAlAs LEDs—820–850 nanometer transmission systems on multimode fibers
- InGaAsP LED—1330 nanometer transmission systems on multimode fibers

Lasers

Semiconductor lasers provide a higher power and a better directivity than LEDs. These characteristics are inherent to their different design, although they use the same principle of photon emission by electron excitation.

When an electron moves from one energy level to a lower level, it emits a photon. This is called spontaneous emission, and can be controlled and triggered by a photon hitting the electron, which then emits a photon with the same wavelength as the first one. This is the *light amplification by stimulated emission of radiation* effect (or the *laser* effect).

Laser diodes are made of a semiconductor crystal, one side of which is coated by a reflective layer (to form the reflecting mirror), while the emission side is left bare.

The laser effect in a semiconductor is triggered above a given threshold, when the initial stimulation has produced enough photons for the process to be maintained and amplified. Below this threshold, the laser diode functions as an LED. Beyond this value, a small increase in the control current will generate a very high increase in the emitted power.

The two common types of lasers used in fiber systems are as follows:

- *Multimode lasers*—Classical laser diodes (LDs), called *Perot-Fabry*, for 1300-nanometer transmission systems on conventional single-mode fibers and 1550-nanometer transmission systems on dispersion-shifted single-mode fibers
- *Single-mode lasers*—Monochromatic LDs, called *DFB* for 1550-nanometer transmission systems on conventional single-mode fibers.

Cabling

The first protective layer for an optical fiber is the coating. The next layer of protection is a *buffer* layer. The buffer is typically extruded over the coating to further increase the strength of the single fibers. This buffer can be of either a *loose tube* or *tight tube* design. Most data communication (datacom) cables are made using either one of these two constructions. A third type, the ribbon cable, is frequently used in the telecommunications work and may be used for data communication applications in the future. It uses a modified type of tight buffering.

After the buffer layer, the cable contains a *strength member*. Most commonly, the strength member is Kevlar fabric (the material that bulletproof vests are made from). The strength member not only protects the fiber, but is used to carry the tensions of pulling the cable. (You can never pull the fibers themselves.) In many cases, however, there will be additional *stiffening members* that also increase the cable's strength and durability.

The final layer of protection is the cable's jacket.

Many different configurations of optical cables are made. Several are shown in Figure 14-7.

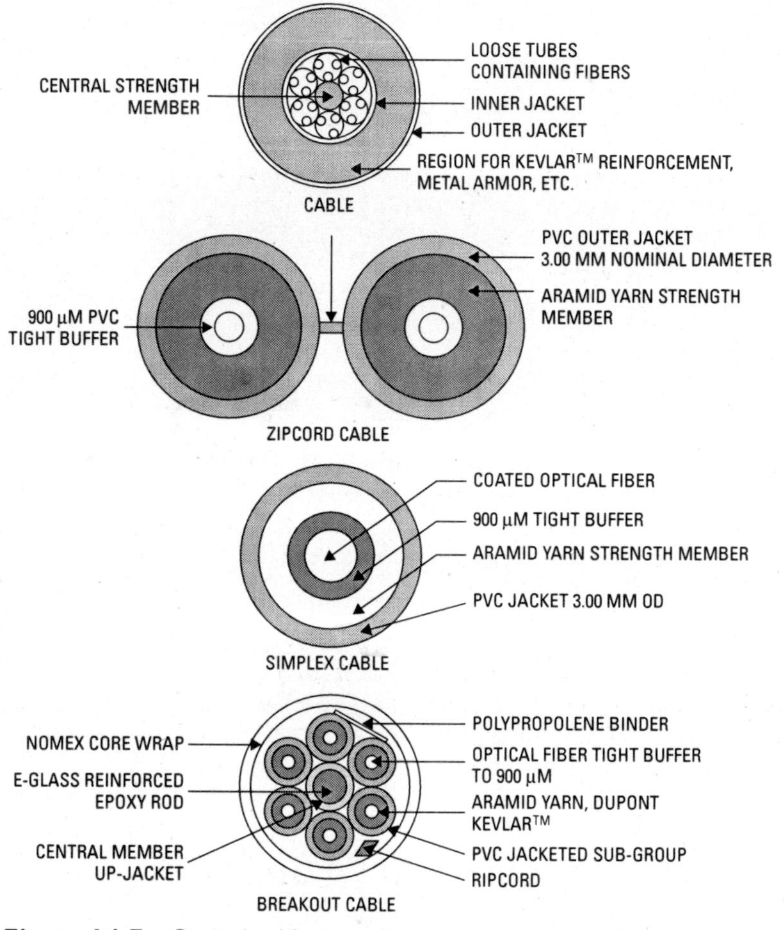

Figure 14-7 Optical cables.

Connectors

Because optical fibers have small diameters, they must be held rigidly in place and properly aligned to mate with other fibers, sources, or detectors.

Terminating a fiber involves installing a connector and polishing the face of the fiber. Figure 14-8 shows several common fiber connectors.

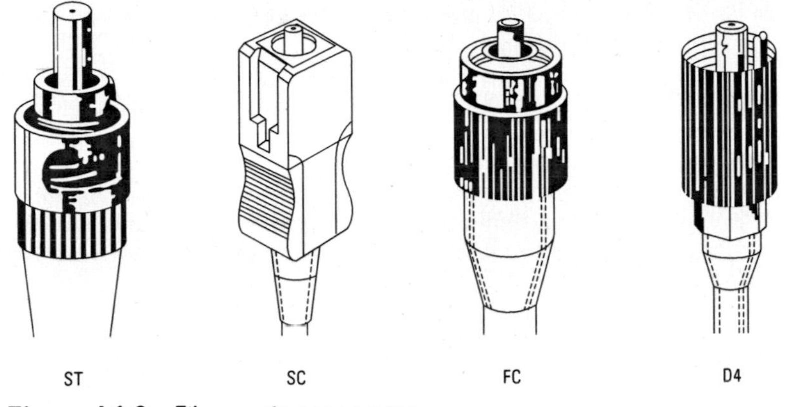

Figure 14-8 Fiber-optic connectors.

A variety of glues and crimping techniques are used to hold fibers firmly and permanently inside of connectors.

Splices

Splices are used to join two ends of fiber to each other permanently. This is done either by *fusion* (that is, by melting the pieces of glass together) or by mechanical means. The critical factors in splicing are that the joint passes light without loss and that the joint is mechanically secure.

Single-mode fiber is almost always fusion-spliced. While fusion splicing offers lower loss and better performance, this is not the only reason to use this method of splicing. Mechanically splicing single-mode fiber, with its very small core, is often problematic.

Multimode fiber, with its complicated core structure, does not always fusion-splice easily, so mechanical splices can give equal performance at a lower overall cost.

Fusion splicers use an electric arc to ionize the space between prepared fibers to eliminate air and to heat the fibers to proper temperature (2000°F). The fibers are then fed into the splicing machine as semi-liquids, and melted together. One drawback to fusion splicing is that it most generally must be performed in a controlled environment (such as a splicing van or trailer) and should not be done in open spaces because of dust and other contamination. Another drawback is that fusion splicers can be expensive.

Mechanical splices are widely used for multimode fiber. They do not require a controlled environment other than a reasonable level of dust control. The strength of a mechanical splice is better than most connectors, although fusion splices are stronger.

Mechanical splices use either a V groove or tube-type design to obtain fiber alignment. The fibers are then glued or crimped into place. Mechanical splices use some type of index matching gel for proper light transmission between the surfaces of the two fiber ends being joined. The V groove is probably the oldest and still most popular method, especially for multifiber splicing of ribbon cable. This type of splice is either crimped or snapped to hold the fibers in place.

Completed splices, whether fusion or mechanical, are placed into splicing trays that are designed to accommodate the particular type of splice in use. Splicing trays then fit into splice organizers and, in turn, fit into a splice closure.

Receivers

The detectors most used in optical-signal transmission are phototransistors, photodiodes, and avalanche photodiodes (APDs). *PIN-FET* diodes (combining a PIN photodiode and a field effect transistor) and phototransistors are used in numerous applications because of their low cost. Photodiodes are used in a variety of systems. APDs are used where operations at low signal strengths are necessary.

Testing

When installing a fiber system (the whole system of optical fibers is often called a *cable plant*), it must be tested to ensure that it will perform. The purpose of testing is to ensure that light will pass through the system properly.

While continuity testing is very simple, both power testing and Optical Time Domain Reflectometry (OTDR) testing require training. We will explain the basics in this book, but to actually perfom these tests reliably requires specialized training.

Continuity Testing

A continuity test is a simple visible light test. Its purpose is to ensure that the fibers in your cables are continuous (that is, that they are not broken). This test is done with a modified type of flashlight device and the naked eye, and it takes only a few minutes to perform.

Power Testing

Power testing accurately measures the quality of optical-fiber links. A calibrated light source puts infrared light into one end of the fiber, and a calibrated meter measures the light arriving at the other end of the fiber. The loss of light in the fiber is measured in decibels.

The most basic (and most commonly performed) fiber-optic test is called an *insertion-loss* test. Insertion loss is the loss caused by the insertion of a component (such as a splice or connector) in an optical fiber. Figures 14-9 and 14-10 show two basic methods of insertion-loss testing.

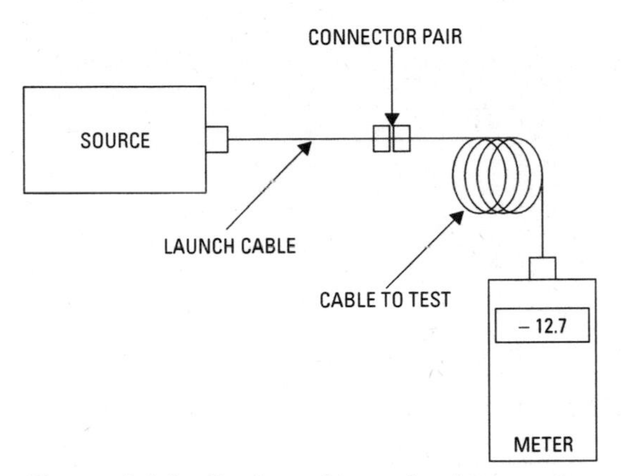

Figure 14-9 Single-end insertion-loss testing.

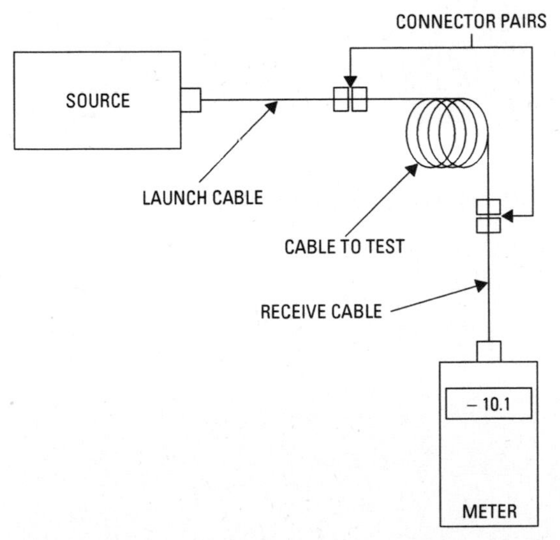

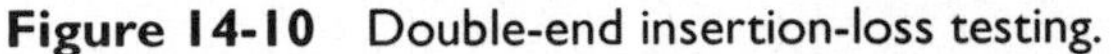

Figure 14-10 Double-end insertion-loss testing.

Table 14-1 System Performance Data

Link Type	*Source/Fiber Type*	*Wavelength (nm)*	*Transmit Power (dBm)*	*Receiver Sensitivity (dBm)*	*Margin (dB)*
Telecom	Laser/SM	1300	+3 to −6	−40 to −45	34 to 48
		1550	0 to −10	−40 to −45	40 to 45
Datacom	LED/MM	850	−10 to −20	−30 to −35	10 to 25
		1300	−10 to −20	−30 to −35	10 to 25
CATV (AM)	Laser/SM	1300	+10 to 0	0 to −10	10 to 20

Table 14-2 System Fiber Type and Bandwidth Figures

Fiber Type	Core/Cladding Diameter (μm)	Attenuation Coefficient (dBkm) 850 nm	1300 nm	1550 nm	Bandwidth (MHz-km)
Step index	200/240	6		NA	50 @ 850 nm
Multimode	50/125	3	1	NA	600 @ 1300 nm
Graded-index	62.5/125	3	1	NA	500 @ 1300 nm
	85/125	3	1	NA	500 @ 1300 nm
	100/140	3	1	NA	300 @ 1300 nm
Single-mode	9/125		0.5	0.3	High
Plastic (POF)	1 mm	(0.2 dBm @ 665 nm)			Low

Table 14-3 Optical Networking Parameters

Network	***IEEE802.3 FOIRL***	***IEEE802.3 10base F***	***IEEE802.5 Token Ring***	***IEEE802.12 100base F***	***ANSI X3T9.5 FDDI***	***ESCON IBM***
Bit rate (MB/s)	10	10	4/16	100	100	200
Architecture	Link	Star	Ring	Star	Ring	Branch
Fiber type	MM, 62.5	MM, 62.5	MM, 62.5	MM, 62.5	MM/SM	MM/SM
Link length (km)	2	2	—	2	2/60	3/20
Wavelength (nm)	850	850	850	1300	1300	1300
Connector	SMA	ST	FDDI	ST	FDOI	ESCON

Virtually all optical-fiber systems (also called *fiber-optic cable plants*) are tested for end-to-end insertion loss to confirm that the required signal will be permitted to pass. Loss is typically measured at 850- and 1300-nanometer wavelengths for multimode links. Single-mode fiber is always tested at 1300 nanometers, and may also be tested at 1550 nanometers.

OTDR Testing

Optical time domain reflectometer (*OTDR*) uses light backscattering to analyze fibers. The OTDR device sends a high-powered pulse into the fiber and measures the light scattered back toward the instrument. In essence, it takes a snapshot of the fiber's optical characteristics. The OTDR can be used to locate fiber breaks, splices and connectors, as well as to measure loss. However, the OTDR may not give the same value for loss as a source-and-power meter, because of the different methods of measurement.

Optical System Parameters

There are many types of optical-signal transmission systems. Table 14-1 shows typical system performance figures. Table 14-2 shows fiber types and bandwidth figures. (*Bandwidth* means the amount of signal transmitted. This is an old radio industry term.) Table 14-3 shows typical optical network parameters.

Note that although all the data shown in Table 14-1, Table 14-2, and Table 14-3 are very common, specific systems may vary.

Summary

Optical fibers can transmit huge quantities of data and do so at very low costs and very high reliability. This technology is becoming ever more important for the electronics industry. Although the signals used in these systems are light, not electricity, electronic components are used on both ends of such communications links, as well as in the middle of many optical links as well. Fiber-optic systems are especially important because of the general increase in communication signal usage worldwide. Not only are worldwide communications growing rapidly, but many systems that were formerly implemented using only copper wiring are now being implemented with fiber. Telephone systems in particular are rapidly changing over to fiber, as are cable television, local area networks, and many other means of communication.

Light begins in the outer electron shells of atoms. The essential particle of light is the photon. Photons are expelled from the outer shells of electrons when an electron moves from one energy level to another, as if the excess energy of the electron is thrown off in the form of a photon.

The color of light is measured in wavelength. Although wavelength also translates into frequency, wavelength is the standard measurement of the color of light, and is usually expressed in nanometers. The intensity of light is measured in milliwatts or microwatts.

There are three concentric layers to an optical fiber. Light travels only through the glass core of the fiber. The cladding serves as a barrier to keep the light within the core, functioning much like a mirrored surface. The coating has nothing to do with light transmission and is used only for mechanical strength and protection.

Optical fibers are very thin strands of ultrapure glass. Light stays in the center of the fiber, not because there is a physical opening there, but because the cladding glass reflects any escaping light back to the core.

To suitably protect glass fibers, we package them in cabling. Actual fiber itself (not the cable, but only the thin fiber) is surprisingly flexible and will not break easily. Fibers are, however, susceptible to internal cracking if twisted and/or pulled with too great a tension.

Fiber-optic cables are not fragile. Optical cables encase the glass fibers in several layers of protection.

The simple weakening of an optical signal as it passes through a fiber is called attenuation.

The most serious problem in optical signal transmission is signal integrity. When a signal degrades, we say that it is dispersed. This results in a phenomenon called pulse spreading. Because of dispersion, clearly defined signals (square pulses) spread out. If this happens to a significant degree, the detectors on the far end of the link will not be able to distinguish between zeros and ones, and the signal will be unusable.

Chromatic dispersion occurs because various wavelengths (that is, colors) of light travel at different speeds through a glass fiber. Over a long run, certain wavelengths will arrive sooner or later than others. This tends to spread the pulse, turning it from a square wave to a much flatter, more rounded wave.

Modal dispersion is caused by the path of various light rays in a fiber. Rays of light that bounce from end to end rather than traveling directly through the middle of an optical fiber will arrive at the far end of the fiber later than the ray that goes straight. This tends to spread the pulse, turning it from a square wave to a much flatter, more rounded wave.

Optical fiber functions well for signal transmission because of the principle of total internal reflection. When light goes from one material to another of a different density, the light's path will bend. When the light bends at a certain angle (and this angle is different for different types and densities of materials), all of it is reflected, and none passes through the boundary between the two materials.

Single-mode fiber allows only one light wave ray to be transmitted down the core. The core is extremely small, usually between 8 and 9 microns. Single-mode fiber has an advantage over all other types in that it can handle far more signal over far greater distances.

Graded-index fibers contain many layers of glass, each with a lower index of refraction as it moves outward from the center. Since light travels faster in the glass with lower indexes of refraction, the light waves refracted to the outside of the fiber are speeded up to match those traveling in the center. Graded-index fiber allows for high-speed data to be transmitted over a reasonably long distance. Multimode fibers are used with LED light sources, which are less expensive than the laser light sources used for single-mode.

Step-index fibers are used far less than the other types, having a far lower capacity. They have a relatively wide core (like multimode, graded-index fibers), but since they are not graded, the light put through them bounces wildly through the fiber and exhibits high levels of modal dispersion.

The size of an optical fiber is referred to by the outer diameter of its core and cladding. For example, a size given as 62.5/125 indicates a fiber that has a core of 50 microns and a cladding of 125 microns. The coating is not typically mentioned in the size.

Transmitters and *decoders* are the devices that change electricity into light pulses and change light pulses back into electrical pulses.

Optical transmissions are classified as simplex, half-duplex, or full-duplex. Simplex transmission is in one direction only. Half-duplex systems send signals in both directions, but not both directions at the same time. Full-duplex transmission uses two fibers to communicate in both directions at the same time.

In optical systems, all energy and power levels are expressed in decibels rather than in watts. All transmission calculations and measurements on a fiber are almost always made as comparisons against a reference. The decibel is a comparative measurement. Almost all optical power is specified in dBm. This signifies that the reference level of 0 dBm corresponds to 1 milliwatt of power. A difference of 3 decibels equals a doubling or halving of power.

Passive components are those that do not change the light signal, but merely pass it. Passive optical components would include couplers, connectors, splices, and similar components. Active components are those that alter, produce, or use the light signals in some way. Active components include light sources, receivers, amplifiers, and so on.

LEDs are the most commonly used light emitters. They are inexpensive and work well with multimode fibers.

Semiconductor lasers provide a higher power and better directivity than LEDs. They are more expensive than LEDs and are used with single-mode fiber systems.

The first protective layer for an optical fiber is its coating. The next layer of protection is a buffer layer. The buffer is typically extruded over the coating to further increase the strength of the single fibers. This buffer can be of either a loose tube or tight tube design. After the buffer layer, the cable contains a strength member. Most commonly, the strength member is Kevlar fabric. The final layer of protection is the cable's jacket.

Because optical fibers have small diameters, they must be held rigidly in place and properly aligned in order to mate with other fibers, sources, or detectors. Terminating a fiber involves installing a connector and polishing the face of the fiber.

Splices are used to join two ends of fiber to each other permanently. This is done either by fusion (that is, by melting the pieces of glass together) or by mechanical means. The critical factors in splicing are that the joint passes light without loss and that the joint is mechanically secure.

The detectors most used in optical-signal transmission are phototransistors, photodiodes, and avalanche photodiodes.

Optical-fiber systems must be tested to ensure that they will properly perform. While continuity testing is very simple, both power testing and OTDR testing require training.

In the next chapter, we will begin our examination of specific electronic applications with a look at radio transmissions.

Review Questions

1. What is throughput?
2. What is a photon?
3. How do photons behave?
4. What do we use as the measurement of color?
5. What is the diameter of a single-mode fiber's core?
6. Name a typical core diameter for a multimode fiber.
7. What is attenuation?

8. What is modal dispersion?
9. What is chromatic dispersion?
10. What is pulse spreading?
11. What is total internal reflection?
12. What does a decoder do?
13. Why are decibels used in the measurement of optical power?
14. What would be signified by a signal loss of 6 decibels?
15. Describe half-duplex transmission.
16. Why are LEDs commonly used as light sources?
17. Why are lasers used with single-mode fibers?
18. What material is used to make strength members?
19. What type of optical testing uses a flashlight type of device?
20. What are the critical factors in splicing?

Exercises

1. Draw a single-mode optical-transmission system. Label the parts.
2. Draw a multimode optical-transmission system. Label the parts.

Chapter 15

Radio Transmission

No technology has been more important to the electronics industry than radio. It was the first major electronic technology, and, for all practical purposes, it created the electronics industry. The introduction of radio was far bigger in its time than the introduction of the Internet in our time. All of the early electronics pioneers started in radio.

The term radio is a shortening of the original radiotelegraphy. The word radio is derived from the Latin *radius*, which means "a ray," as in a ray of light. In the early days, the term wireless telegraphy (or simply wireless) was usually applied to this technology.

The idea of broadcasting never occurred to the inventors and businessmen who developed radio. They saw it as an improvement over the telegraph—a telegraph system requiring no wires. Only later did anyone see the possibilities of broadcasting nonpersonalized signals to a very large number of people with home receivers.

Early Development

There are endless arguments, even today, over who really invented radio. The name that appears in most textbooks is Guglielmo Marconi, a young Italian inventor. And, without question, Marconi was the first person to demonstrate long-distance communication using Hertzian waves, which we now call radio waves. He did this in December 1901, when he sent the letter "S" in Morse code between Cornwall, England, and St. John's, Newfoundland. (The spot in St. John's where this took place is called *Signal Hill* to this day.) This was a transmission covering 2100 miles, and it caught the attention of the world. This was really the founding moment of radio, even though transmissions of this time were simply coded pulses of static.

Marconi went on to set up business interests (his family had extensive business interests and contacts, especially in England), to improve wireless technology, and to set up numerous transmitting and receiving stations.

While Marconi certainly deserves credit for sparking the industry, all the elements of his system were invented by others before him. Heinrich Hertz discovered the propagation of electromagnetic waves (which we will explain shortly) in 1888. In fact, a number of people were transmitting wireless signals prior to Marconi's 2100 mile transmission of 1901. Inventors named Lodge, Stone, Stubbefield, and Popov were definitely doing such work.

The person with the best claim to have beaten Marconi to radio was Nikola Tesla. While we will not take time here to cover Tesla's life story, suffice it to say that he was the prototype mad genius and invented an amazing number of the electrical and electronic devices we use today. I highly recommend the amazing biography, *Tesla: Man Out of Time* (New York: Simon and Schuster, 2001).

Tesla was reportedly demonstrating remote-controlled boats at the World's Fair in Chicago in 1893 and in St. Louis in 1894. He definitely made such demonstrations at an engineering conference in New York's Madison Square Garden in 1898. In fact, when an associate informed Tesla that Marconi had beaten him to trans-atlantic transmission, Tesla replied, "Marconi is a good fellow. Let him continue. He is using 17 of my patents."

But regardless of Tesla, Lodge, Stone, and the others, Marconi was the first to actually put all of the pieces together and to send a trans-Atlantic signal. The final key was in adding tuning (in this case, a variable inductor) to the resonant circuit in the receiver. Marconi took the developments of all the others, incorporated them together, and took them to an effective conclusion. He had, as his daughter later wrote, "a clear objective in mind, a definite goal to reach which he considered so important that he devoted all his intellectual, moral, and material resources to its pursuit."

In the end, Marconi may not deserve sole credit for developing effective long-range wireless transmission, but his name is deservedly at the top of the list. Yes, it is almost certain that Tesla could have done it first if he would have focused solely on wireless telegraphy, but he did not do that. Marconi did.

Radio Waves and Propagation

Radio waves were predicted by James Clerk Maxwell in 1864, then created, measured, and described by Heinrich Hertz between 1895 and 1898. So, since that time, it has been known that radio waves *propagate* (that is, they travel). But, to this day, scientists argue over the precise reasons why radio waves propagate and what fundamental mechanisms are involved.

Since this is not a physics text, our explanations here will stick to what is known and avoid arguments about technicalities.

There are two parts that make a propagating wave: the magnetic component and the electrostatic component.

The Magnetic Wave

The magnetic field is the same as we described in the early chapters of this book (especially Chapter 4): the magnetic field that surrounds any conductor when a current flows through it. We said that such a field arises when a current flows through any conductor, grows or contracts according to the amount of current, and collapses altogether when current through the conductor stops. This phenomenon would make radio communication possible within the reach of this field, but no further.

At radio frequencies, this field would expand on either side of the conductor for a number of meters, but no further. The formula for wavelength is as follows:

$$\lambda = \frac{d}{f}$$

The Greek letter lambda (λ) represents wavelength. So, wavelength equals distance (d) per second divided by frequency. And, since distance per second is considered to be equal the speed of light (300,000,000 meters per second), the practical formula is:

$$\lambda = \frac{300{,}000{,}000}{f}$$

So, for a typical radio frequency of 100 megahertz (MHz), the wavelength would be as follows:

$$\lambda = \frac{300{,}000{,}000}{100{,}000{,}000} = 3 \text{ meters}$$

You can see from this that communication at long distances would be impossible for a simple inductive magnetic wave, such as the one shown in Figure 15-1. Current in the antenna gives rise to a magnetic field, which then collapses when the current declines and stops.

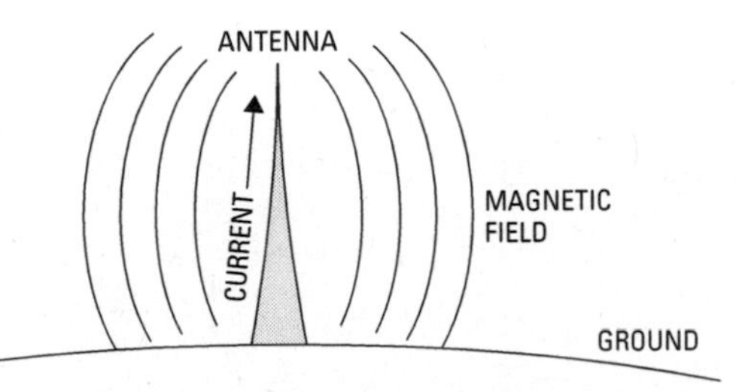

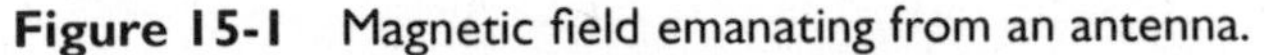

Figure 15-1 Magnetic field emanating from an antenna.

The Electrostatic Wave

We have already covered the electrostatic force to some degree in our coverage of capacitors in Chapter 4. Such a force affects the two plates of a capacitor. Remember that these plates never touch, but charges build up in them. In effect, electrons on one side try to get to the other side but are restrained because of the *dielectric* (insulating) layer in between. This force between the plates is the electrostatic field.

Figure 15-2 shows an electrostatic field from the top to the bottom of an antenna. You can think of this as the top of the antenna being one capacitor plate and the ground at the base of the antenna forming the other plate. The static that builds up between them is the electrostatic field.

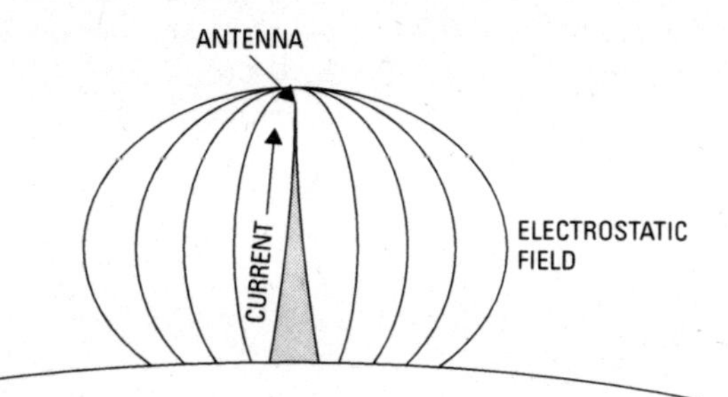

Figure 15-2 Electrostatic field emanating from an antenna.

Figure 15-3 shows the difference between the magnetic and electrostatic fields. Note that the magnetic lines of force are parallel

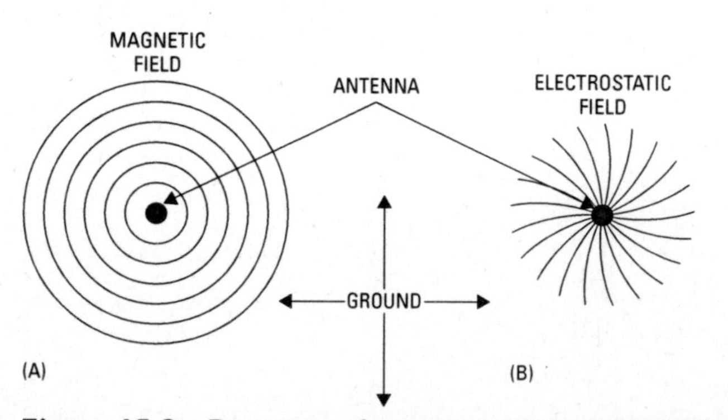

Figure 15-3 Directions of magnetic and electrostatic fields.

with the ground, just like lines of latitude around a globe. The lines of force from the electrostatic field are perpendicular to the ground, like the lines of longitude around a globe.

How Waves Are Formed

Figure 15-4 shows how radio waves are formed. In this example, we will focus on the electrostatic wave, even though the magnetic component is also present. Pay attention to the polarities of the lines of force in these drawings.

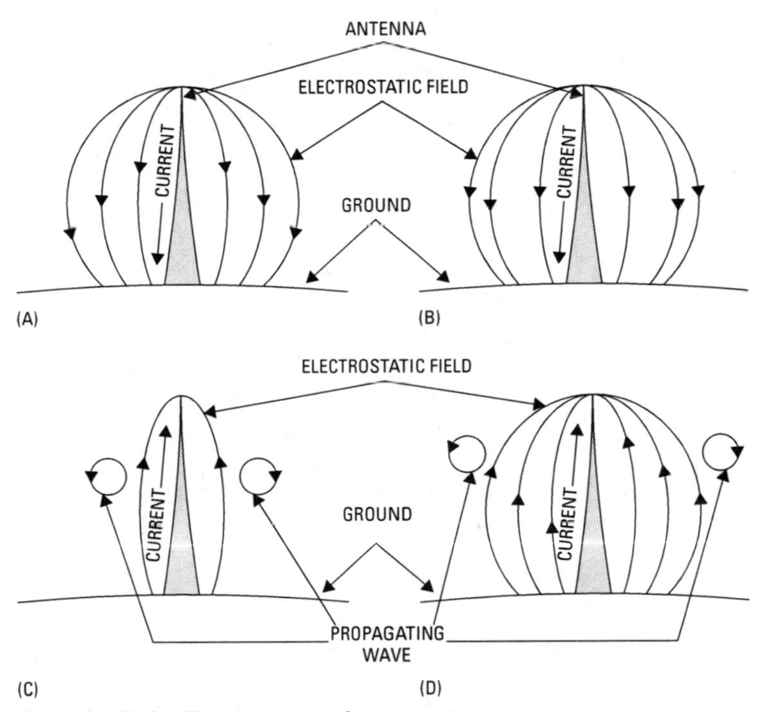

Figure 15-4 The creation of propagating waves.

In Figure 15-4A, an electrostatic field has been created around an antenna, as we have already described. In Figure 15-4B, however, this field has begun to recede because of reduced current in the antenna. Note that the outermost lines of force are pushed out fairly far from the antenna. This is because like charges repel each other. So, the outer layers remain pushed outward while the inner layers have already begun to collapse.

Note in Figure 15-4C that the current in the antenna has reversed. Now the field will quickly reverse also. This happens before the outer layers of the magnetic field can collapse all the way. At this point, as shown in Figure 15-4D, the lines of force form into their own ring. Now, new (reversed) lines of force from the antenna begin to spread, and these push the rings from the previous electrostatic field outward.

These rings of force that are pushed away from the antenna are *radio waves*, and they can travel great distances.

Wave Characteristics

While we are not going to describe every aspect of wave propagation, there are a few other things that are worth stating here.

First is that the magnetic and electrostatic waves are 90 degrees out of phase with each other, as shown in Figure 15-5. Notice that the fields are labeled *H field* for the magnetic field and *E field* for the electrostatic field. These are the standard industry terms.

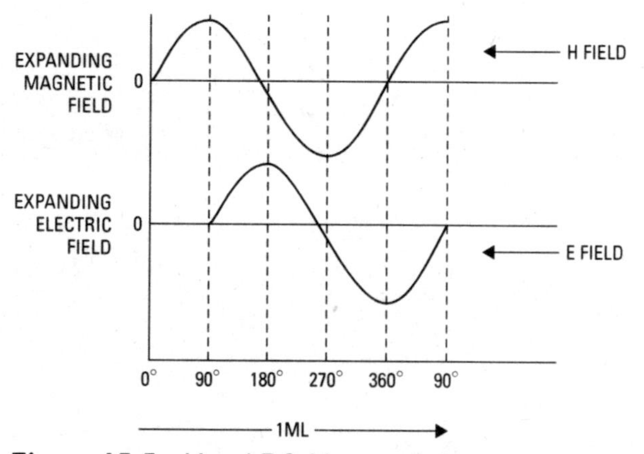

Figure 15-5 H and E fields out of phase.

Figure 15-6 shows a simple way of determining the direction of a propagating wave: the *right-hand rule*. With your right hand held this way, your thumb in the direction of the electrical field, and your index finger in the direction of the magnetic field, your middle finger will point in the direction of propagation.

Tuned Circuits

To send intelligible signals over a distance, tuning the transmitting and receiving circuits was critical.

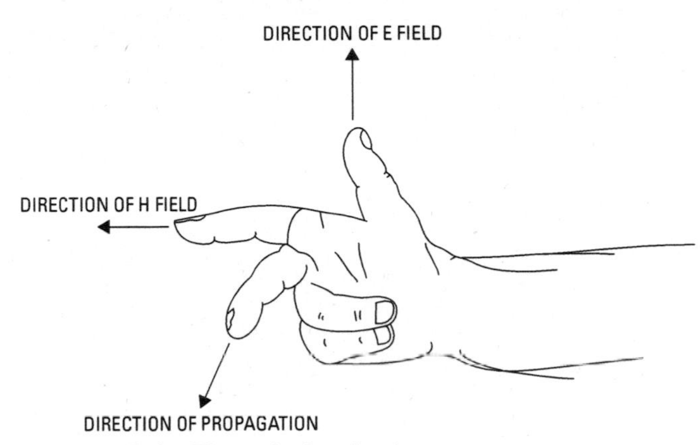

Figure 15-6 The right-hand rule.

The first oscillating circuits were operated with spark-gap devices, as shown in Figure 15-7. If you remember back to Chapter 12, "Oscillators," we said that a tank circuit could be started with a single burst of current and then continue oscillating until resistance diminished the current flow and ultimately stopped it altogether. This is precisely how an oscillating circuit was made before the positive feedback mechanism was created. A capacitor (called a *condenser* at the time) was made to spark across the primary of a transformer, inducing a short current burst into it. The secondary of this transformer would be a resonant circuit.

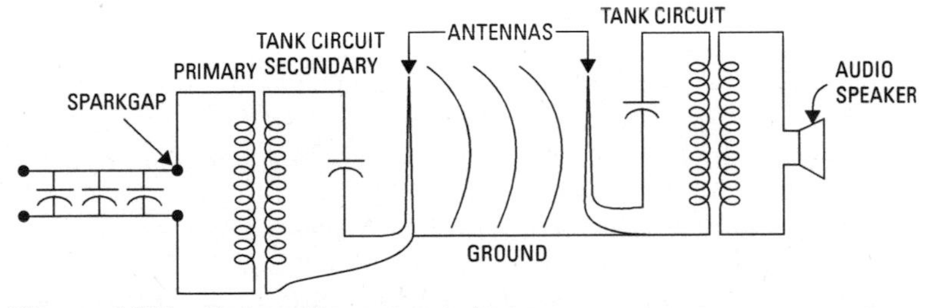

Figure 15-7 Early radio transmission.

A burst of current across the spark-gap device would cause a burst of current in the primary winding then a burst of current into the tuned (tank) circuit of the secondary, which would continue oscillating for some period of time. This oscillating current formed a steadily oscillating magnetic field. Then, if this field were imposed on an antenna, radio waves of a fixed wavelength would be created and would travel a significant distance. At this distance, an antenna would be set up to catch the wave and attached to a circuit tuned to the same frequency. This would induce a current into the circuit, which could be used to make a sound or otherwise affect some indicating device.

Since the early days, this process has been improved greatly, but the need for tuned circuits on each end remains unchanged. This, obviously, is also how we tune-in individual broadcast channels. In the early days, however, tuning was essential to functionality, not to find a specific station.

Oscillators and the Audion

For a decade or two after radio had been put into use, signals consisted mostly of dots and dashes, either on or off. In fact, Morse code was very common. A series of pulses separated by intervals made up the messages.

To discern whether a signal was present or not, early receivers often relied on devices called *coherers*, which were essentially switches that turned on when a pulse excited an antenna, and had to reset themselves before the next pulse arrived.

Two developments improved this early technology: *effective oscillators* and the *audion tube*. These two devices replaced the clumsy (though effective) arrangement shown in Figure 15-7.

As mentioned earlier, waves of a consistent frequency are required for effective transmission. Also, the size of the antenna is critical. We will explain this in some detail later in this chapter, but an effective antenna is usually designed to be one-half of the signal's wavelength for maximum effectiveness. This is another reason to use a fixed frequency.

Since frequency and wavelength are simply different measurements of the same thing, any frequency will always have the same wavelength. For this reason, you may see a specific radio signal being referred to either as a specific frequency or as a specific wavelength. In the example we gave earlier, we said that a signal of 100 megahertz has a wavelength of 3 meters. We could call it a 3-meter signal or a 100-megahertz signal and be equally correct either way.

The development of oscillating circuits with positive feedback gave radio developers the ability to use a continuous signal of a specific frequency. This ended the limitation to dashes and dots.

Next came the audion, the amplifier that we explained in Chapter 7. This made possible not only continuous signals but signals that were both continuous and variable.

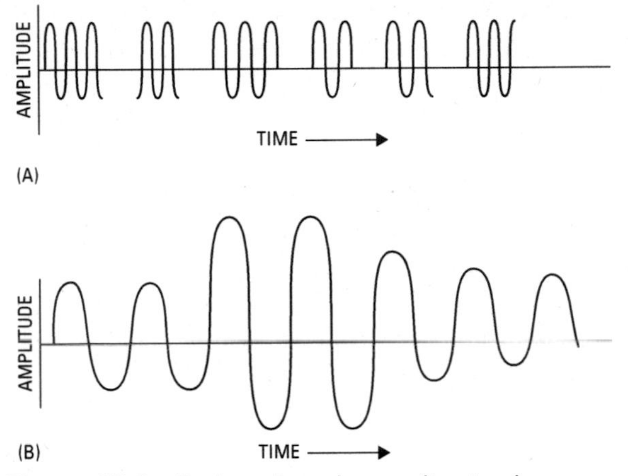

Figure 15-8 Early and modern radio signals.

Figure 15-8 shows the difference between the early radio signals and the more modern radio signals. Figure 15-8A shows what radio signals originally looked like, and Figure 15-8B shows the modern *AM* (*amplitude modulation*) signal made possible by the oscillator and the audion. *Amplitude* means size or strength. The greater the amplitude is, the stronger the signal will be.

Basics

Having covered the development of radio technology fairly well, let us simply state the basic facts.

Radio refers to sending and receiving electromagnetic waves in the range of frequencies lying between 1 hertz and a few gigahertz (GHz). This is shown in Table 15-1. Notice that above LF range, all the frequency ranges are multiples of ten.

Table 15-1 Radio Frequency Classifications

Classification	*Abbreviation*	*Frequency Range*
Very low frequency	VLF	9 kHz and below
Low frequency (longwave)	LF	30 kHz to 300 kHz
Medium frequency	MF	300 kHz to 3 MHz
High frequency (shortwave)	HF	3 MHz to 30 MHz
Very high frequency	VHF	30 MHz to 300 MHz
Ultra-high frequency	UHF	300 MHz to 3 GHz
Microwaves		3 GHz and above

Electromagnetic radiation occurs in waveform (that is, in a series of regularly rising and falling strengths). Distance from one crest to the next makes up one wavelength. The number of crests going by in 1 second is called the *frequency* of the wave. This is shown in Figure 15-9.

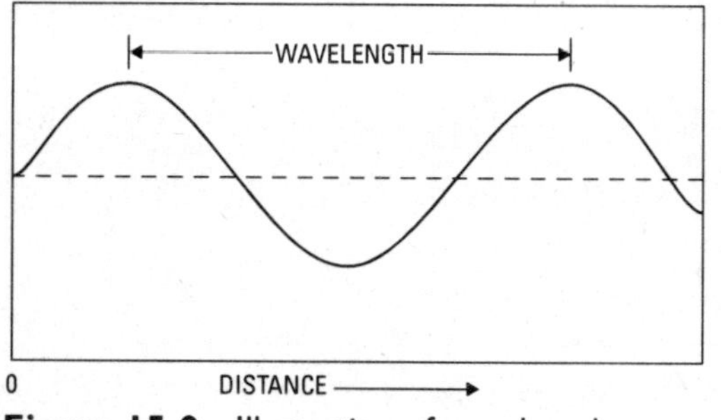

Figure 15-9 Illustration of wavelength.

The height of a wave is its amplitude, which is usually expressed in volts.

Radio communication requires a transmitter that produces, amplifies, and radiates power at a useful radio frequency, while at the same time incorporating some kind of information into its signal. It also requires a receiver that can detect the selected frequency, separate the signal's information content, and somehow indicate it.

A normal electromagnetic wave has the shape of a sine wave. A sine wave has a complete cycle of 360 electrical degrees. Starting from zero, the first crest occurs a quarter of the way through the cycle, at 90 degrees. The wave passes through zero again, at 180 degrees, before cresting in the opposite direction, at 270 degrees. Reaching zero for the second time, at 360 degrees, the cycle repeats. Expressed in degrees, any point in a wave's cycle is called a *phase angle*.

When several waves meet, they add together. If two waves of the same height happen to be in phase (reaching the same crests at the same time), the result is a wave of identical frequency but of twice the amplitude. Should they crest in opposite directions at the same time (180 degrees out of phase), they cancel exactly and disappear. This is shown in Figure 15-10. Figure 15-10A and Figure 15-10B show the two original waves. Figure 15-10C shows what happens if the waves are in phase with each other, and Figure 15-10D shows what happens if they are 180 degrees out of

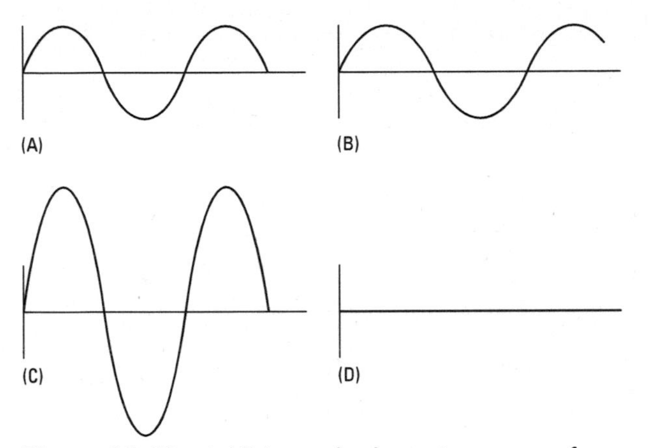

Figure 15-10 Additive and subtractive nature of waves.

phase. In real-world cases, waves of many frequencies, amplitudes, and relative phases are always being thrown together, resulting in complex waveforms. As a practical matter, it is almost always possible to separate an irregular, jumbled form into its pure sine-wave components.

A number of things can affect a radio wave as it propagates from one place to another. In the simplest case, two antennas (a transmitter and a receiver) are located so that a direct path connects them. Waves at most radio frequencies will experience no difficulties in transit beyond a normal attenuation as the wave spreads itself through space and with the usual losses from the resistance of the transmission medium (which, in this case, is air).

However, Earth is a radio environment covered with obstacles. Firstly, Earth has an atmosphere that thins with altitude, topped with an electrically charged layer on top of it (the ionosphere). The atmosphere is further changeable with varying clouds, rain, and hot and cold air. The surface is covered with irregularities, such as mountains, buildings, and the curvature of the planet itself. In addition, the various land and sea areas have differing conductive and reflective properties. Fortunately, the effects of these obstacles on propagating waves are quite dependent on frequency.

Longer, or lower, frequency waves generally get around obstacles quite well. Like ocean waves, they simply flow around obstacles that are significantly smaller than they are. (Remember, the lower the frequency is, the longer the wavelength is.) Long waves do quite well over the curve of Earth as well. This is because they spread widely enough to stay in contact with Earth's surface as they move outward. Earth absorbs higher frequencies rapidly, however, so waves that are to be received deep in a *shadow zone* (as shown in Figure 15-11), the area would need to be about a half-mile long, about 300 kilohertz, which is too low, say, for good voice transmission. To some degree, Earth conducts these waves, as well as changing their properties over distance.

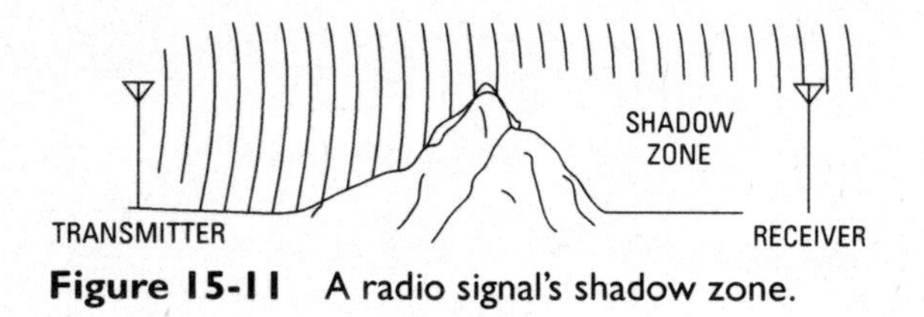

Figure 15-11 A radio signal's shadow zone.

Waves below 300 kilohertz do not penetrate the ionosphere at all and, as a result, are trapped in a zone between the surface and the charged layer above. They are conducted by this natural waveguide around Earth, losing energy with distance. However, small signals can be measured where the wave meets and reinforces itself on the opposite side of the globe.

Most of the spectrum chosen for communications of any sort are frequencies that spread just enough over the horizon (as in AM broadcasting), must have line-of-sight connection (as in FM), or reflect back from the ionosphere the *sky waves* of shortwave radio (illustrated in Figure 15-12). The shorter the wavelength is, however, the more likely it will experience scattering or absorption by obstacles, moisture, or Earth itself. Satellites beaming broadcasts back to the surface use efficient, line-of-sight, ultrashort wavelengths—but rain or heavy fog can absorb them.

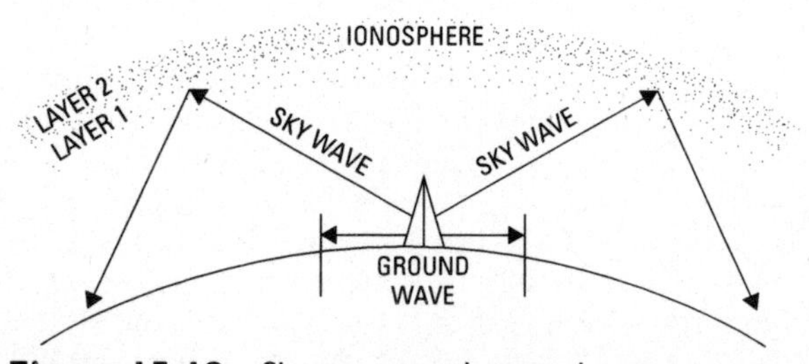

Figure 15-12 Sky waves and ground waves.

Modulation

To send or receive a radio signal, you must use circuits that resonate at a particular frequency, and you must send radio waves that have a certain frequency. However, the sound of a person's voice, or music, varies usually in the range of 100 hertz to a few thousand hertz. The problem, then, is how to use a fixed frequency of wave to carry a voice signal. We do this by several techniques that are collectively called *modulation.*

Amplitude Modulation

The first type of modulation used in radio was amplitude modulation (AM). This, obviously, was used to convert a voice or other signal into electrical waveforms. Voice and music are rather low frequencies (in the 100- to 10,000-hertz range). They are added together with a *carrier* wave of single frequency. For example, 1400 kilohertz would be a common AM carrier wave frequency. This frequency would be produced by a basic oscillator circuit. The result is a very dense sequence of radio frequency waves whose amplitudes vary over time in a way that mirrors the amplitudes of the much longer voice wavelengths that have been imposed over them. This is shown in Figure 15-13. Notice that the carrier wave remains at 1400 kilohertz. The signal that can be derived from it (the modulating waveform) is of a much lower frequency.

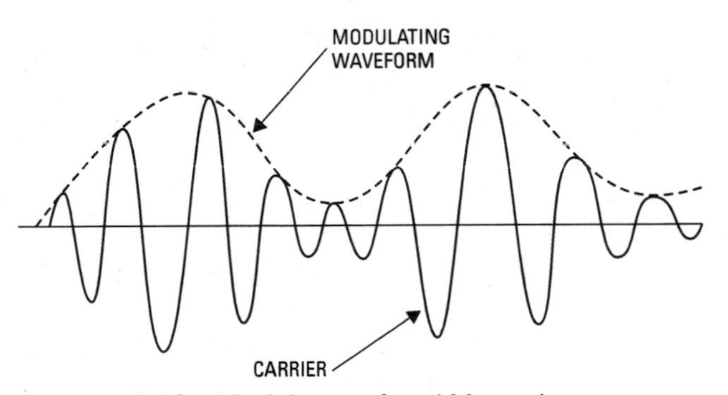

Figure 15-13 Modulation of an AM signal.

Figure 15-14 shows a circuit for creating the kind of modulated wave that is shown in Figure 15-13. Notice that the carrier wave frequency (here noted "RF in") connects to the base of the transistor. The voice signal is added to the circuit through a transformer and passed through the transistor, where its strength is added to the RF signal, increasing it or decreasing it from moment to moment.

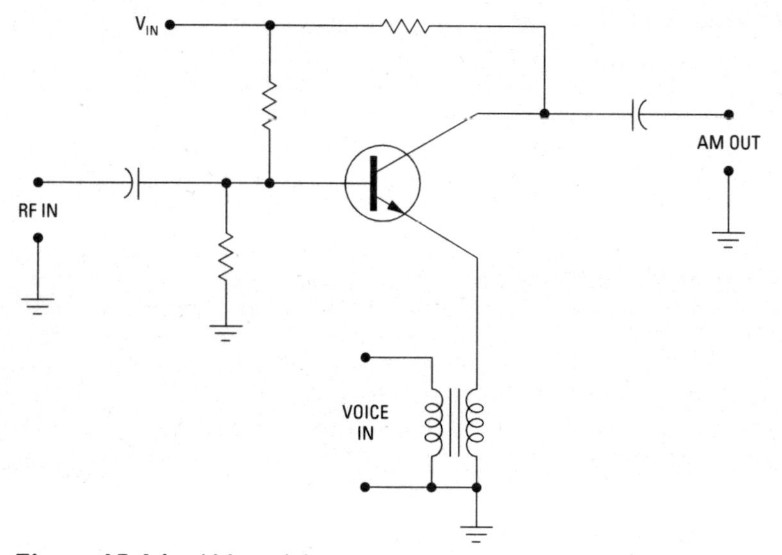

Figure 15-14 AM modulation circuit.

The result is a signal using amplitude modulation to carry a lower-frequency signal.

In the receiver's tank circuit, tuned to the radio frequency, 1400 kilohertz resonates, but other frequencies do not. Once captured the signal's radio component is stripped away (the tank circuit sends 1400 kilohertz resonances in this case to ground) and only the audio frequencies are left, which are amplified and reconverted into sound by other components in the receiver.

Demodulation

Once an amplitude-modulated signal arrives at a receiver, it has to be demodulated in order to reproduce the sound of the voice or music that was combined with the carrier wave.

Figure 15-15 shows a demodulation circuit, also called a *detector* circuit. Demodulation is done in a few steps. You will first notice that this figure shows a resonant circuit on the left, and that from there the signal goes through a diode. This cuts off one-half of the sine wave, simplifying the process, yet still leaving enough signal for high-quality reproduction. Continuing, you will notice that the signal must pass a capacitor. This capacitor filters out the carrier wave by being appropriately sized; that is, the capacitor is large enough to hold a charge for one carrier wave cycle but not so large as to smooth out the slower cycles of the modulating signal. The circuit shown in Figure 15-15 (called an *envelope detector*) is commonly used for both AM radio and for television.

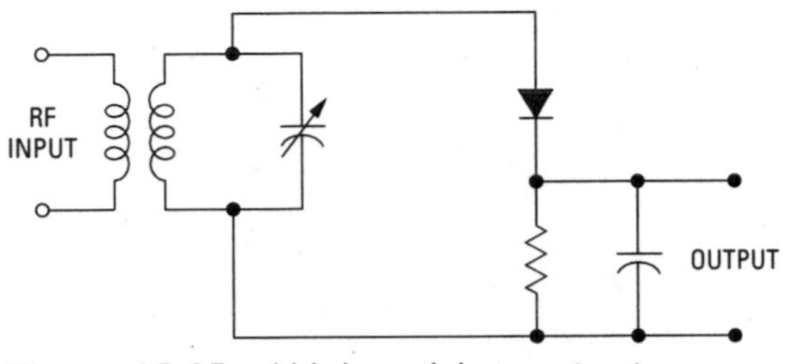

Figure 15-15 AM demodulation circuit.

Frequency Modulation

While the original modulation technology was amplitude modulation, *frequency modulation* (*FM*) has become very widely used since the 1970s. In the FM method, sound sources are electrically impressed upon carrier waves whose frequency (not amplitude) is made to change up and down at rates that match the audio signal. This gives a somewhat cleaner signal than an AM signal. There are many disturbances to a radio signal, both natural and artificial, and these distort the amplitudes of radio waves. (Remember how Figure 15-10 showed frequencies either adding to or subtracting from each other.) On the other hand, very few outside sources can alter a signal's frequency. For this reason, FM signals come in cleaner, with less static and noise.

FM signals are designed so that the carrier wave increases in frequency to show greater amplitude, as shown in Figure 15-16. Figure 15-16A shows the audio signal, Figure 15-16B shows the carrier wave, and Figure 15-17C shows the final FM signal. Notice that the signal bunches-up when the voltage of the audio signal is most positive and spreads out when the signal is most negative. This deviation from the carrier wave frequency encodes the audio wave onto the carrier wave.

Figure 15-17 shows another aspect of FM transmission. Notice that this figure is very similar to Figure 15-16. However, in Figure 15-17, the amplitude of the audio signal is greater. In other words, the audio input is louder. Now look at Figure 15-17C. Notice that the final FM signal bunches-up more during the peaks than the milder signal from Figure 15-16. So, how much the signal bunches up is used to encode the amplitude of the audio wave.

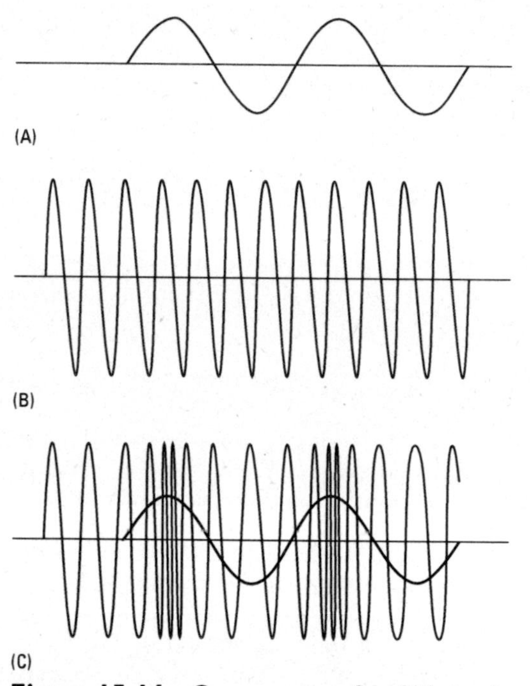

Figure 15-16 Components of an FM signal.

We now have two things we are showing with the FM signal: where the tops and bottom of the audio signal should be, and how strong or weak they should be.

Notice that the amplitude of the final FM signal does not change like an AM signal. The amplitude of an FM signal is constant.

Figure 15-18 shows a common method of combining an audio signal with a carrier to obtain an FM signal. This circuit uses *reactance modulation* to create the complete FM signal. The critical component in this circuit is a *varactor diode*. This type of diode is useful in a reverse-biased position. As you will remember, connecting a diode in a reverse bias causes the barrier layer (or *depletion zone*) to expand. You can also think of this depletion zone as a dielectric, an insulator. In effect, the varactor diode can function in this way as a capacitor. Now, since the amount of capacitance is determined by the thickness of the dielectric layer (the thinner the dielectric, the greater the capacitance; but the thicker the dielectric, the lower the capacitance), by widening or narrowing the depletion zone, we can vary the device's capacitance. Thus the varactor, when properly

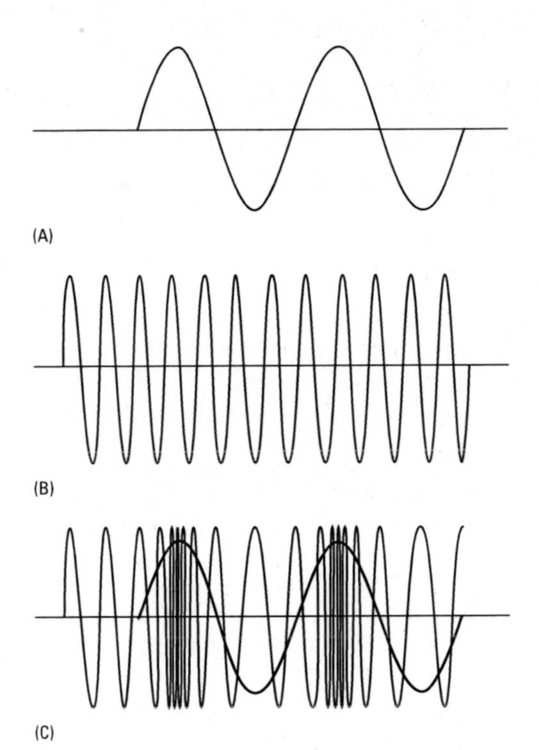

Figure 15-17 Components of an FM signal, showing greater amplitude.

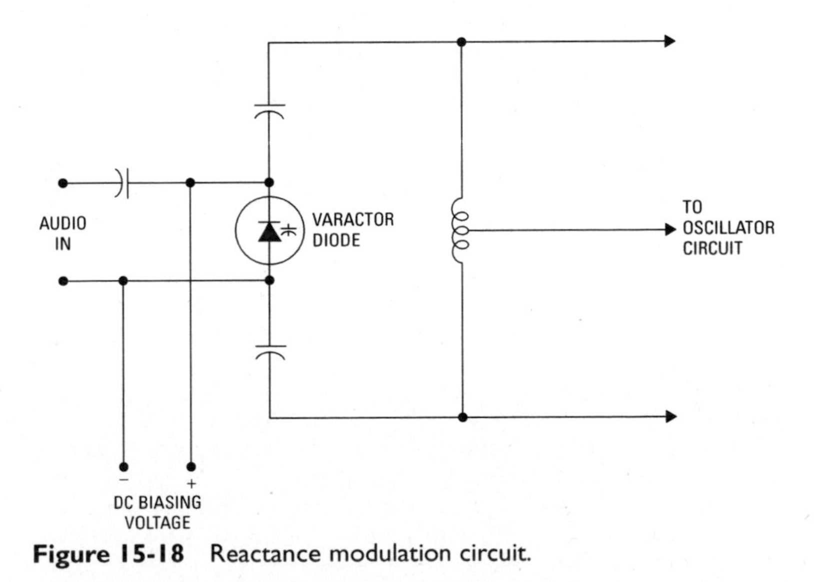

Figure 15-18 Reactance modulation circuit.

biased and connected, can function as a variable capacitor with very fast response times. In other words, we can vary the capacitance of this device millions of times per second simply by connecting it to a variable voltage.

Other methods have been developed to modulate FM (and AM) signals, but our purpose here is not to cover every detail of modern radio circuitry. Keep in mind, however, that many specialized methods for all of these processes have been developed over the years.

Figure 15-19 shows a circuit used to decode FM signals. This circuit is called a *ratio detector*. This circuit has two basic operations. First, it divides the FM waveforms into two parts with the diodes. Because these two halves are 180 degrees out of phase, any change of amplitude will cancel itself out (the difference being of the opposite polarity in the other half of the circuit). Next, remember that a change in frequency causes a change in a circuit's resonant frequency. Because of this relationship, changes in the frequency of the FM signal will cause the impedance of the resonant circuits to change, which, in turn, changes the voltage made available at the output in exactly the frequency of the original audio signal.

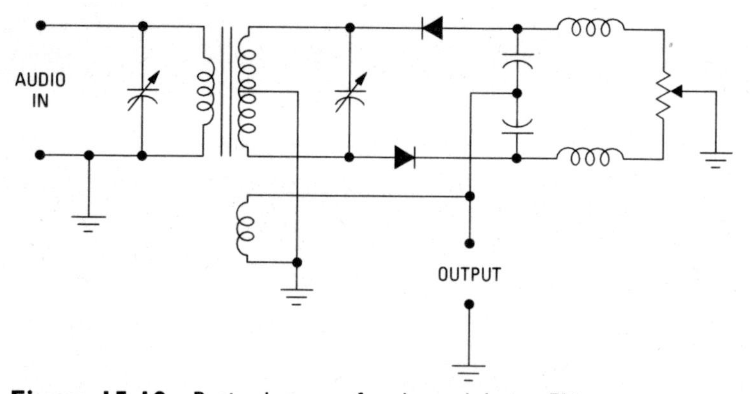

Figure 15-19 Ratio detector for demodulating FM.

Pulse Modulation

The term *pulse modulation* is really a misnomer. Pulse modulation is simply representing a signal with a series of short pulses. There is no carrier wave, and nothing is modulated.

Figure 15-20 shows the three usual forms of pulse modulation (again, representing signals with pulses). Notice that *pulse amplitude modulation* (*PAM*) uses pulses of greater or lesser strength (amplitude) to represent the sine wave signal. *Pulse width modulation*

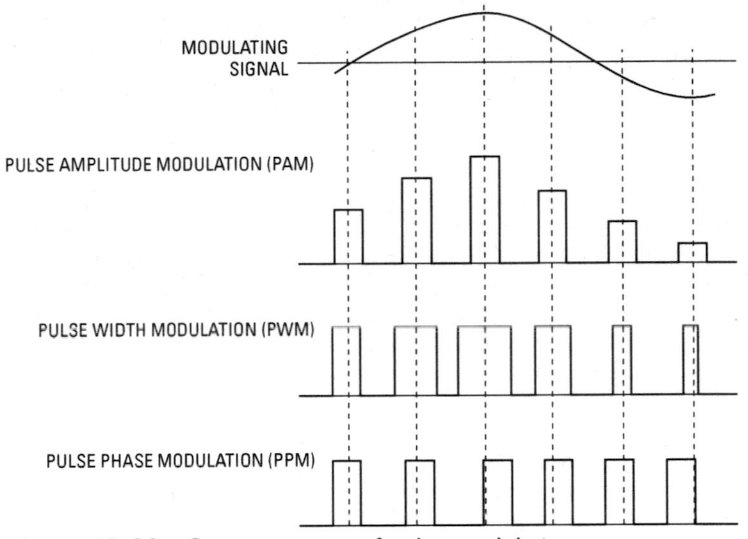

Figure 15-20 Common types of pulse modulation.

(*PWM*) uses wider or narrower pulses to do the same thing. *Pulse phase modulation* (*PPM*) uses pulses that are shifted to one side or another of the basic time interval to indicate the amplitude of a signal.

Multiplexing

One of the interesting things that can be done with pulse modulation is called *multiplexing*. Multiplexing is sending more than one stream of information in the same signal. For example, look at the pulse amplitude representation in Figure 15-20. In addition to using the amplitude of the pulses to convey information, we could use the width of the pulses to convey information at the same time. Considering that there would be at least thousands of pulses per second, quite a bit of information could be encoded in the pulse widths. So, we could send music with amplitude modulation and send any other form of data with pulse width modulation at the same time.

While Figure 15-20 shows the most common forms of multiplexing, be aware that there are others.

Antennas

All radio transmitters require tuned antennas with a length or shape that permits the oscillating currents fed into it to radiate their energy efficiently into the atmosphere. At the receiving end (where much-weakened waves crossing through a receiver's antenna induce very small currents), tuning is not as necessary. A receiver will pick up frequencies within a whole range, or broadcast band, such as the ranges shown in Table 15-1.

An antenna is used to couple with either the electrostatic or the magnetic component of a radio wave. (You will remember that radio waves are composed of both electrostatic and magnetic components.)

An antenna designed to couple to the electrostatic component of the wave is designed as a simple length of conductor through which the electrical charge can move. This is the type of antenna you are probably most familiar with, such as a car antenna.

An antenna designed to couple with the magnetic component of a radio wave is usually a coil of wire that forms an electromagnet.

Transmitting Antennas

Antennas (especially transmitting antennas) are tuned to the frequencies they are to be used with. This is especially critical in the transmitting antenna and its circuit. Not only does the antenna need to be the proper length (usually one-half or one-quarter wavelength), but the circuit run to a remote antenna can also be critical. This is a discipline all its own, but there is one phenomenon that is worth explaining briefly.

When a high-frequency signal moves from a conductor of one impedance level to a conductor of a higher impedance level, the signal will tend to reflect back to the source. This *back reflection* is objectionable (or an undesirable effect). Obviously, it reduces the signal that reaches the antenna, but in addition, it creates waves that bounce back and forth in the conductor. This creates standing waves in the conductor, which reduce and distort the signal to be transmitted. For this reason, the impedance of cables and components must be carefully matched.

Transmitting antennas are usually connected to circuits on the far side of a capacitor, as shown in Figure 15-21. Notice in this drawing that the antenna is on the left and connects to the circuit at the far capacitor plate. With the antenna being electrically joined to the capacitor, it experiences all the rises and falls in voltage and current that the capacitor experiences.

Polarity

Transmitting antennas may transmit their signals in either a vertical or a horizontal polarity, as shown in Figure 15-22. For antennas that will be positioned near the ground, a vertically polarized

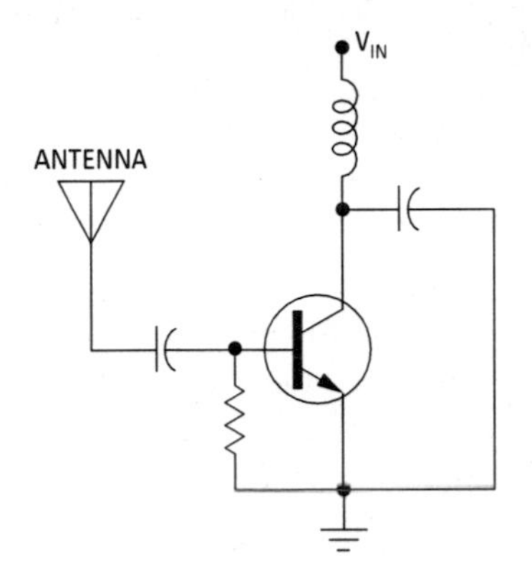

Figure 15-21 Antenna connected to capacitor.

configuration is almost always used, because it provides greater signal strength along Earth's surface. For an antenna that will be at an altitude, a horizontally polarized configuration provides better signal strength at ground level.

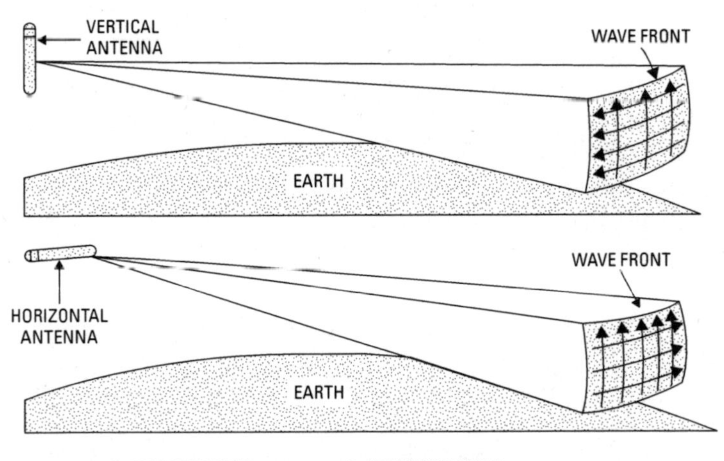

Figure 15-22 Horizontally and vertically polarized antennas.

Receiving Antennas

Receiving antennas come in a wide variety of types. However, the most important criterion for them is that they be sized according to the wavelength or the signal they are to receive. This is not as critical for a receiving antenna as it is for a transmitting antenna, but it is an important factor.

Figure 15-23 shows a *dipole antenna*, along with its connection to a circuit. Notice that the two halves of the antenna are connected to the two conductors of a coaxial cable. Coaxial is generally the preferred type of cable to use with antennas, especially in larger systems. (You are not likely to find coaxial cable inside a small home radio.)

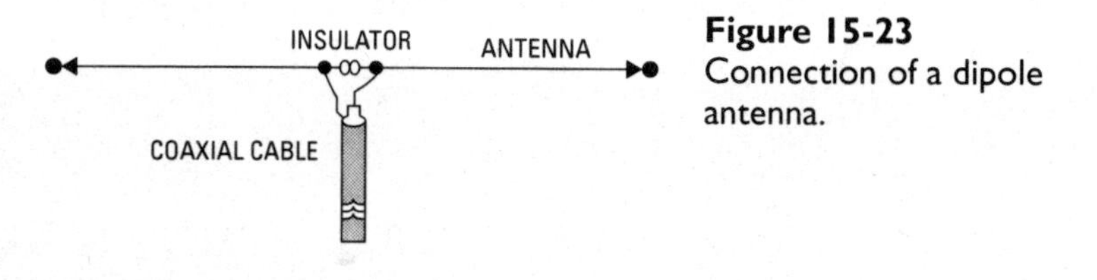

Figure 15-23 Connection of a dipole antenna.

Wave Paths

The signals that emanate from transmitting antennas can travel across Earth's surface in several different ways.

One of these paths is called the *ground wave* or *surface wave*, as illustrated in Figure 15-24. Surface waves tend to follow the surface of Earth. Because of diffraction, they bend around obstructions and stay very close to Earth's surface. But also because of diffraction, Earth tends to absorb much of the signal's strength, limiting the distance at which a ground wave will remain useable.

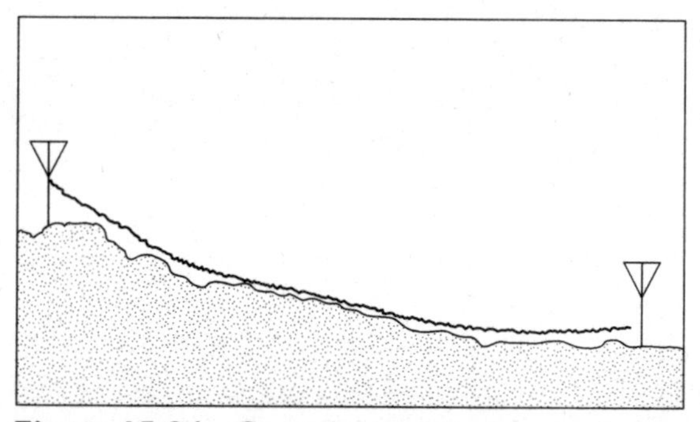

Figure 15-24 Ground wave travel.

Radio signals, especially from tall antennas, can be reflected off of the ground, as shown in Figure 15-25.

Finally, radio signals can bounce off of high levels in the atmosphere, as is illustrated in Figure 15-26. (This was discussed briefly in the earlier section, "Basics," and shown in Figure 15-12.)

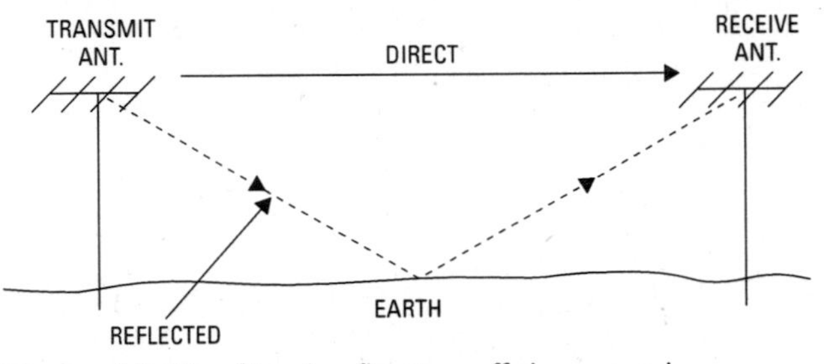

Figure 15-25 Signal reflection off the ground.

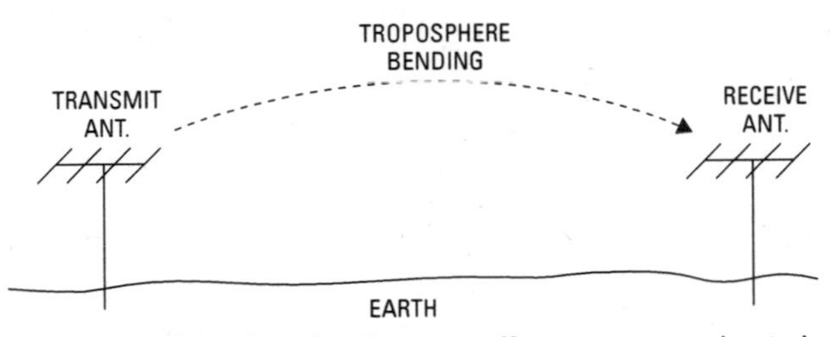

Figure 15-26 Signal reflection off upper atmospheric layers.

Summary

The term radio is a shortening of the original word, radiotelegraphy. In the early days, the term wireless telegraphy (or simply wireless) was usually applied to this technology. *Radio* refers to sending and receiving electromagnetic waves in the range of frequencies lying between 1 hertz and a few gigahertz.

Guglielmo Marconi was the first person to actually put all of the pieces together and send a trans-Atlantic signal. The final key to creating the radio was in adding tuning (a variable inductor) to the resonant circuit in the receiver. Marconi took the developments of others, incorporated them together, and took them to an effective conclusion.

Radio waves were predicted by James Clerk Maxwell in 1864, then created, measured, and described by Heinrich Hertz between 1895 and 1898. Since that time, it has been known that radio waves *propagate* (travel). But, to this day, scientists argue over the precise reasons why radio waves propagate and what fundamental mechanisms are involved. There are two parts to a propagating wave: the magnetic component and the electrostatic component.

An electrostatic field forms from the top to the bottom of an antenna. The top of the antenna acts similarly to one capacitor plate, and the ground at the base of the antenna acts as the other plate. The static that builds up between them is the electrostatic field.

The magnetic and electrostatic fields of a radio wave are configured differently. The magnetic lines of force are parallel with the ground, just like lines of latitude around a globe. The lines of force from the electrostatic field are perpendicular to the ground, like the lines of longitude around a globe.

Radio waves form when rings of force from a collapsing field are pushed away from the antenna.

To send intelligible signals over a distance, tuning the transmitting and receiving circuits is critical.

Frequency and wavelength are simply different measurements of the same thing. Any frequency will always have the same wavelength. A signal of 100 megahertz has a wavelength of 3 meters. We could call it a 3-meter signal or a 100-megahertz signal and be equally correct either way.

When several waves meet, they add together. If two waves of the same height happen to be in phase, reaching the same crests at the same time, the result is a wave of identical frequency but of twice the amplitude. Should they crest in opposite directions at the same time (180 degrees out of phase), they cancel exactly and disappear.

A number of things can affect a radio wave as it propagates from one place to another. Waves at most radio frequencies will experience no difficulties in transit beyond a normal attenuation as the wave spreads itself through space and with the usual losses from the resistance of the transmission medium (which, in this case, is air).

To send or receive a radio signal requires you to send radio waves that have a certain frequency. However, the sound of a person's voice, or music, varies, usually in the range of 100 hertz to a few thousand hertz, which is significantly below the range of frequencies that are used in radio transmission. To carry both the steady radio wave and a voice signal, we use several techniques that are called modulation.

Amplitude modulation (AM) adds audio signals together with a carrier wave of single frequency. The result is a very dense sequence of radio frequency waves whose amplitudes vary over time in a way that mirrors the amplitudes of the much longer voice wavelengths that have been imposed over them. The signal that can be derived from this (the modulating waveform) is of a much lower frequency than the carrier wave.

In frequency modulation (FM), sound sources are electrically impressed upon carrier waves whose frequency (not amplitude) is made to change up and down at rates that match the audio signal. This gives a somewhat cleaner signal than an AM signal. FM signals are designed so that the carrier wave increases in frequency to show greater amplitude.

Pulse modulation is simply representing a signal with a series of short pulses. There is no carrier wave.

Multiplexing is sending more than one stream of information in the same signal.

All radio transmitters require tuned antennas with the antenna's length or shape permitting the oscillating currents fed into it to radiate their energy efficiently into the atmosphere. At the receiving end, where much-weakened waves cross through a receiver's antenna and induce very small currents, tuning is not as necessary. A receiver will pick up frequencies within a whole range or broadcast band.

An antenna is used to couple with either the electrostatic or the magnetic component of a radio wave. An antenna designed to couple to the electrostatic component of the wave is designed as a simple length of conductor through which the electrical charge can move. An antenna designed to couple with the magnetic component of a radio wave is usually a coil of wire that forms an electromagnet.

When a high-frequency signal moves from a conductor of one impedance level to a conductor of a higher impedance level, the signal will tend to reflect back to the source. This back reflection is objectionable.

Transmitting antennas may transmit their signals in either a vertical or a horizontal polarity. For antennas that will be positioned near the ground, a vertically polarized configuration is almost always used, because it provides greater signal strength along Earth's surface. For an antenna that will be at an altitude, a horizontally polarized configuration provides better signal strength at ground level.

Chapter 16 continues our look at specific electronic applications with an examination of audio systems.

Review Questions

1. Describe the first radio signals.
2. What was the original purpose of radio?
3. Who do you consider to be the most important developers of radio?
4. What do we call the traveling of a wave?
5. What are the components of a radio wave?

6. What is the formula for wavelength?
7. Describe the orientation of a radio signal's magnetic field.
8. Describe the orientation of a radio signal's electrostatic field.
9. Describe the formation of a radio wave.
10. What is the right-hand rule?
11. What is the frequency range in the VHF band?
12. What is amplitude?
13. What is amplitude modulation?
14. What is frequency modulation?
15. What is pulse modulation?
16. What is multiplexing?
17. What is a surface wave?
18. What is a sky wave?
19. Where might you use a horizontally polarized antenna?
20. What is back reflection?

Exercises

1. Draw a transmitting antenna and circuitry for a 500-megahertz signal. Show as many components as you can. Label the components. Show your wavelength calculations.
2. Show an antenna and receiver sized for a 250-megahertz signal. Show as many components as you can. Label the components. Show your wavelength calculations.

Chapter 16

Audio

Obviously, audio (sound) systems incorporate a significant number of electronic circuits. We have already covered many specific types of circuits that are used for audio equipment (such as amplifiers, oscillators, modulators, and signal receivers). We have covered radio, and we will shortly cover television, both of which incorporate audio technology.

This chapter focuses on sound systems and the components that they are made of. As always, bear in mind that we will cover basics here, not every innovation in audio circuitry.

The Nature of Sound

To understand audio systems in general, it is first necessary to understand something of sound itself.

Elements of Sound

Sound is nothing but vibrations. A sound wave is a series of vibrations passed through the air. (They can be passed through other substances as well, but since air is by far the most common and desirable carrier of sound waves, we will restrict our coverage to that.) Notice that the sound waves in Figure 16-1 are composed of

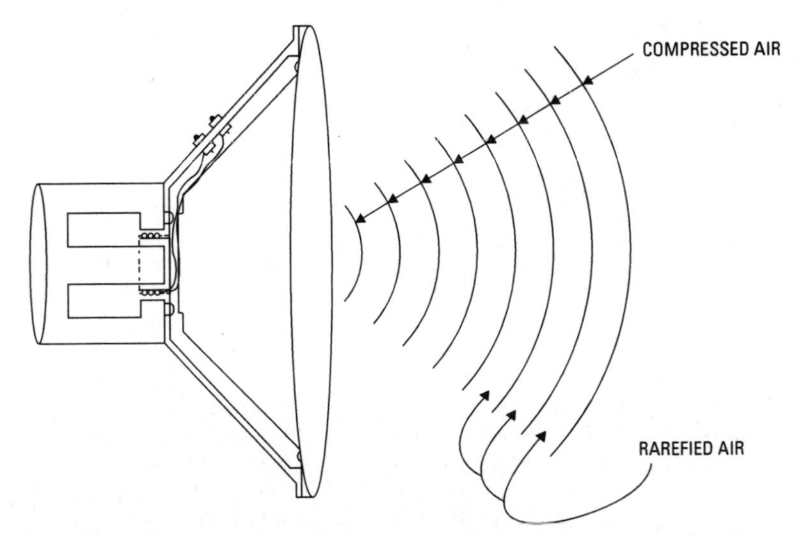

Figure 16-1 Sound waves. Rarified is the space between compression lines.

alternating layers of compressed and rarefied air. In other words, sounds waves are areas of high and low pressure that travel from a source.

Sound waves in dry air travel at 343 meters per second. While this is certainly a high speed, it is much, much slower than the speed of radio waves. Sound takes just under 3 seconds to travel 1000 meters—1 kilometer. Radio waves travel 300,000 times that distance in a second. While we can ignore the slowness of sound for most uses of audio technology, bear in mind that this can become an issue when trying to coordinate sound-producing components in a large space. At 100 meters, an audio delay is almost one-third of a second. That is noticeable and bothersome.

Sound waves have two primary components: frequency and amplitude. *Frequency* is determined by how many compressions and rarefactions pass a given point per second. *Amplitude* is determined by the strength of the compressions and rarefactions. In other words, it is determined by the number of compressions and rarefactions per second, and how strong they are.

The comparative measurement method of *decibels* (dB) was created to represent how the human ear perceives sound. (We covered this in Chapter 14.). You will almost always see sound measured in decibels. The threshold of hearing (the quietest sounds that we can notice) is identified as 0 dBSPL (*SPL* stands for sound pressure level). The minimal audible level for sound becomes the base reference point for decibels. Sounds above 85 decibels begin to damage the ear, and those above 130 decibels generally cause pain. Referring back to the fact mentioned in Chapter 14 that every 3 decibels connotes a doubling of the base signal, you can see that 85 decibels would be an enormous increase in a sound's amplitude.

The sounds we hear are almost never as simple as one single sine wave. Rather, sounds are composed of many frequencies and variations. This is why you can strike the same note on a piano, a guitar, and an organ and still easily differentiate between them. The base frequency of vibration is the same for all three instruments, but the extra signals and the qualities of the signals make a very significant difference.

Figure 16-2 shows a major factor in this quality differentiation. This figure shows *harmonics* (frequencies that are related to, and sometimes caused by, the primary signal). Note that the second harmonic has a frequency that is exactly twice that of the fundamental frequency. In musical terms, we call this an *octave*. For example, if you press the A key on a piano, then press the next A key above it, you would note the similarity of the sounds but also recognize that

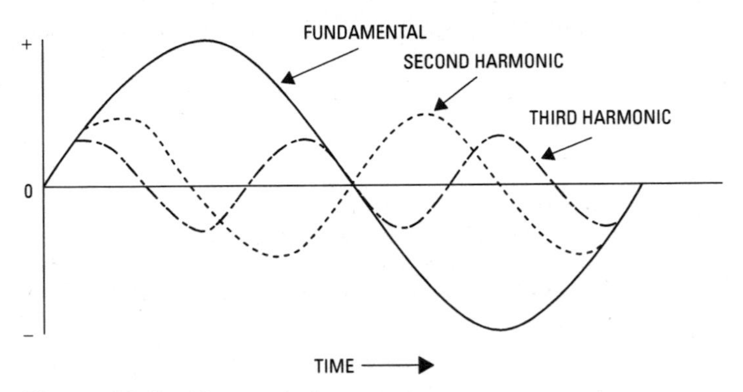

Figure 16-2 Harmonic frequencies.

the second note is of a higher pitch than the first. This is an interval of one octave. The third harmonic has exactly three times the frequency of the fundamental frequency. In musical terms, we would say that this is an octave plus a fifth higher than the fundamental frequency. This would be from the lower A, past the next A, to the next E. If you were to play these notes, you might say that they sound good together but that they are not the same as the octaves were.

We can continue with the example of music to point out that *harmonies* in musical performance are notes that complement a fundamental note. Second and third harmonics (octaves and fifths) are generally the most common kind of harmony notes. Following the musical example, we come to one more interesting fact. Playing harmonic notes (such as the A and E mentioned previously) can induce other harmonic notes to sound. If you were to play an A and an E on a guitar, for example, you might soon notice another E would begin sounding, even though you had not plucked that string. This happens because the A and E that were plucked put off harmonics that induce vibrations into all the guitar's strings. And, if there is a string that has an oscillatory frequency that is related to the A and E, small oscillations from the air-transmitted signal will build up in the string and will cause it to sound.

Harmonic tones, even if not prominent, give the flavor to many of the sounds we hear and allow us to distinguish between them.

The human ear picks up sounds in the vibratory range of approximately 20 hertz to 20,000 hertz. This varies from one individual to another and generally is diminished with age, especially on the high end. Sounds at about 3500 hertz are perceived as the loudest, with higher or lower frequencies not being perceived as loud even if the true measurement of amplitude is the same. Our ears do not have an even response curve.

Acoustics

It is helpful to have a basic understanding of how sound behaves in a room and how it is received by the human ear. The study of such things is called *acoustics*. We will not attempt to go through a great deal of acoustics in this chapter, but a few of the more important points are as follows:

- Sound waves spread out from their source in the same way that waves of water spread out in a pond.
- Sound waves readily bounce off of hard surfaces (such as plaster walls) and are readily absorbed by soft, flexible surfaces (such as carpeting or fabric).
- The angle at which the sound waves bounce off of hard surfaces is equal to, and opposite of, the angle at which they strike the surface.
- Sound waves that reach the human ear from the side are perceived as sounding better than waves coming from the front or from the rear.
- Because sound waves travel relatively slowly, delay units must be used in large buildings. If they are not used, signals from distant speakers will reach the listener later than sound from closer speakers. This makes the sounds jumbled and unpleasant.

Audio Systems

Before we explain specific components and their operations, we will briefly explain the main types of sound systems themselves.

There are two main divisions of sound systems: high-fidelity systems and facility sound systems. *High-fidelity systems* include everything from a portable radio to the sound system for an outdoor concert. *Facility sound systems* are the typical sound systems that are used in grocery stores, offices, and similar locations. They are used not only to transmit sound, but also for paging and public address uses. At times, the line of differentiation between these two types of sound systems begins to blur. This is primarily because of technological improvements that have made some facility systems almost equal to high-fidelity systems.

The three main types of sound system equipment are as follows:

- Signal generators
- Signal processors
- Sound generators

In addition to these three types of equipment, all sound systems require some method of signal transmission (usually in the form of copper wires). Radio waves can also be used, but wires are almost always preferable.

Signal generators are the source of the audio signal. The normal items we consider to be signal generators are tape decks, compact disc players, turntables, and similar devices. These are the devices that feed the signals into the audio system.

We also categorize radio receivers and microphones as signal-generating equipment, even though they are not the true sources of the signals. A radio receiver, for example, does not originate as a signal; it merely receives a wireless signal and translates it into an electronic form. Because of the way the receiver is used with other audio equipment, however, it is treated as a signal generator.

Signal processors are the intermediate step through the signal generators and the sound generators. These devices normally include amplifiers, preamps, switches, noise-reduction circuits, equalizers, filters, delay units, mixers, monitors, and similar items. These devices boost or reduce a signal, cut off part of it, color it, or in some other way modify it.

In designing sound systems, a great deal of time is spent considering signal-processing equipment. Each combination of signal processors makes the final sound a little different than the other combinations.

Sound generators are simply devices that change electrical signals into audio signals. The signal generators we described earlier generate electronic signals. These signals are then processed and sent to the sound generator, which changes the electronic signals into audio sound waves.

By far the most common types of sound generators we use are speakers. There are others, such as buzzers, but almost all sound generation revolves around a basic speaker-type design.

Audio Devices

We will begin coverage of audio devices with signal generators, continue through signal processors, and finish with sound generators.

Microphones

Microphones convert sound waves into electronic signals. They do this with some sort of diaphragm or disk that vibrates when it is hit by sound waves. Then, by one of several mechanisms, the mechanical movement of the diaphragm is used to create electrical signals.

Microphones are rated by the frequency range they respond to, their impedance, and their sensitivity. One of the key factors in producing a good-quality sound is the microphone's response. It should be relatively flat throughout any critical frequency range. That is, it should put out the same amplitude of signal for all frequencies and not respond differently to various pitches.

The impedance of a microphone should closely match the circuitry that drives it. Typically microphones are rated as high- or low-impedance units.

Figure 16-3 shows one method of doing this: the carbon microphone. Notice that the diaphragm is attached to a plate that pushes down on the carbon granules when pressure is applied to it. This pushes the granules of carbon closer together and makes them conduct electricity better. So, the carbon granules function as a variable resistor. If the pressure is greater, the resistance is lower, and more current passes through the circuit. This, obviously, raises the output voltage and produces a louder sound at the end of the process. Also, the higher the frequency of sound waves striking the diaphragm is, the higher the pitch of the final output will be. The transformer in this circuit serves to raise the voltage of the output to an appropriate level.

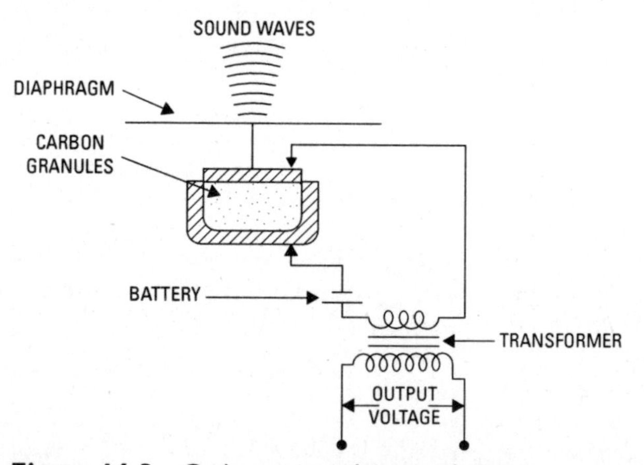

Figure 16-3 Carbon microphone and circuit.

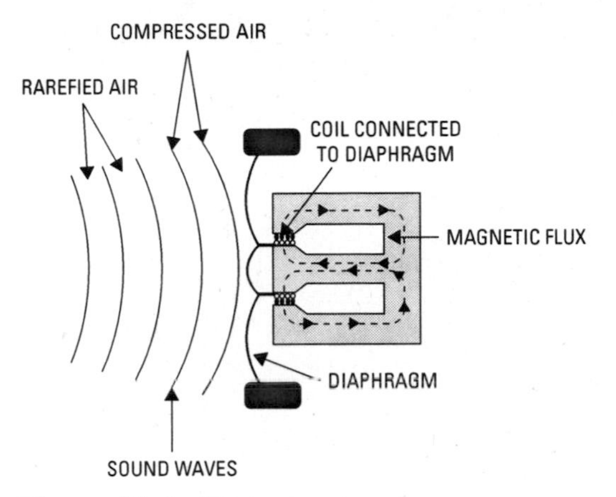

Figure 16-4 Dynamic microphone.

Carbon microphones are relatively inexpensive and can provide a high output. They are also compact and rugged. However, they do not have especially good sensitivity or an even response over a wide range of frequencies. (They can over-respond to lower frequencies.) They are commonly used in telephones and in other applications where high fidelity is not a serious issue.

Figure 16-4 shows a *dynamic microphone*. The term dynamic comes from the fact that the coil of wire used in this design moves. Notice that the coil of wire is attached to the diaphragm and moves with it through the magnet's field. This induces voltage into the coil. As the coil moves in one direction, one direction of current is pushed through the coil. As the coil moves the opposite direction, current moves in the opposite direction. More intense sound waves will cause it to move farther and to induce a larger current into the coil, and so on.

Figure 16-5 shows the very simple circuit for this. Note that no external power source is required. The dynamic is, in fact, a miniature generator (actually, an alternator). Output could be either to sensitive circuitry or to an amplifying transistor or a step-up transformer.

Figure 16-6 shows a different view of a dynamic microphone. Note the air chambers in this design. Captive air in these chambers acts to dampen the movement of the diaphragm so that it doesn't move too dramatically. This helps keep the output and response of the device within design limits.

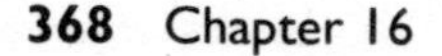

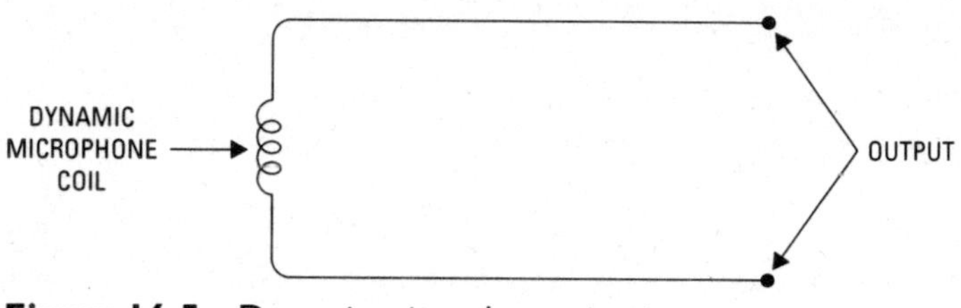

Figure 16-5 Dynamic microphone circuit.

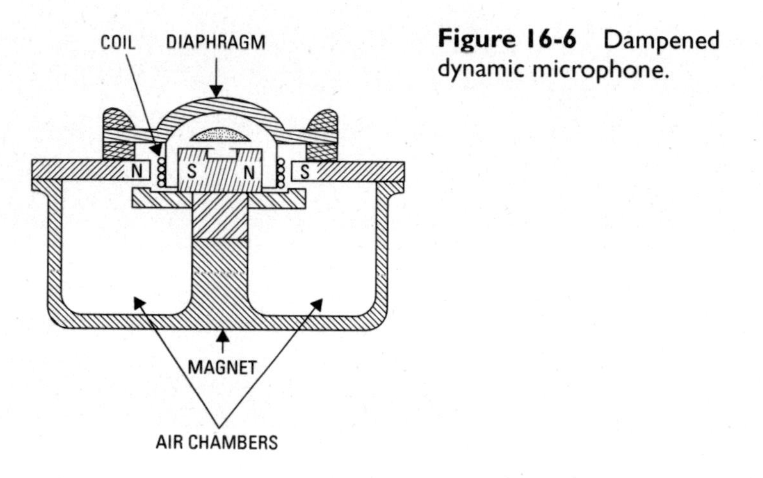

Figure 16-6 Dampened dynamic microphone.

Dynamic microphones have excellent sensitivity. In addition, they are generally small, rugged, and do not require a separate power source. As you will see later on, they can also function as speakers.

Figure 16-7 shows a *crystal microphone*. You will remember from Chapter 12 that a crystal is used to generate a piezoelectric voltage—a voltage generated by physical force. We can also use this

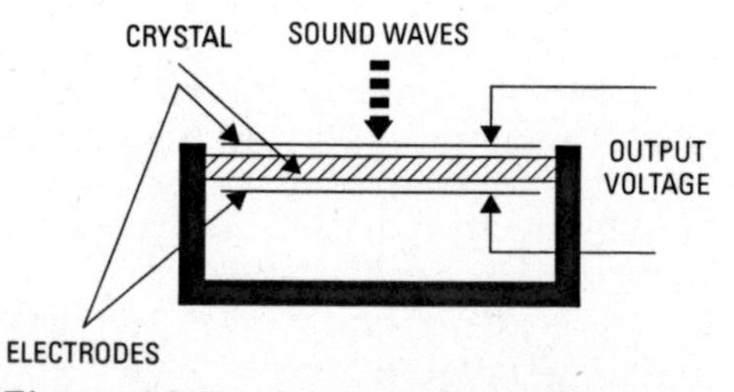

Figure 16-7 Crystal microphone.

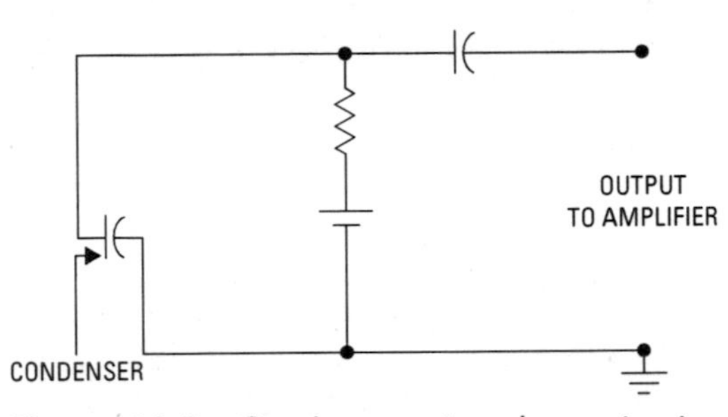

Figure 16-8 Condenser microphone circuit.

effect to turn the force of sound waves into an electrical signal. This figure shows sound waves directly striking the crystal and generating electricity. In the case of small crystal microphones, a diaphragm with a larger physical area than the crystal may also be used. Crystal microphones do not have especially good audio characteristics, as do dynamic microphones, but they are durable, easy to use, and relatively inexpensive.

Another microphone with excellent audio characteristics is the *condenser microphone*. You will remember that condenser is the old term for capacitor. So, a condenser microphone is simply a specially designed capacitor. It is built and mounted so that sound waves will force the plates on the condenser closer together, thus increasing their capacitance and lowering their capacitive reactance. Figure 16-8 shows such a capacitor connected in a circuit. Note that variations in the circuit's capacitance will cause proportionate variations in the circuit's output to an amplifier. As mentioned earlier, condenser microphones have excellent audio characteristics and are used a great deal for recording.

Tape Recorders

Tape recorders of all types store representations of sound in the form of permanent magnetism on metal-coated tapes. Microphones are used to turn sound waves into electrical pulses, which are then used to drive electromagnets. Tape passed underneath these magnets at a carefully controlled speed will be magnetized in the same pattern as the sound waves. Then, if the tape is passed at the same speed underneath a magnetic sensor, the same pattern will be picked up by the sensor's circuitry and can be passed on to a speaker, which will reproduce the original sound. Obviously, the reproduction can be done at a great distance and at a different time.

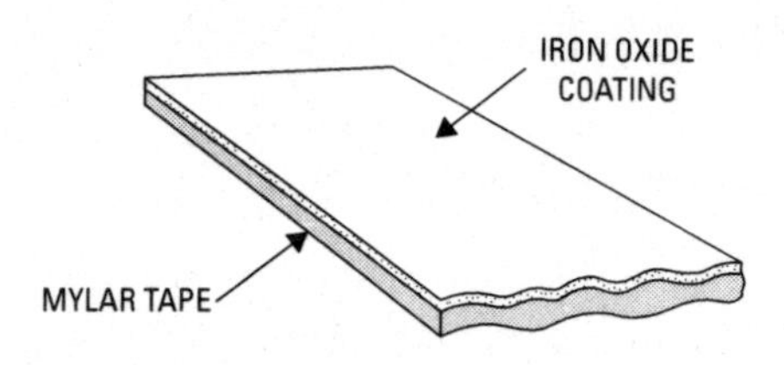

Figure 16-9 Magnetic tape.

Figure 16-9 shows a cross-sectional view of a basic magnetic tape. The mylar base of magnetic recording tape is typically between 0.0005 and 0.0015 inch. The iron oxide coating is usually between 0.00035 and 0.00065 inch. Iron oxide is used for its magnetic properties (being mainly composed of iron), and because its material strength and characteristics allow it to be successfully used in this type of application.

Figure 16-10 shows an electromagnet being used to impress magnetic fields on a moving tape. Such a magnet (which is usually mounted in some type of protective holder) is called a *recording head*. Recording heads, *playback heads,* and *erase heads* all use a basic coil design to interact with the magnetic tape. In the case of the playback head, it is used to create a variable current based upon the magnetic field of the tape's surface inducing current into the head's coil.

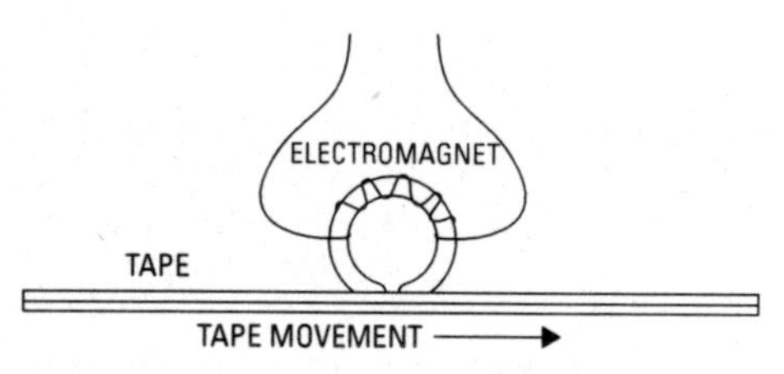

Figure 16-10 Recording head.

Figure 16-11 shows the magnetization of a tape's iron oxide coating at various spots along a tape's surface. Notice that these spots are also associated with specific signal levels of an audio signal, as indicated by the sine wave. Note also the opposite polarities of the iron oxide magnets that are magnetized at the positive and negative peaks of the signal.

Figure 16-12 shows the general arrangement of a typical tape recording system. This arrangement is approximately the same for either small cassette recorders or for large studio models.

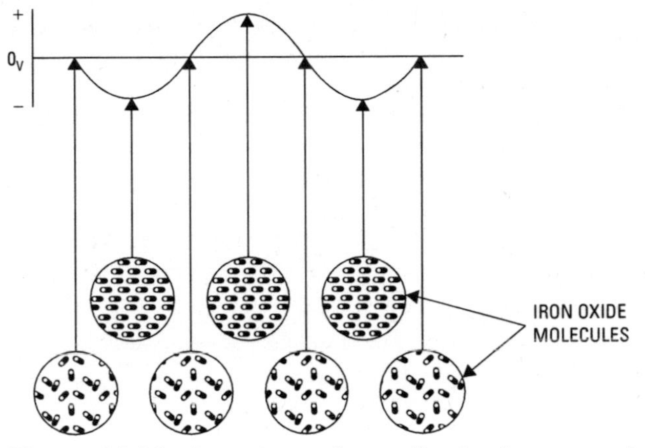

Figure 16-11 Imposition of an audio signal to magnetic tape.

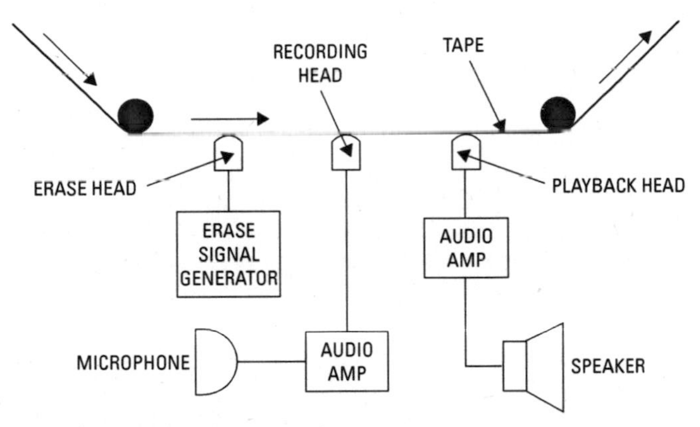

Figure 16-12 Tape recoding system.

Equalizers

No sound system or even any single piece of audio equipment can reproduce sound without some type of distortion or alteration. Not only is this a problem related to audio equipment, it also arises because of the architectural characteristics of any room. Typically, certain frequencies are too loud or too soft. An amplifier, or microphone, or the acoustic characteristics of a room, for example, may cut back the higher frequencies.

In such cases, a device called an *equalizer* is used to attenuate specific frequencies or ranges of frequencies and thus balance the final output more appropriately.

Equalizers come in two primary types: graphic equalizers (which affect several ranges of frequencies) or parametric equalizers (which can be used to affect only a very narrow band of frequencies).

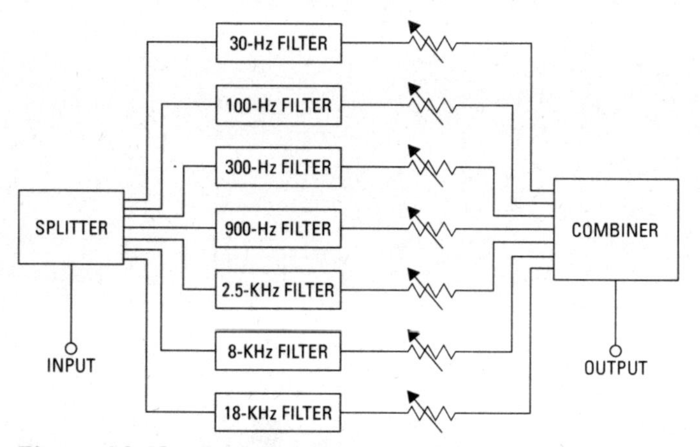

Figure 16-13 An equalizer.

Figure 16-13 shows a typical equalizer. The audio signal enters a splitter, which simply creates seven parallel paths for the current to take. Each parallel leg passes through a filter set to a different frequency. (We covered filters in Chapter 10.) From the filter, the parallel legs pass through variable resistors, which are used to control the amount of current (and, therefore, power) that passes through each leg. By modifying these potentiometers (the variable resistors), the amplitudes of all these frequencies can be adjusted, producing a high-quality final sound.

Amplifiers

Amplifiers were discussed in Chapter 11, so we will not go over them again here. However, two common amplifier circuits for driving speakers are shown in Figure 16-14.

Speakers

Speakers change electrical signals into audible sound. They do this by running an electronic audio signal through an electromagnet. This magnet pulls a flexible cone in toward the magnet and allows

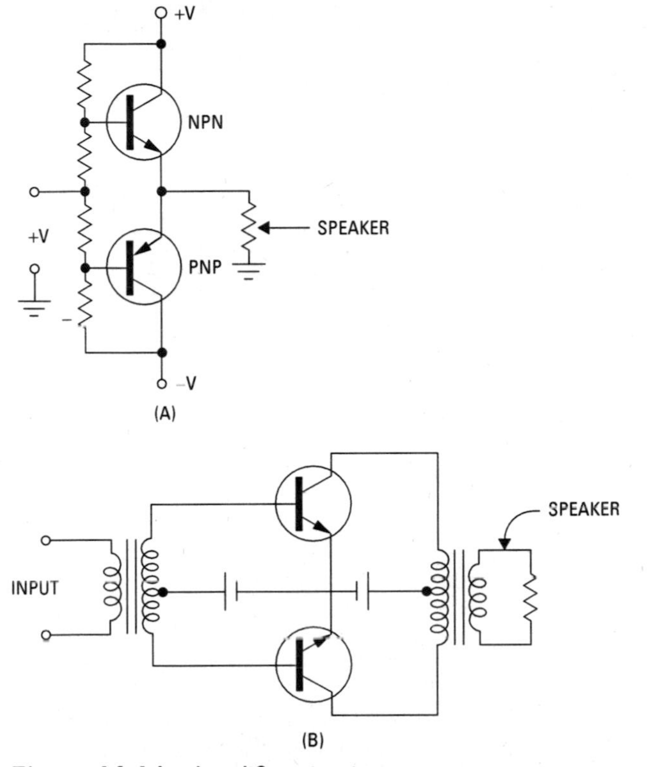

Figure 16-14 Amplifier circuits.

it to move in or out in the same pattern as the audio signal. This displaces air adjacent to the cone, producing sound waves.

Figure 16-15 shows a typical audio speaker. Large or small, nearly all speakers share this basic design, which is called a *permanent magnet* design. The cone is typically made of a heavy grade of paper. The *spider* is a flexible circular piece of material that holds the cone in the round inner frame.

Speakers are rated according to their impedance, and it is necessary to match them to the ratings of the amplifiers that drive them. Audio amplifiers can produce only a limited amount of current to drive speakers. If too many speakers are connected in parallel (which is the normal connection) to an amplifier, it will be overloaded, and either function poorly or be damaged.

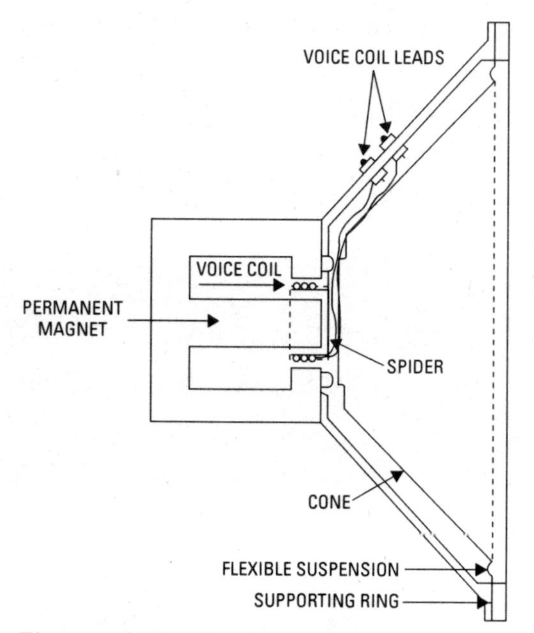

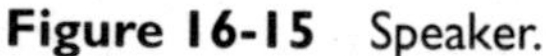
Figure 16-15 Speaker.

Audio amplifiers have their output limits measured in ohms. For example, an amplifier may have an output rating of 3 ohms (Ω). This indicates that the circuit you connect to the output terminals (the speaker connections) must have a total impedance (total resistance, including inductive and capacitive reactance) of no less than 3 ohms.

Most speakers are rated at 8 ohms. Thus, two such speakers connected in parallel would give the circuits an impedance of 4 ohms (8 ohms per branch, divided by the number of equal branches). Three such speakers connected to a circuit would yield an impedance of 2.67 ohms (8 ohms divided by 3 branches). In these calculations, we are ignoring the resistance value of the conductors, which is negligible except when very long runs are required.

If you compare the speaker in Figure 16-15 with the microphone in Figure 16-4, you can see that they share the same basic design of a coil of wire moving back and forth past a permanent magnet. Review these two drawings and notice that the speaker could function as a microphone (if connected properly) and that a microphone could act as a speaker. In actual use, this is seldom done, primarily because we design these devices for maximum efficiency in a specific use, but a very generally designed device would work well as either a speaker or as a microphone.

Autotransformers

While there are a few amplifiers that can handle loads with impedances as low as 1.5 ohms, they are far from common. In most cases, if you wish to connect more than two 8-ohm speakers on a circuit, you must install an autotransformer to adjust the impedance of multiple speaker pairs and present a sufficiently high impedance to the amplifier. You will often find these audio autotransformers sold as impedance-matching transformers.

Volume controls are often autotransformers (they can also be potentiometers) and can be used to reduce the level of signal sent to the speakers, while still presenting the same circuit impedance to the amplifier.

Crossovers

Crossovers are filters that route specific frequencies of signals to specific speakers. You are probably familiar with speakers that are designed for sounds of a specific pitch, such as *woofers* for low frequencies and *tweeters* for high frequencies.

Figure 16-16 shows a typical crossover circuit. This circuit passes low frequencies for the 12-inch (woofer) speaker, and passes high frequencies to the 4-inch (tweeter) speaker.

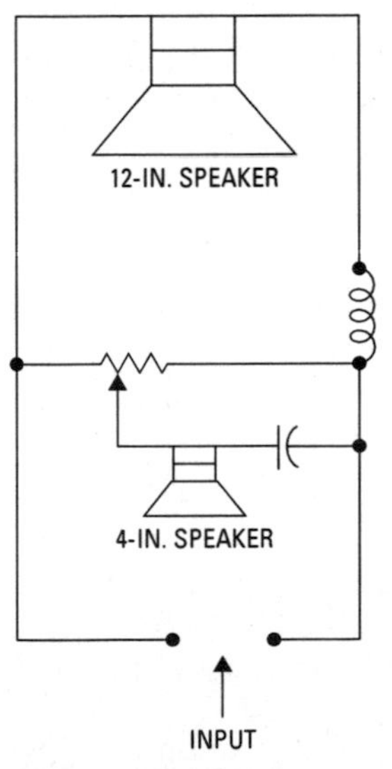

Figure 16-16 Crossover circuit.

Summary

A sound wave is a series of vibrations passed through the air. Sound waves in dry air travel at 343 meters per second and have two primary components: frequency and amplitude. Frequency is determined by how many compressions and rarefactions pass a given point per second, and amplitude is determined by the strength of the compressions and rarefactions.

The comparative measurement method of decibels was created to represent how the human ear perceives sound. Sound is measured in decibels. The threshold of hearing is identified as

0 dBSPL (SPL stands for sound pressure level). So, the minimal audible level for sound becomes the base reference point for decibels. Sounds above 85 decibels begin to damage the ear, and those above 130 decibels generally cause pain.

The sounds we hear are almost never as simple as one single sine wave. Sounds are rather composed of many frequencies and variations.

Harmonics are frequencies that are related to (and sometimes caused by) the primary signal. A second harmonic has a frequency that is exactly twice that of the fundamental frequency. A third harmonic has exactly three times the frequency of the fundamental frequency. Harmonic tones, even if not prominent, give the flavor to many of the sounds we hear, and allow us to distinguish between them.

The human ear picks up sounds in the vibratory range of approximately 20 hertz to 20,000 hertz. This varies from one individual to another and generally is diminished with age, especially on the high end. Sounds at about 3500 hertz are perceived as the loudest, with higher or lower frequencies not being perceived as loud even if the true measurement of amplitude is the same.

Sound waves spread out from their source in the same way that waves of water spread out in a pond. They bounce off of hard surfaces (such as plaster walls) and are readily absorbed by soft, flexible surfaces (such as carpeting or fabric). The angle at which the sound waves bounce off of hard surfaces is equal to, and opposite of, the angle at which they strike the surface.

Microphones convert sound waves into electronic signals. They do this with some sort of diaphragm or disk that vibrates when it is hit by sound waves. Then, by one of several mechanisms, the mechanical movement of the diaphragm is used to create electrical signals.

Microphones are rated by the frequency range they respond to, their impedance, and their sensitivity. One of the key factors in producing a good-quality sound is the microphone's response. It should be relatively flat throughout any critical frequency range. That is, it should put out the same amplitude of signal for all frequencies and not respond differently to various pitches.

The impedance of a microphone should closely match the circuitry that drives it. Typically, microphones are rated as high- or low-impedance units.

Tape recorders of all types store representations of sound in the form of permanent magnetism on metal-coated tapes. Microphones are used to turn sound waves into electrical pulses, which are then used to drive electromagnets. Tape passed underneath these magnets at a carefully controlled speed will be magnetized in the same pattern as the sound waves. Then, if the tape is passed at the same speed underneath a magnetic sensor, the same pattern will be picked up by the sensor's circuitry and can be passed on to a speaker, which will reproduce the original sound.

Equalizers are used to attenuate specific frequencies or ranges of frequencies and, thus, balance the final output more appropriately. Equalizers come in two primary types: graphic equalizers (which affect several ranges of frequencies) or parametric equalizers (which can be used to affect only a very narrow band of frequencies).

Speakers change electrical signals into audible sound. They do this by running an electronic audio signal through an electromagnet. This magnet pulls a flexible cone in toward the magnet and allows it to move in or out in the same pattern as the audio signal. This displaces air adjacent to the cone, producing sound waves.

Crossovers are filters that route specific frequencies of signals to specific speakers. You are probably familiar with speakers that are designed for sounds of a specific pitch, such as woofers for low frequencies and tweeters for high frequencies.

In the next chapter, we will examine the role of electronics in the world of television.

Review Questions

1. What is rarefied air?
2. What is 0 dBSPL?
3. What is the speed of sound?
4. Describe a third harmonic.
5. What is an octave?
6. What is the approximate range of hearing of a healthy human ear?
7. What is meant by the term acoustics?
8. Why are delay units used in large buildings and in large rooms?
9. Name three types of signal processors.
10. Compare a carbon microphone to a simple electronic device.
11. What would happen if you used a dynamic microphone connected in reverse?

12. What type of microphone uses the piezoelectric effect?
13. What is the key component in a condenser microphone?
14. How are sounds stored on tape?
15. What is the name of the typical speaker design?
16. How are autotransformers used in audio systems?
17. How are crossovers used?

Exercises

1. Design a crossover circuit. Show values and state the frequencies being sent to each speaker.
2. Illustrate impedance matching. Show values.

Chapter 17

Television

The development of the television industry was similar to the development of radio in several ways. First of all, it was chaotic, complete with multiple inventors and multiple patent battles. Financial hopes and arrangements came and went for a number of years until the field stabilized.

The first working television systems were partly mechanical, using rotating disks or systems of mirrors. The first completely electronic television system was invented by a farm boy from Idaho named Philo T. Farnsworth. Farnsworth envisioned the system at age 14, when he was in high school, and finally built and patented the system at age 21. Needless to say, Farnsworth was a rather exceptional young man.

But young Farnsworth was lacking in experience. His patents were contested by RCA, and he ended up getting very little recompense for his efforts. He eventually won his patent suits against RCA, but, by that time, the industry had moved on and he wasn't able to recover the losses. (RCA, obviously, did and does dispute this opinion.) Farnsworth was able to continue working in the television industry and in various laboratories all his life. He was awarded some 300 patents, including important patents in the field of nuclear power.

Television Basics

Probably the easiest way to explain basic television technology is to use the fax machine as an example. Fax machines are properly called *slow-scan television.* They send one television picture frame, in black and white, over telephone lines. We are all familiar with the process: the machine scans one line at a time, interprets that area as black or white space, translates the information into electronic pulses, and transfers the information to a receiver.

Black-and-white television uses the same technology as the fax machine, except at 30 frames per second, and using a camera rather than a scanner. Each line of the image is scanned and translated into a complex video signal. The color process is the same, except that a still more complex signal is used.

Because of the complexity of the television signal, it requires a significant amount of bandwidth. A standard television channel uses 6 megahertz of bandwidth. In other words, the signal requires 6 million cycles per second of signal space to get all the information

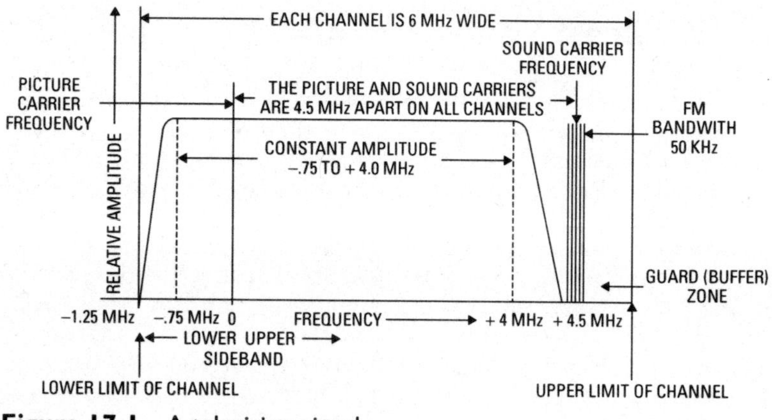

Figure 17-1 A television signal.

sent fast enough. Video images could be sent at lower speeds, but the images would be in slow motion.

Figure 17-1 shows a diagram of a television channel signal. Note the 6-megahertz channel width. The picture and sound frequencies are set very specifically and are 4.5 megahertz apart. Note also that the sound signal requires much less signal than the picture signal. The guard zone, which appears on the far right, keeps a buffer between one channel and the channel next to it.

To use multiple channels, each channel must use separate frequencies or else their signals would all be jumbled together. For example, the Federal Communications Commission (FCC) has assigned channel 2 the frequency range of 54–60 megahertz, channel 3 60–66 megahertz, channel 4 66–72 megahertz, and so on. (UHF channels go up to 890 MHz.) Each channel transmits its programming in that 6-megahertz channel. Cable television (CATV) works the same way, except it sends the signal through coaxial or optical cables, not via radio waves.

Basic 6-megahertz television signals are modulated to match one of the standard channel frequencies.

For example, the TV signal coming out of the camera at your local station operates at about 6 megahertz. If we assume that your station is channel 13, it must send its programs in the 210–216 megahertz slot. So, before sending the signal to the antenna, it must change the 6-megahertz signal to a 210–216 megahertz signal. This is done with a modulator circuit. We covered amplitude and frequency modulation in Chapter 15, so we will not go into a deep discussion of modulation here. However, Figure 17-2 shows a video

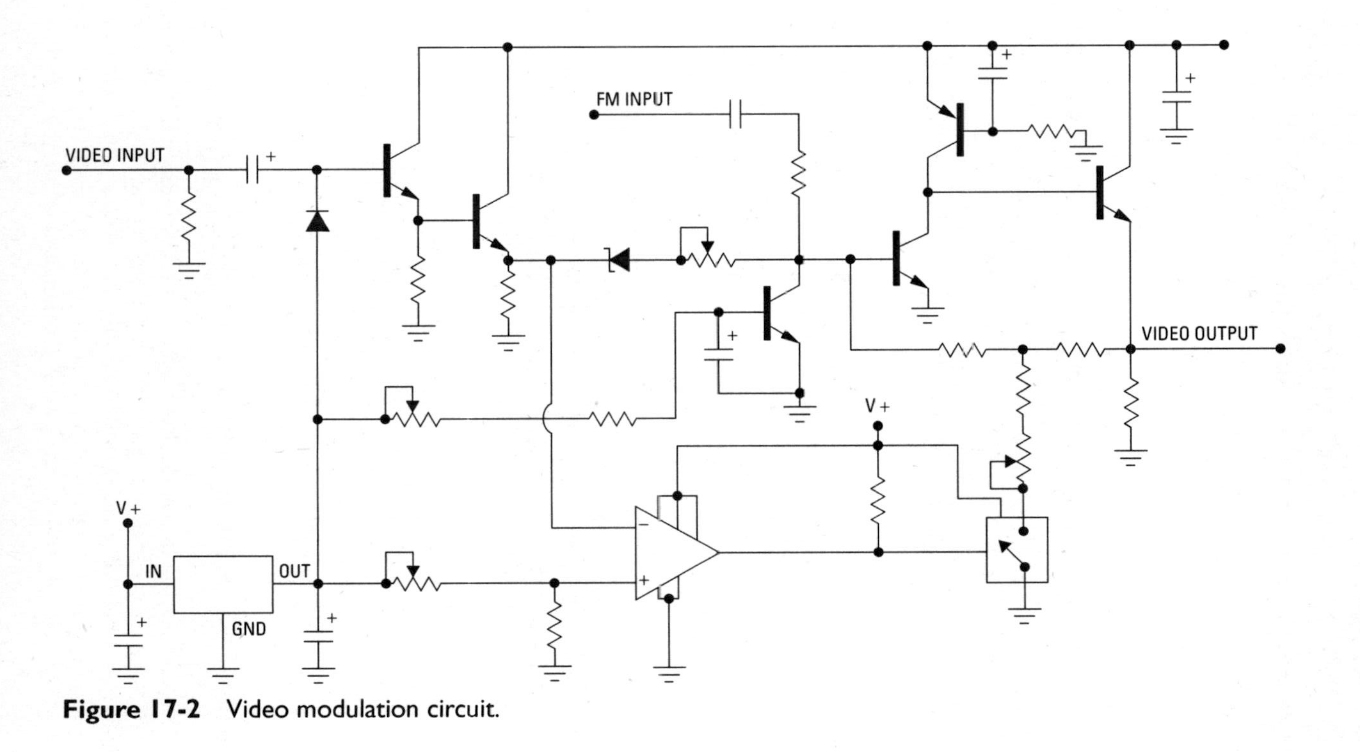

Figure 17-2 Video modulation circuit.

modulation circuit. Notice the complexity of this circuit. It must modulate not only the complex picture signal but the sound signal as well.

The Television Picture

The ability to broadcast and receive a complete television signal at 30 frames per second is really quite amazing. (The fact that a 21-year-old farm boy invented the process in 1927 is even more amazing.)

The basic format of a television signal is called a *raster*. This is essentially just a grid. Figure 17-3 shows a National Television System Commission (NTSC) system raster. (The NTSC is responsible for standardizing the television format used throughout North and Central America, as well as in Japan and South Korea.) An NTSC signal produces of 525 image lines, each with 440 pixels. As mentioned in Chapter 8, pixel is a shortening of *picture element*, or approximately one dot on the screen. Looking at Figure 17-3, you see the raster composed of many horizontal and vertical lines. An actual television picture has 525 horizontal lines from top to bottom. Each line is composed of 440 separate pixels, from left to right. (The new high-definition televisions, or HDTVs, will differ.)

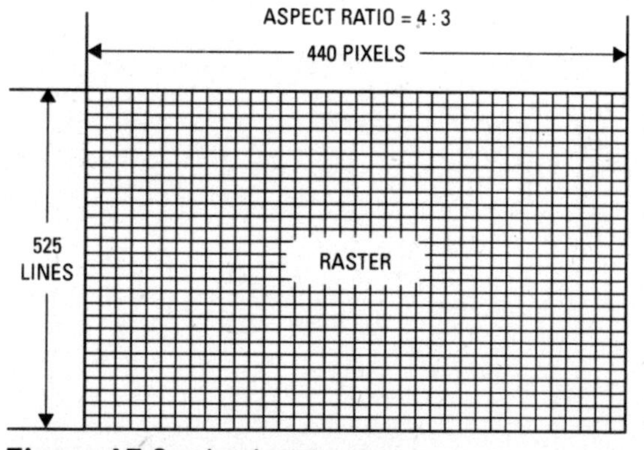

Figure 17-3 A television raster.

A television signal works by sending one line of signal at a time. In between each line of signal are a *blanking* pulse and a *synch* pulse. The blanking pulse tells the circuit to go completely dark upon completion of the line, so that the electron beam that creates the image can be moved back to the other end of the line without leaving an image, and will be set to start over. The synch (short for *synchronizing*) pulse is used to synchronize the signal and the electron beam continually.

The television signal broadcasts one line, goes blank while the *electron gun* (which we will describe shortly) retraces to the left side again, synchronizes, and then begins on the next line. This process is done 525 times in $^1/_{30}$ of a second.

It is worth noting that the NTSC system doesn't go from line 1 through line 525 directly. Rather, it produces the odd lines in one sweep (lines 1, 3, 5, 7, 9, and so on) then produces the even-numbered lines (2, 4, 6, 8, and so on) on the next sweep. This even-and-odd pattern is called an *interlaced* scanning pattern. The interlaced pattern produces a somewhat better overall appearance. So, we could say that the interlaced NTSC signal produces 60 half-frames per second.

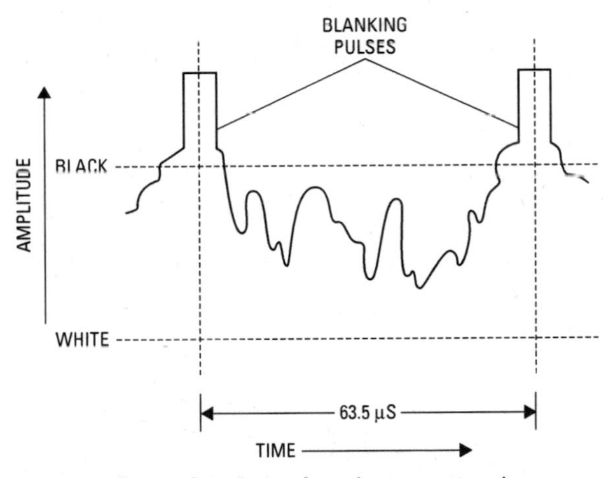

Figure 17-4 One line of a television signal.

Figure 17-4 shows the signal for one line of a black-and-white television broadcast. Note that this figure is showing time from left to right, not bandwidth as in Figure 17-1. Notice the blanking pulses at the beginning and end of each line. Again, this allows for the electron beam to be shut off while the angle of the beam readjusts to send the next line of signal.

The Cathode Ray Tube (CRT)

The device that produces the lines of the screen is called a *cathode ray tube* (*CRT*).

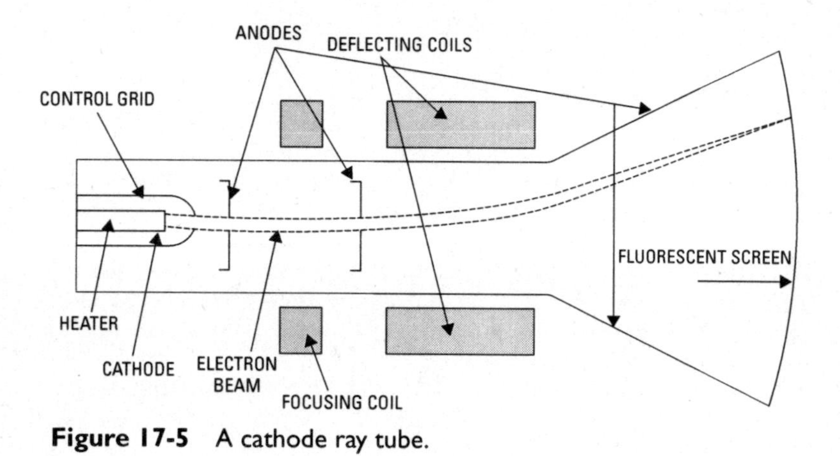

Figure 17-5 A cathode ray tube.

Figure 17-5 shows the construction of a CRT in a cut-away side view. If you look at this carefully, you will see that it resembles the vacuum tubes we covered in Chapter 7. There is a cathode (essentially a heated, negatively charged, metal plate) that emits electrons. There are also anodes (positively charged plates). Notice, however, that the anodes have holes in them. They draw the electrons toward themselves but let them pass through the holes. In other words, the electrons accelerate and shoot right past the anodes.

Another key feature of the CRT is the focusing coil. It was discovered early on that magnetic fields can affect the path of electrons as they move from cathode to anode. This *focusing coil* keeps the electrons in a straight beam, rather than being deflected to one side as the beam passes through the hole in the first anode.

After flying past the first anode, the electron beam is attracted to the next anode, and passes through the hole in that anode as well. Now the electron beam passes the *deflecting coils*. These coils are the ones that align the electron beam to trace a path back and forth across the fluorescent screen at the end of the tube. When electrons strike this chemically coated surface, they cause the fluorescent (also called *phosphorescent*) coating to emit light. As we mentioned previously, this is done one dot (that is, one pixel) at a time, and by causing the electron beam to very quickly trace a raster, a picture can be created.

So, the electron beam is created by the cathode and anodes, and controlled by the focusing and deflection coils. The picture is produced by the fluorescent effect of electrons striking the screen's surface.

An early version of the CRT was originally invented by Karl Ferdinand Braun, a German physicist who shared the Nobel Prize for the development of wireless telegraphy with Marconi. The modern version of the CRT, however, was invented by Philo T. Farnsworth, and was obviously critical to the invention of the all-electronic television.

Color Television

Up to this point we have been describing black-and-white television. Color television works in exactly the same way (which is why you can see the same broadcasts on either black-and-white or color sets), except that color television adds more information to the broadcast signal.

Instead of electrons hitting shades-of-gray pixels (as in black-and-white television), color television is arranged to have electrons hitting red, blue, or green spots. By combining these three colors, almost any color can be created on the television screen.

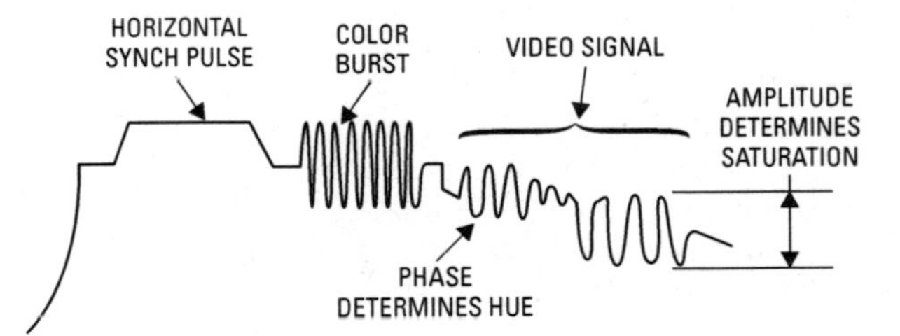

Figure 17-6 Signal for one line of a TV transmission.

Figure 17-6 shows one line of a color television signal. Again, the color signal matches the black-and-white signal, with extra information added for color.

Television Systems

The internal design of a modern television set is very complex, incorporating dozens of specialized circuits developed over decades. We will not attempt to analyze them here—that is a complete subject of its own. Broadcasting systems are just as complex.

Closed-Circuit Television

Closed-circuit television (CCTV) systems are the most common application of television technology after broadcasting. In fact, CCTV systems share a great deal with the CATV systems that are slowly replacing broadcasting.

The principal elements of a CCTV system are these:

- Cameras
- Lenses
- Mountings
- Switching and synchronizers
- Monitors
- Communications
- Video recorders
- Video motion detectors

In addition to these, some form of *communications media* is required. *Communications media* simply refers to the method of bringing the video signal from one place to another (that is, the wiring method).

Copper coaxial cable has long been the industry standard, although it is slowly being replaced by optical-fiber cables, especially for long runs. Microwave transmission is used only where required, because its cost is significantly more than the other two methods. Using a high-quality coaxial cable is essential. For installations of less than 1000 feet, the RG59U type is fine. But for distances of 1000 to 2000 feet, RG11U should be used. Installations of more than 2000 feet in length require the use of amplifiers to keep the signals at usable levels, or the use of optical fiber as the communication media.

Video Cameras

For many years, video cameras consisted of specialized tube cameras such as the *vidicon*. But these have been almost completely replaced by the charge-coupled device (CCD) that we covered in Chapter 8.

The sensing function of the CCD is performed by an array of photodiodes, each serving as one pixel. These are arranged in a grid (rows and columns). This entire gridwork is mounted on a silicon substrate. The individual photosensitive elements store up electrical charges that are proportional to the amount of light that strikes them. If a large amount of light strikes an element between discharges, it will accumulate a relatively high charge. If it receives little or no light between discharges, the electrical charge in the element will be very small. The CCD generates a television signal by discharging the individual elements in the proper order (the same order as the tube-scanning process). The discharging process can be done either directly or through *shift registers*, but the final result is the same—the conversion of photoelectric charges into a television signal.

The proper scientific name for the CCD is *charge-coupled device imager*, and it specifically refers only to the component that performs the transfer function and not to the photosensitive elements. Nonetheless, the entire imager is called a CCD.

The most difficult function of the CCD is the method of converting a charge pattern on an array of photosensors into a complete video signal. This is also frequently referred to as the CCD's *architecture*, rather than method of charge transfer. There are two architectures commonly used: *frame transfer* (*FT*) and *interline transfer* (*IT*).

Cable TV Channels

Since CATV signals are not radiated, a cable television system can use frequencies that are assigned to radio services without interference. These frequencies (between 88 and 174 megahertz) include 88–108 megahertz for the FM radio broadcast band, plus various marine and aircraft communication services. Cable television channels generally do not, however, use the FM radio frequencies. The standard configuration for cable television provides 55 channels, with the highest channel frequency just below the UHF band.

Additional cable channel frequencies can extend farther upward toward or into the UHF band. To accommodate these higher frequencies, with their greater losses (reactance being proportional to frequency), the cable providers have been forced to install better cables, use more amplifiers, and use optical-fiber cables for trunk and feeder lines.

Summary

The first working television systems were partly mechanical, using rotating disks or systems of mirrors. The first completely electronic television system was invented by Philo T. Farnsworth.

Black-and-white television uses the same technology as the fax machine, except at 30 frames per second, and uses a camera rather than a scanner. Each line of the image is scanned and translated into a complex video signal. The color process is the same, except that a still more complex signal is used.

Because of the complexity of the television signal, it requires a significant amount of bandwidth. A standard TV channel uses 6 megahertz of bandwidth. In other words, the signal requires 6 million cycles per second of signal space to get all the information sent fast enough. Video images could be sent at lower speeds, but the images would be in slow motion.

The picture and sound frequencies are set very specifically and are 4.5 megahertz apart. The sound signal requires much less signal than the picture signal.

To use multiple channels, each channel must use separate frequencies. Each channel transmits its programming in its own 6-megahertz channel. Cable television (CATV) works the same way, except that it sends the signal through coaxial or optical cables, not via radio waves. Basic 6-megahertz television signals are modulated to match one of the standard channel frequencies.

The basic format of a television signal is called a raster. This is essentially just a grid. An NTSC signal produces of 525 image lines, each with 440 pixels. An actual television picture has 525 horizontal lines from top to bottom. Each line is composed of 440 separate pixels, from left to right.

In between each line of signal are a blanking pulse and a synch pulse. The blanking pulse tells the circuit to go completely dark upon completion of the line, so that the electron beam can be moved back to the other end of the line without leaving an image on the screen. The synch pulse is used to synchronize the signal and the electron beam continually. A television signal broadcasts one line, goes blank while it retraces to the left side again, synchronizes, then begins on the next line. This process is done 525 times in $^{1}/_{30}$ of a second.

An interlaced NTSC signal produces 60 half-frames per second.

The device that produces the lines of the screen is called a cathode ray tube (CRT).

A CRT is a type of vacuum tube. Its cathode emits electrons. The CRT's anodes have holes in them. They draw the electrons toward themselves but let them pass through the holes and into the screen. A key feature of the CRT is the focusing coil. Magnetic fields are used to control the path of electrons as they move from cathode to anode. The focusing coil keeps the electrons in a straight beam, rather than being deflected to one side as the beam passes through the hole in the first anode.

The CRT's deflecting coils align the electron beam to trace a path back and forth across the fluorescent screen at the end of the tube. When electrons strike this chemically coated surface, they cause the coating to emit light. This is done one pixel at a time, and by causing the electron beam to very quickly trace a raster, a picture can be created.

Color television works in the same way as black-and-white television, except that color television signals contain more information than the black-and-white signal. Instead of electrons hitting shades-of-gray pixels, as in black-and-white television, color television is arranged to have electrons hitting red, blue, or green spots. By combining these three colors, almost any color can be created on the television screen.

Closed-circuit television (CCTV) systems are the most common application of television technology after broadcasting. CCTV systems are quite similar to CATV systems.

Copper coaxial cable has long been the industry standard communications medium for CCTV and CATV systems, but it is being replaced by optical-fiber cables, especially for long runs.

For many years, video cameras consisted of specialized tube cameras, such as the vidicon. These have now been replaced by the charge-coupled device (CCD). The sensing function of the CCD is performed by an array of photodiodes, each serving as one pixel. These are arranged in a grid (rows and columns). This entire gridwork is mounted on a silicon substrate. The individual photosensitive elements store up electrical charges that are proportional to the amount of light that strikes them. The CCD generates a television signal by discharging the individual elements in the proper order. The discharging process results in the conversion of photoelectric charges into a television signal.

Since CATV signals are not radiated, a cable television system can use frequencies that are assigned to radio services without interference.

In the next chapter, we will examine radar systems.

Review Questions

1. Describe the earliest television systems.
2. How is slow-scan television used?
3. What is the size of a television signal?
4. What is a raster?
5. Describe the NTSC system.
6. How many full frames per second are broadcast under the NTSC format?
7. Why are television signals modulated?
8. What is a pixel?
9. What is the purpose of a blanking pulse?

10. What is unique about the design of CRT anodes?
11. What is the purpose of a CRT's focusing coil?
12. What is the purpose of a CRT's deflection coil?
13. Describe the configuration of a color television signal.
14. How are multiple colors created in a color television picture?
15. How do cable television and broadcast television channel usage differ?

Exercises

1. Make a drawing of how an NTSC signal creates a television picture. Label parts and show where blanking pulses should be.
2. Draw the cathode and anodes of a CRT. Show as much detail as possible.

Chapter 18

Radar

Radar is an acronym for *radio detection and ranging*. As you most likely know, radar is used to detect and determine the location of large objects such as airplanes. (*Ranging* means "to determine the distance to.")

The development of radar includes a cast of characters very similar to the developers of radio. It begins, again, with James Clerk Maxwell predicting the existence of radio waves and with Heinrich Hertz not only measuring radio waves but noting that they were reflected off certain surfaces. Following this, a German engineer named Christian Huelsmeyer invented a very basic spark-gap and antenna device to detect boats (and prevent collisions) on the high seas. The idea did not catch on.

Next came Nikola Tesla, who published an article in 1917 that stated the principles of modern radar in detail, even including a fluorescent display screen. Using Tesla's principles, and based upon the findings of Hertz, a number of primitive radar systems began to be constructed in the 1920s and 1930s. These were primarily for military uses, especially in locating enemy equipment.

Since the radar development surge during World War II, uses have continued to be primarily military, though notable (and frequently hated) devices such as the police radar gun have come into broad use.

One of the great advantages of radar is that even very small echoes returning from an object can be amplified greatly by the electronic circuit that receives it. Radar is far more effective at detecting objects at great distances than light or sound signals are. That being said, radar does not work in water, and sound-based (*sonar*) systems are used for such applications.

Radar Basics

Radar operates by reflection. Radar systems generate strong radio waves, then wait for the waves to reach distant objects and reflect back toward the radar receiver, and then detect the echoes of the waves when they return. The distance between the radar transmitter and the object is determined by the amount of time passing between the sending of the signal and its echo back. Since radio waves travel at a known and fixed speed, determining distance requires little more than accurate time measurements and simple calculations.

Pulse radar (which will receive our primary coverage here) operates by sending short pulses of radio waves via antenna then measuring the time interval between the sending and the receiving back of the signals after they bounce off of some distant object. The *pulse width* (duration of the high-frequency pulse) of the signals is usually between 10 and 50 microseconds (μs).

The typical wavelength of these radar signals is between 1 centimeter and 1 meter. This corresponds to frequencies of between 300 megahertz and 30 gigahertz. These pulses are generally composed of a small range of frequencies centered around the frequency chosen for the system. Table 18-1 shows the typical radar frequency bands and their common uses.

The reflected signals that are used to produce radar images are called *backscatter*, and are also called the radar *cross-section*. The scientific and trade measurement of backscatter is called *nominalized radar cross-section* and is noted as *sigma 0* (measured in decibels). Typical values for sigma 0 are from +5 decibels (bright images) to – 40 decibels (very dark images).

Shorter wavelengths are better than longer wavelengths at detecting small objects.

The radio waves used by radar systems are frequently polarized either horizontally or vertically. (We covered this in Chapter 15, and the same principles apply here.)

Measuring Distance

The distance between any radar transmitter and the object being detected is half the round-trip time, divided by the speed of the signal. The formula is as follows

$$d = \frac{\text{time} \times 300{,}000}{2}$$

where the following is true:

d is distance (measured in kilometers)

300,000 is the speed of radio wave propagation (in kilometers per second)

t is time (measured in seconds)

So, if our radar system picked up an object, and if the round-trip time for the signal was 56 microseconds (μs), the calculation would look like this:

$$d = \frac{0.000056 \times 300{,}000}{2} = 8.4 \text{ km}$$

Table 18-1 Radar Band Designations and Uses

Band	*Frequency Range*	*Wavelength Range*	*Use and Notes*
HF	3–30 MHz	10–100 m	High-frequency band.
P	< 300 MHz	1 m +	The "P" stands for "previous"; applied retrospectively to early systems.
VHF	50–330 MHz	0.9–6 m	Very long range; ground penetrating.
UHF	300–1000 MHz	0.3–1 m	Very long-range (ballistic early warning); ground penetrating.
L	1–2 GHz	15–30 cm	Long-range air traffic control and surveillance.
S	2–4 GHz	7.5–15 cm	Terminal air traffic control; long-range weather.
C	4–8 GHz	3.75–7.5 cm	The "C" stands for "compromise" between X and S bands; used for weather.
X	8–12 GHz	12.5–3.75 cm	Missile guidance; marine radar; weather; in the U.S., the narrow range 10.525 GHz (± 25 MHz) is used for airport radar.
K_u	12–18 GHz	1.67–2.5 cm	High-resolution mapping; satellite altimetry.
K	18–27 GHz	11.11–1.67 cm	Used only for detecting clouds (because of absorption by water vapor).
K_a	27–40 GHz	0.75–1.11 cm	Mapping; short range; airport surveillance.
mm	40–300 GHz	1–7.5 mm	Millimeter band.
V	40–75 GHz	0.4–0.75 cm	
W	75–110 GHz	0.27–0.4 cm	

Note that, since radar signals travel so fast, pulse width becomes an issue. Earlier we said that standard pulse widths range from 10 and 50 microseconds. You can see from the previous calculation that picking up objects closer than about 8 kilometers would be a problem at 50 microseconds, where the echo signal would be returning before the pulse was finished sending. (Think back to what we said in Chapter 15 about waves combining to either enhance or diminish each other.) A mix of waves of the same frequency coming and going at the same time would make any usable result almost impossible to obtain. So, to range close objects, you must use especially short wavelengths, or high frequencies.

Modern radar systems can determine distances to within a centimeter or so.

Measuring Speed

Radar systems can sense the speed of an object at a distance by the *Doppler effect*. However, this measurement is of the object's speed relative to the radar transmitter. So, if the object is moving toward or away from the transmitter, the Doppler effect can be used. But if the object is moving parallel to the radar station, movement is much more difficult to sense.

The Doppler effect is the change in the frequency of a received signal because of the relative motion between the signal generator and the observer. For example, you no doubt have noticed the change in pitch of an ambulance siren as it passes you at a fairly high speed. Since sound waves travel slowly (roughly 340 meters per second), an ambulance driving at 35 miles per hour would be traveling at roughly 3 percent of the speed of sound. So, as the ambulance approaches you, the sound wave comes toward you at 103 percent of the speed of sound; and once the ambulance passes you and moves away, the sound it generates reaches your ears at 97 percent the normal (standing) speed. As the ambulance comes near you, more vibrations per second reach your ears; and as it goes away from you, fewer waves per second reach your ears. You perceive this as a change of pitch.

The same thing happens with radar waves bouncing off of distant objects. If the radar signal is sent at one frequency but returns at a lower frequency, we know that the object is moving away from the transmitter. And if we have good measuring equipment (to precisely measure the difference between the two frequencies), we can determine how fast the object is moving away. Conversely, if the echo pulse returns at a higher frequency than what it was sent, we can be certain that the object is moving toward the transmitter. Again, if we have precise measuring equipment, we can easily calculate the speed of the object toward the transmitter.

Modern radar systems can sense speeds as low as a few centimeters per second.

Measuring Position

Position measurement can be difficult with a single radar antenna. Radar waves spread out from a transmitter. In addition, radar signals bounce off distant objects unequally. Objects with different shapes and characteristics may bounce signals at odd angles.

The critical factor is being able to determine the angle of the signal as it returns to the receiver. Once this has been determined, the position of the object is certain.

One way that radar systems determine position is by using a moving (usually rotating) antenna, which will generally receive maximum power from the echo when facing the object directly. So, when the return signal is the strongest, the antenna is facing the object most directly. This requires either careful observation or power-monitoring equipment.

Another method of measuring position is by use of a technique called *phased array radar*. This method uses a large number of transmitters, operating together. By controlling the interference between the signals, the resultant signal can be steered without moving the antenna.

Waveguides

At very high frequencies (such as those used in radar systems), the use of copper cables becomes problematic. Remember that inductive reactance is a function of frequency. The higher the frequency is, the greater the reactance is. So, for ultra-high frequencies, it is easier to move waves from place to place than it is to move high-frequency currents.

Waveguides are used to move short-wavelength radar waves from place to place. This need not be in a straight line. A waveguide can move waves around corners with very close to zero loss.

A waveguide is generally constructed of a hollow metal conductor. This conductor may be round or rectangular. The metal raceway literally guides the wave to its intended point of use. The short waves travel through the waveguide in much the same way as pulses of light travel through optical fibers.

Cavity Resonators

If you think back to Chapter 1, you will remember that we said that electricity has different characteristics at different frequencies. As you can see from the previous explanation of waveguides, this is true.

This is also true as it relates to creating resonance at very high frequencies. You will no doubt remember from previous chapters that creating resonance requires the use of a capacitance and an inductance, either in series or in parallel.

At very high frequencies, coils are problematic. Remember, the reactance of a coil is proportional to frequency. So, at a frequency of 10 gigahertz (10,000,000,000 cycles per second), a coil that operates nicely at AM frequencies has an incredible resistance to current flow. So, at high frequencies we must minimize inductance. Of course, to maintain any specific resonant frequency, we must reduce capacitance in the same proportion.

In reducing both inductance and capacitance, we go from many turns of wire in a coil to less than one turn. Also, in reducing capacitance, we end up eliminating the capacitor entirely and simply using the trace of capacitance in any type of conductors that are near each other. For example, a half-turn of wire would have a very small amount of capacitance between one end and the other at 180 degrees opposite. And at several gigahertz, this is more than enough.

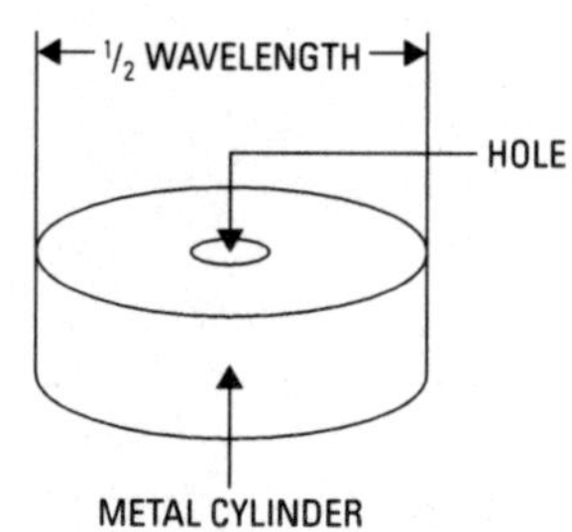

Figure 18-1
Resonant cavity.

Ultimately it was found that a closed metal cylinder (such as shown in Figure 18-1) works quite well to create resonance in such circuits. (You can see from this that, at very high frequencies, the borders between current and waves and conductors and waveguides begin to blur.) Note that the size of the resonant cavity is a function of wavelength, with one-half of a wavelength being the chosen size. Again, this makes such devices useful at very high frequencies. At 1 megahertz, such as device would be hundreds of feet across. At 10 gigahertz, however, it would be less than an inch in diameter.

Resonant cavities are nearly always used at frequencies of 3 gigahertz or greater.

Resonant cavities tend to amplify waves as well, much in the way that lasers amplify light. We will, however, leave that phenomenon unexplored here.

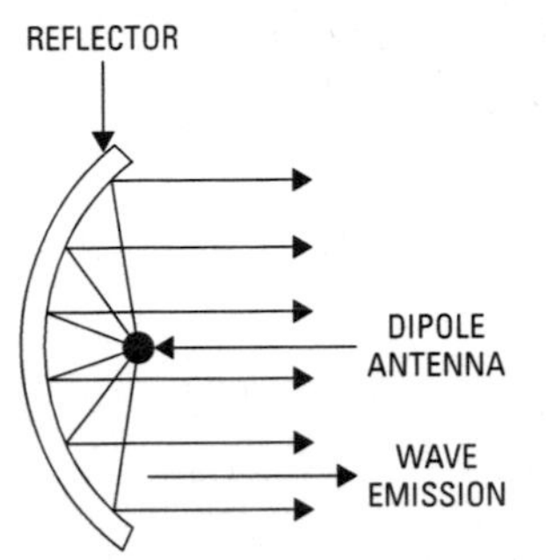

Figure 18-2 Parabolic radar antenna.

Antennas

The typical radar antenna is a *parabolic* antenna, as shown in Figure 18-2. Note that this is an arrangement of an antenna and a reflector. The shape of the reflector is called a *parabola*, which gives the entire arrangement its name. The parabolic shape allows for the waves emanating from the antenna to all be reflected out in a uniform direction, and this directionality ensures that maximum power will be transmitted in one area at a time, providing better results. Note that very short wavelengths tend to act much like light in terms of reflecting off surfaces, such as a parabolic reflector.

The antenna itself is a dipole of one-half wavelength. Remember that, in this case, it is quite a short antenna.

Radar systems use ground wave transmission exclusively. The sky wave is not useful, since very short waves tend to pass through the upper layers of atmosphere without reflecting back, as do longer wavelengths. In addition, the irregularity of the atmosphere's various layers and temperatures make it unsuitable, at least for ground-based systems (as opposed to systems in aircraft.)

Reciprocity

Reciprocity is the characteristic of being reciprocal, or operable in both directions. Radar antennas are almost always *reciprocal antennas* (that is, the same antenna is used both for transmitting and receiving). Note, however, that the operation is not sending and receiving at the same time. If it did, the waves would interact with each other and destroy the usefulness of the signals, as we have mentioned elsewhere.

Since radar signals are short pulses, measured in microseconds (millionths of a second) there is little trouble with overlapping send and receive signals. Since radar signals travel at 300,000 kilometers per second, they return to the antenna in a very short time, requiring very little delay before a second burst may be transmitted.

Analyzing Radar Images

Interpreting radar signals can be something of an art, requiring considerable experience to do it properly. Some of the large, advanced systems minimize the difficulty, but experience still helps quite a bit. Radar waves bounce differently from objects and are affected

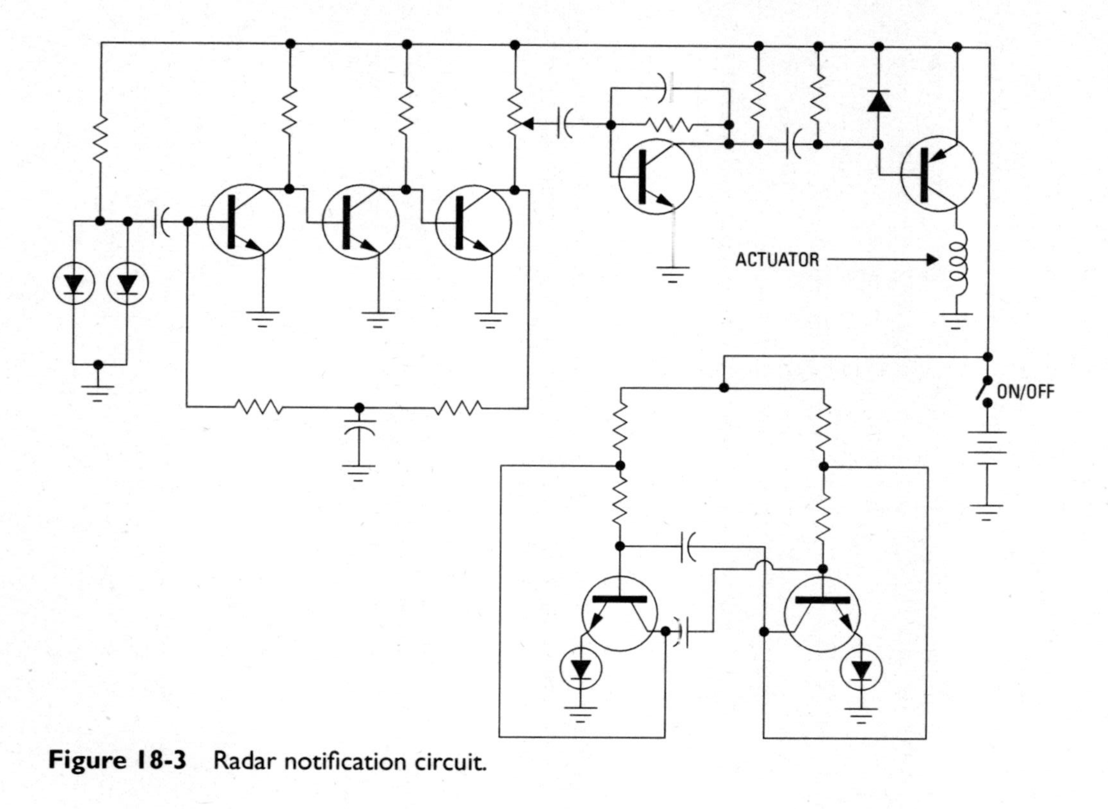

Figure 18-3 Radar notification circuit.

not only by the composition of the object being struck by the wave but by its composition, the angle of incidence, and sometimes by atmospheric conditions.

Backscatter from radar signals is sensitive to the target's electrical properties, including its water content. Wetter targets appear brighter and drier targets appear darker. One exception is a calm body of water. Radar waves tend to bounce off of smooth surfaces of this type without creating backscatter. (Think back to our discussion of internal reflection in optical fibers in Chapter 14. This is the same principle.)

Figure 18-3 shows a radar receiver circuit. The *actuator* is any device that is actuated (that is, activated, by the system). This could be any type of light, buzzer, or similar device that notifies the operator that the system they are overseeing is being struck by radar waves.

Radar Systems

There are obviously many types of radar systems. As noted earlier, our coverage in this chapter was of pulse radar systems, which are the most common type. But there are many other specialized systems, such as *phased-array* systems, which use a large number of transmit-and-receive antennas operating together as a unit.

Specialized radar systems have been developed for searching, targeting, sensing proximity (closeness), weather forecasting, navigation, and others.

Summary

Radar is an acronym for radio detection and ranging. As you most likely know, radar is used to detect and determine the location of large objects such as airplanes. (Ranging means "to determine the distance to.")

The development of radar includes a cast of characters very similar to the developers of radio.

Radar systems generate strong radio waves, wait for the waves to reach distant objects, reflect back toward the radar receiver, and detect the echoes of the waves when they return. The distance between the radar transmitter and the object is determined by the amount of time passing between the sending of the signal and its echo back. Since radio waves travel at a known and fixed speed, determining distance requires little more than accurate time measurements and simple calculations.

Pulse radar operates by sending short pulses of radio waves. The duration of these pulses is usually between 10 and 50 microseconds.

The typical wavelength of these radar signals is between 1 centimeter and 1 meter. This corresponds to frequencies of between 300 megahertz and 30 gigahertz.

Shorter wavelengths are better than longer wavelengths at detecting small objects. The radio waves used by radar systems are frequently polarized either horizontally or vertically.

Since radar signals travel so fast, pulse width becomes an issue. Picking up objects closer than about 8 kilometers would be a problem at 50 microseconds, where the echo signal would be returning before the pulse was finished sending. To range close objects requires especially short wavelengths, or high frequencies.

Radar systems can sense the speed of an object at a distance by the Doppler effect. The Doppler effect is the change in the frequency of a received signal because of the relative motion between the signal generator and the observer. If the radar signal is sent at one frequency but returns at a lower frequency, we know that the object is moving away from the transmitter. Conversely, if the echo pulse returns at a higher frequency than that at which it was sent, we can be certain that the object is moving toward the transmitter.

A waveguide is generally constructed of a hollow metal conductor. This conductor may be round or rectangular. The metal raceway literally guides very high-frequency waves to their intended point of use.

Resonant-cavity devices are used to replace coils and capacitors in creating resonance in extremely high-frequency circuits. They generally consist of a covered metal cylinder of one-half wavelength diameter.

The typical radar antenna is a parabolic antenna, which focuses radar waves in the same way that a reflector focuses light beams.

Radar antennas are used for both transmitting and receiving, a characteristic called reciprocity.

Radar systems use ground wave transmission exclusively. The sky wave is not useful.

Chapter 19 concludes our discussion of electronic applications by taking a look at computers.

Review Questions

1. Who do you think should get the most credit for the invention of radar? Why?
2. What does radar stand for?
3. Name several applications of radar.
4. Describe the pulses used in radar.

5. What is the Doppler effect used for?
6. Describe the Doppler effect.
7. Describe how radar systems perform ranging.
8. What is a waveguide?
9. What is a cavity resonator?
10. What is significant about the shape of a radar antenna?
11. What is reciprocity?

Exercises

1. Draw a radar pulse from an antenna to a remote object and back. Show as much detail as possible.

Chapter 19

Computers

Since it would be superfluous to describe the importance and impact of computers in the modern world, we will simply pass over such material. If you were ignorant of this fact, it is most unlikely that you'd ever make your way to a book on basic electronics.

Computing machines date back to the abacus, which was invented in about 3000 B.C., probably in Babylonia. However, more complex machines did not appear for a long, long time. (However, there is a reasonable argument that some forgotten ancient Greek genius may have created a mechanical computer, which has recently been discovered. As of this writing, this is still conjecture.)

The first complex computing machines in modern times (and maybe ever) were created by English professor and polymath Charles Babbage in the 1800s: the *difference engine* and the *analytical engine*. Earlier work along these lines was done by French mathematician Blaise Pascal, Gottfried Wilhelm von Leibniz, and Wilhelm Schickard.

The first electronic computer was the Atanasoff-Berry Computer that was built at Iowa State College between 1939 and 1942 by John Vincent Atanasoff and Clifford Berry. Other notable machines, such as Colossus and ENIAC, followed shortly thereafter, with ENIAC being a direct result of the builder's meetings with Atanasoff. These machines were built using vacuum tube technology. Each weighed several tons and was the size of at least one fairly large room.

Throughout the 1950s and 1960s, computers decreased in size and weight, and increased in computing power. Nonetheless, they remained very large and very expensive, finding limited application, primarily in government, for institutional research, and at central banks.

As you are no doubt aware, the smaller/faster trend led to the development of the personal computer in the late 1970s and to the entry of desktop computers into widespread business use in the 1980s. The 1990s saw the introduction of the Internet to wide popular use.

Needless to say, the computer, and especially the Internet, will influence our futures to a very significant degree. It is quite unlikely that these changes are anywhere near having played themselves out. In other words, there are probably many more changes to come. Two books, *The Third Wave* (New York: William Morrow, 1980) and *The Sovereign Individual: Mastering the Transition to the*

Information Age (New York: Touchstone, 1997) are highly recommended reading along these lines.

The computer industry is currently the largest field of application for electronics, and people from other electronic fields have been rapidly brought into the computer business. This is largely because a lot of money has been made in the computer and Internet industries. But, in addition, computer-company executives actively recruit people with a background in electronics, knowing that they are capable of ordered, concentrated thought. This makes them valuable and reliable as programmers, where precisely this type of skill is highly valued.

Computer Basics

All modern computers use binary mathematics and logic gates, which we cover in detail in Chapter 13. These are the technological foundations of the computer. To recap:

- Using digital pulses to represent digits allows us to transmit, modify, and codify data far better and easier than do analog electronics. Digital electronic systems use voltages to represent digits (that is, numbers). The presence of a sufficient voltage represents the number 1, and voltage below a certain level represents the number 0.
- *Binary* is a way of counting using a *base 2* method, rather than our usual *base 10*. Under our normal system, we move one space to the left, adding a new digit when we reach 10. Under a base 2 system, we move to the left and add a digit when we reach 2.
- Binary mathematics uses two digits only. Normally we use 10 (0–9). The value we gain from digital electronics and binary numbering is certainty. Differentiating between "yes or no," "zero or one," and "all or nothing" is much easier than differentiating between 10 levels of voltage.
- One type of digital signal transmission is *series transmission*. A series transmission is a stream of voltage pulses, with low voltage representing 0 and higher voltages representing 1. Another type of digital signal transmission is called *parallel transmission*, which consists of sending several streams of pulses at the same time.
- A *bit* is a single binary digit (a zero or a one). A *byte* is a group of eight bits. Computers and many microprocessors use bytes as a standard binary group.

- Logic circuits output signals in certain conditions. The basic conditions are AND, OR, and NOT. We build circuits called gates that will put out a high voltage. If both A and B are ones, this is called an AND gate. If either A or B are ones, this is called an OR gate. If A is a one, no voltage is output, but if A is a zero, a one will be put out. This is called a NOT gate. The NAND gate incorporates the operations of both NOT and AND gates. NOR gates incorporate the operations of both NOT and OR gates. Listings of exactly how gates operate are called *truth tables*. As logic circuits get more complex, truth tables become a necessity.
- Transistors can be used as switches as well as for amplification. Both bipolar and field-effect transistors are used in logic gates.
- Logic circuits are considered sequential or combinational. *Combinational circuits* do the same thing every time. After each operation they return to their original state. *Sequential logic circuits* are different; they react based upon their previous state. They do not automatically return to their original state. Combinational logic circuits are usually built entirely of NAND and NOR gates.
- Sequential logic circuits act based upon their previous state. They do not automatically return to their original state, and they use this characteristic to accomplish certain goals. Data (pulses) generally proceed through a sequential logic circuit step by step (in other words, one pulse sets a device, or puts it in a specific state of operation). The device then remains in that position until it is acted upon. In this way, we store data in a sequential logic circuit. Then, when the next bit of data arrives, it will encounter the stored data, and a specific reaction to both pieces of information will occur. The final output will reflect this.

Integrated Circuit Chips

A modern integrated circuit (IC) is a thin chip of silicon upon which are thousands or millions of semiconductor devices. These devices are primarily transistors, though there are usually a large number of resistors as well. Most chips are approximately 1 centimeter square, though many other sizes exist. The most important types of IC chips are microprocessors and memory chips. IC chips are just large-scale combinations of the devices and circuits that have been previously covered in this text.

In the construction of a pure silicon IC, the insulation layer on top is silicon dioxide (SiO_2). Layers of P and N semiconductor material are embedded in the silicon. Running on top of the silicon dioxide (and penetrating through to the P and N layers at times) is a layer of aluminum conductor. From this combination of materials, resistors, diodes, and transistors are constructed and connected together.

The Flip-Flop Circuit

The essential computer circuit is the *trigger* or *flip-flop* circuit, which was originally invented by W. H. Eccles and F. W. Jordan in 1919. You may also see it referred to as a *multivibrator* circuit.

This circuit, shown in Figure 19-1, is used for automatic counting—obviously critical for a computer. In this figure, Q_1 and Q_2 (Q being a common designation for transistors) rely on each other for base current. If Q_1 is conducting, negative voltage from its collector is transferred through R_4 to the base of Q_2. This reduces conduction through Q_2, which results in Q_1 becoming fully on and Q_2 fully off. At this point the circuit will continue as is. However, if a pulse of current is applied to the base of Q_2, it will start conducting, and the process (in reverse) will turn off Q_1.

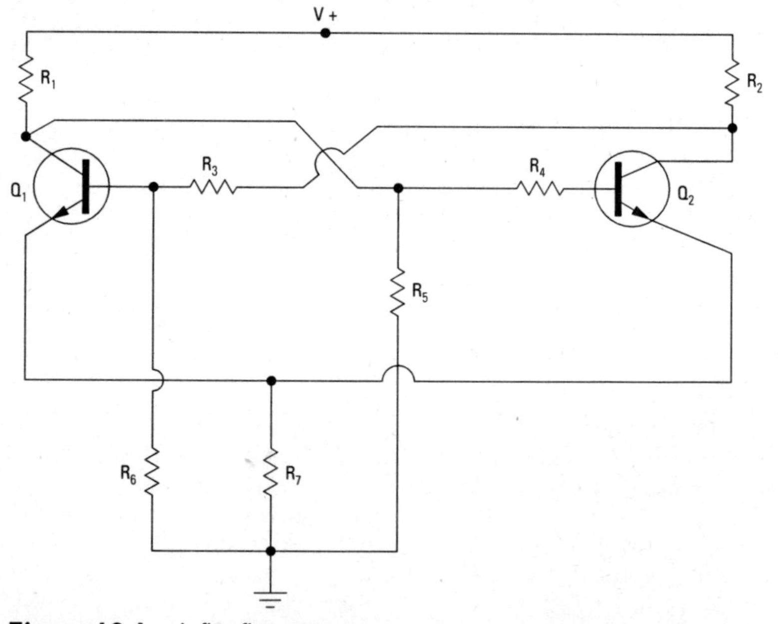

Figure 19-1 A flip-flop circuit.

This circuit thus develops a type of memory. It will stay in one state until acted upon, then change state until acted upon again, and so on.

Personal Computers

We will restrict our coverage here to personal computers (PCs), which are currently the standard for computing. Certainly, very large computers exist, but few of us have ever really seen one. Very small computers, on the other hand, while not the standard, may become very important in the near future. As communication techniques improve, massive computing power becomes less necessary and sharing information more important.

In our discussions here, we will leave out peripheral devices (such as printers) and even monitors and keyboards, focusing instead on activities inside the machine itself.

The CPU

The *central processing unit* (*CPU*) is the device that performs all the calculations made by the computer. It interprets control signals and data and then transfers them from one part of the computer to another.

Central processors are essentially just a very intricate IC chip.

BIOS

The *BIOS* is the *basic input and output system*. This is a chip with data stored on it. This data tells the computer how to start (called booting, as in pulling on your boots), how to read the keyboard, and how to process certain types of data.

The BIOS is a read-only memory (ROM) chip, meaning that the information in it cannot be altered. Under any normal circumstances, a computer user has nothing to do with the BIOS; it is strictly internal.

Memory

The *internal memory* in a personal computer is called *random-access memory* (*RAM*). This is measured in bytes, kilobytes, or megabytes. It can be both read and written to. It is also lost if power to the computer is lost. (ROM, on the other hand, is permanent, whether the power is on or not.) RAM stores various types of data being used by the computer, software programs, and their associated data.

Memory chips can generally be added or removed from computers very easily.

Operating System

The *operating system* (*OS*) governs the operating of the computer. You are doubtless familiar with operating systems such as Linux, Windows, and Mac OS. The OS regulates how software is used, how information is stored, and how data can be arranged, copied, and so on.

I/O Interfaces

Input/output (*I/O*) interfaces provide a way for signals to enter and leave the central computer. These interfaces are necessary to operate disk drives, printers, modems, and other peripheral devices.

Power Supply

The *power supply* for a PC is a metal box containing electrical and electronic components. Much like the power supply shown in Figures 9-20 and 9-21, a PC power supply takes 120 volts of AC power and changes it to the proper voltage or voltages required by the internal components of the computer, which is typically 5 volts DC.

Bus Systems

Bus systems are merely electrical connections between the various parts of the computer system. These are the wires that connect all of the various parts together. There are several types of buses (for different types of signals) that are sent to different parts of the system.

Disk Drives

Hard drives and *disk drives* are, technically, peripheral devices. Nonetheless, they are critical components in any modern computer.

These drives connect to the processor through I/O interfaces and send data to and from the processor. Both hard drives and floppy disks use magnetic disks to store data. The hard drives and floppy disks differ from each other in that hard drives use a much sturdier disk and store far more information than floppies. They are also more expensive and are almost always installed inside the computer, with no access unless the computer is disassembled.

Summary

The computer industry is currently the largest field of application for electronics, and people from other electronic fields have been rapidly brought into the computer business.

All modern computers use binary mathematics and logic gates as the foundations of their operations.

A computer's CPU is the IC device that performs all the calculations made by the computer. It interprets control signals and data and transfers them from one part of the computer to another.

A computer's basic input and output system is a specialized IC chip and the data stored on it. This data tells the computer how to start, how to read the keyboard, and how to process certain types of data.

The internal memory in a personal computer is called RAM. It is measured in bytes, kilobytes, or megabytes. It can be both read and written to. RAM is lost if power to the computer is lost.

The operating system governs the operating of the computer, regulating how information is stored, how data can be arranged, copied, and so on.

I/O interfaces provide a way for signals to enter and leave the central computer. These interfaces are necessary to operate disk drives, printers, modems, and similar peripheral devices.

Bus systems are merely electrical connections between the various parts of the computer system. These are the wires that connect all of the various parts together.

Both internal hard drives and disk drives connect to the processor through I/O interfaces. They are used to send data to and from the processor and to permanently store data.

Review Questions

1. Describe a logic gate.
2. Describe binary mathematics.
3. Describe series transmission.
4. Describe parallel transmission.
5. Describe an IC chip.
6. What is BIOS?
7. What device is used for long-term storage of data?
8. Name two types of operating systems.
9. How does a disk drive connect to a CPU?
10. How does a hard drive differ from a floppy disk?

Exercises

1. Draw a block diagram of the basic components of a PC. Show how they connect with each other. Show as much detail as possible.

Appendix A

Required Mathematics

Although electronic engineering requires significant training in mathematics, basic electronics does not require all that much.

This appendix discusses some of the basic mathematic skills that are necessary for electronic work. We will presume that you know how to add, subtract, multiply, and divide. If you lack these basic skills, put everything else aside and get them immediately, then come back to this book.

Transposing Formulas

Electronic work involves many formulas. So, working with them knowledgeably is a very useful skill.

One of the most appealing things about formulas is that you can change them around to isolate different parts of any given formula.

Remember that formulas are statements of truth. For example, if we say that $x = 3 + 2$, we can take that formula as an absolutely true statement, manipulate it accordingly, and confidently say that $x - 2 = 3$, or, $x - 3 = 2$. This is called *transposing* a formula.

You no doubt noticed in this example that $x = 5$. You probably examined the transpositions, compared them with the known value of x (5), and verified that the transpositions were, in fact, true statements. This is how you can check any of the transpositions we'll cover here. Simply plug in simple numbers and test them.

Transposition Technique #1: Adding or Subtracting from Both Sides

Formulas are balanced statements of equality. To change one side and not the other would throw the formula out of balance. But to change both sides the same way does not throw the formula out of balance.

We can use this fact to isolate one part of a formula. For example, note how we solve for y in the following formula:

$x = y + 12$

subtracting 12 from each side yields:

$x - 12 = y + 12 - 12$

since $12 - 12 = 0$, we can simply say:

$x - 12 = y$

We can use the addition method to transpose a formula as follows:

$x = y - 4$

adding 4 to each side yields:

$x + 4 = y - 4 + 4$

since $-4 + 4 = 0$, we can say:

$x + 4 = y$

This technique can be used in any formula at any time, so long as it is applied to both sides equally.

Transposition Technique #2: Multiplying or Dividing Both Sides

In the same way as with adding and subtracting, we can also multiply and divide both sides. But, again, this must be done equally to both sides. For example, here is how we can use the multiplication technique:

$x = y \div 3$

multiplying both sides by 3, we get:

$3x = y \div 3 \times 3$

since $\div 3$ and $\times 3$ cancel each other, we can say:

$3x = y$

Note in this example that instead of writing "$3 \times x$" we simply write $3x$. This is done to simplify and to avoid confusion, such as having two x-like symbols next to each other. Here are some the simplifications that you will see used:

$2 \times x = 2x$

or

$2 \times x = 2 \cdot x$

$$10 \div 4 = \frac{10}{4}$$

And here is an example of the division technique:

$x = 5y$

dividing each side by 5, we get:

$$\frac{x}{5} = \frac{5y}{5}$$

since $\times 5$ and $\div 5$ cancel each other, we can say:

$$\frac{x}{5} = y$$

Transposition Technique #3: Performing Operations on Both Sides

In addition to adding, subtracting, multiplying, or dividing, we can perform other operations to each side of a formula without throwing it out of balance.

For example, we can square each side of an equation without ruining its effectiveness as a true statement. Here is an example of this:

$$x = \sqrt{y}$$

squaring each side, we get:

$$x^2 = y$$

The Pythagorean Theorem

The theorem of Pythagoras (an ancient Greek mathematician) states the relationship of the sides of a right triangle. A right triangle always has one 90-degree angle. A typical right triangle is shown in Figure A-1.

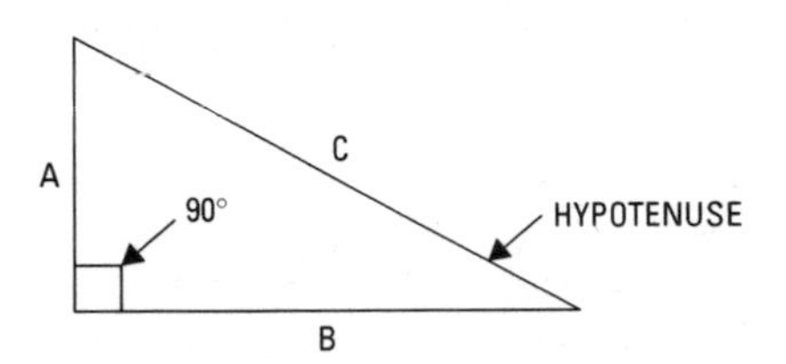

Figure A-1 A right triangle.

Notice that the longest side of a triangle is always called the *hypotenuse*.

The Pythagorean theorem states that

$$C^2 = A^2 + B^2 \qquad \text{or,}$$

$$C = \sqrt{A^2 + B^2}$$

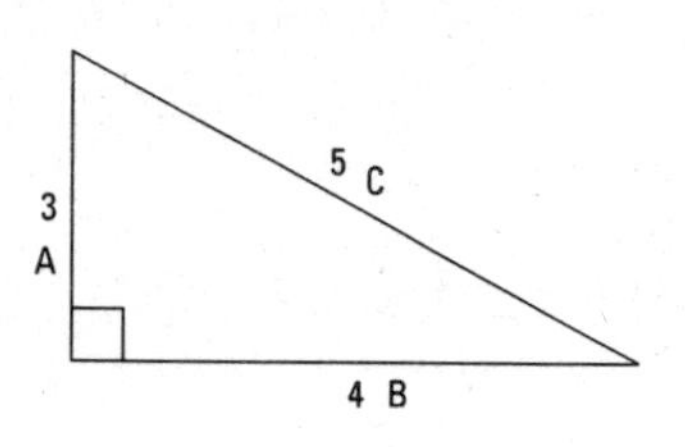

Figure A-2 A 3-4-5 right triangle.

This formula is true for every right triangle, and it is very useful. The simplest triangle (from a calculation standpoint) is one with side measurements of 3, 4, and 5. Figure A-2 shows this 3-4-5 right triangle.

Square Roots

A *square* is any number times itself. The reason it is called a square is because of the way area is measured. Figure A-3 shows why we say that 3 times 3 is the square of 3.

We note a square with a superscript. So, we write "6 squared" as 6^2. The square of 6 is 6 × 6, or 36. In addition to squares (which can also be called "to the second power"), we can raise numbers to higher powers as well. Thus, "6 to the third power" would be written 6^3, and would equal 6 × 6 × 6, or 216.

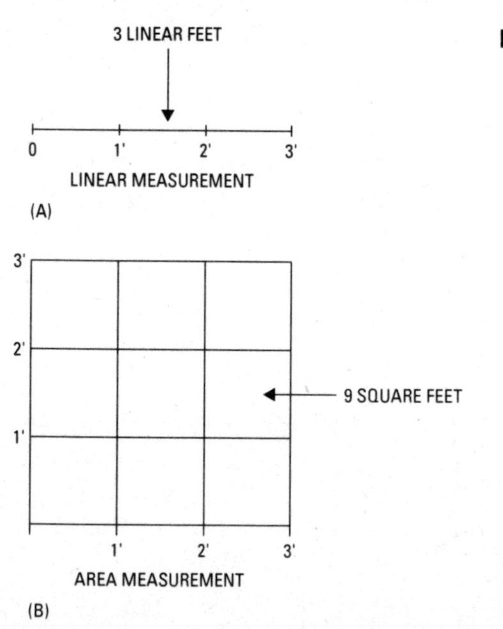

Figure A-3 Squaring 3.

A square root is the number times itself that equals the number in question. We use the $\surd$ symbol to denote square root.

A solution for the square root of 36 looks like this:

$$\sqrt{36} = 6$$

this can be verified by squaring the result:

$$6 \times 6 = 36$$

While calculating square roots can be done by hand, it is a difficult process. It is much easier to use a calculator, a table of square roots, or even a slide rule.

Fractions

Fractions are actually statements of division. The top portion of the fraction is called the numerator, and the bottom portion is called the denominator, as is shown here:

$$\frac{\text{numerator}}{\text{denominator}}$$

For the following fraction

$$\frac{3}{4}$$

we can say "3 over 4," or we could even say "3 divided by 4."

A solution to this fraction would be as follows:

$$\frac{3}{4} = 3 \div 4 = 0.75$$

Adding and Subtracting Fractions

Adding fractions that have the same denominator is simple. The denominators stay the same, and we simply add or subtract the numerators. In the following examples, we are simply counting the number of thirds:

$$\frac{2}{3} + \frac{6}{3} = \frac{8}{3}$$

$\frac{8}{3}$ can also be written as $2\frac{2}{3}$

$$\frac{6}{3} - \frac{1}{3} = \frac{5}{3} \quad \text{or} \quad 1\frac{2}{3}$$

If the denominators are not equal, we have to multiply first to create identical denominators.

Multiplying Fractions

Multiplying fractions is fairly simple: You simply multiply the numerator by the numerator and the denominator by the denominator, as shown here:

$$\frac{6}{4} \times \frac{a}{b} = \frac{6a}{4b}$$

Dividing Fractions

Dividing fractions is exactly the same as multiplication, except you invert the second fraction before multiplying, as shown here:

$$\frac{2}{3} \div \frac{a}{b} = \frac{2}{3} \times \frac{b}{a} = \frac{2b}{3a}$$

Decimals

Decimals are actually a type of fraction. However, decimals always have a multiple of 10 in their denominator. Here are several examples:

$$\frac{36}{10} = 3.6$$

$$\frac{2}{100} = 0.02$$

$$\frac{16}{1000} = 0.016$$

$$\frac{251}{1000} = 0.251$$

$$\frac{1000}{100} = 10.0$$

Note also that in multiplying numbers by factors of 10, we simply move the decimal point to the right, as shown here:

$$42.16 \times 1 = 42.16$$

$$42.16 \times 10 = 421.6$$

$$42.16 \times 100 = 4216.0$$

$$42.16 \times 1000 = 42160.0$$

In dividing by factors of 10, we move the decimal point to the left, as shown here:

$$21.43 \div 1 = 21.43$$

$$21.43 \div 10 = 2.143$$

$$21.43 \div 100 = 0.2143$$
$$21.43 \div 1000 = 0.02143$$

Geometric Functions

The primary geometric functions that are used in electronics are sine, cosine, and tangent. These are all formulas that state the relationship between the sides of a right triangle. The formulas are as follows:

$$\text{sine} = \frac{\text{opposite}}{\text{hypotenuse}}$$
$$\text{cosine} = \frac{\text{adjacent}}{\text{hypotenuse}}$$
$$\text{tangent} = \frac{\text{opposite}}{\text{adjacent}}$$

The simplest way to remember this is the silly phrase "Otto Has A Heap Of Apples." The Initials of this phrase are "O H A H O A." ("O" stands for opposite, "A" for adjacent, and "H" for hypotenuse.) They are the first letters of our formulas, as shown here:

$$\text{sine} = \frac{O}{H}$$
$$\text{cosine} = \frac{A}{H}$$
$$\text{tangent} = \frac{O}{A}$$

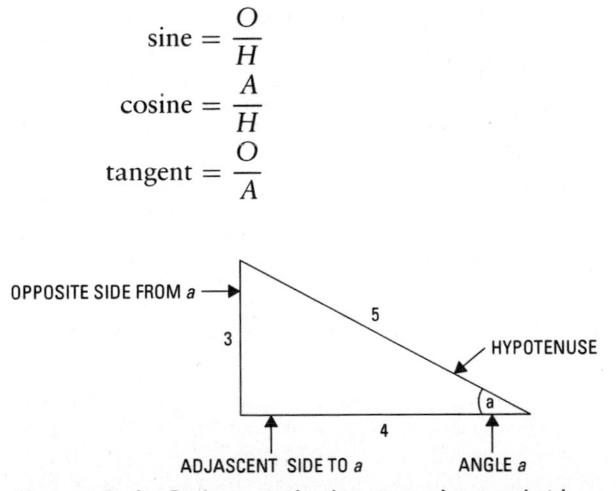

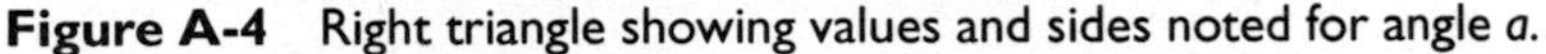
Figure A-4 Right triangle showing values and sides noted for angle *a*.

So, for angle *a* of the triangle shown in Figure A-4, the values of sine, cosine, and tangent for angle *a* would be calculated as follows:

$$\text{sine} = \frac{3}{5} = 0.60$$

$$\text{cosine} = \frac{4}{5} = 0.80$$

$$\text{tangent} = \frac{3}{4} = 0.75$$

Table A-1 shows the sine, cosine, and tangent values for various angles.

Table A-1 Trigonometric Functions

Angle	*Sine*	*Cosine*	*Tangent*	*Angle*	*Sine*	*Cosine*	*Tangent*
0°	0.000	1.000	0.000	30°	0.500	0.866	0.577
1°	0.018	1.000	0.018	31°	0.515	0.857	0.601
2°	0.035	0.999	0.035	32°	0.530	0.848	0.625
3°	0.052	0.999	0.052	33°	0.545	0.839	0.649
4°	0.070	0.998	0.070	34°	0.559	0.829	0.675
5°	0.087	0.996	0.088	35°	0.574	0.819	0.700
6°	0.105	0.995	0.105	36°	0.588	0.809	0.727
7°	0.122	0.993	0.123	37°	0.602	0.799	0.754
8°	0.139	0.990	0.141	38°	0.616	0.788	0.781
9°	0.156	0.988	0.158	39°	0.629	0.777	0.810
10°	0.174	0.985	0.176	40°	0.643	0.766	0.839
11°	0.191	0.982	0.194	41°	0.656	0.755	0.869
12°	0.208	0.978	0.213	42°	0.669	0.743	0.900
13°	0.225	0.974	0.231	43°	0.682	0.731	0.933
14°	0.242	0.970	0.249	44°	0.695	0.719	0.966
15°	0.259	0.966	0.268	45°	0.707	0.707	1.000
16°	0.276	0.961	0.287	46°	0.719	0.695	1.036
17°	0.292	0.956	0.306	47°	0.731	0.682	1.072
18°	0.309	0.951	0.325	48°	0.743	0.669	1.111
19°	0.326	0.946	0.344	49°	0.755	0.656	1.150
20°	0.342	0.940	0.364	50°	0.766	0.643	1.192
21°	0.358	0.934	0.384	51°	0.777	0.629	1.235
22°	0.375	0.927	0.404	52°	0.788	0.616	1.280
23°	0.391	0.921	0.425	53°	0.799	0.602	1.327
24°	0.407	0.914	0.445	54°	0.809	0.588	1.376
25°	0.423	0.906	0.466	55°	0.819	0.574	1.428
26°	0.438	0.899	0.488	56°	0.829	0.559	1.483
27°	0.454	0.891	0.510	57°	0.839	0.545	1.540
28°	0.470	0.883	0.532	58°	0.848	0.530	1.600
29°	0.485	0.875	0.554	59°	0.857	0.515	1.664

(continued)

Table A-1 ***(continued)***

60°	0.866	0.500	1.732	76°	0.970	0.242	4.011
61°	0.875	0.485	1.804	77°	0.974	0.225	4.331
62°	0.883	0.470	1.881	78°	0.978	0.208	4.705
63°	0.891	0.454	1.963	79°	0.982	0.191	5.145
64°	0.899	0.438	2.050	80°	0.985	0.174	5.671
65°	0.906	0.423	2.145	81°	0.988	0.156	6.314
66°	0.914	0.407	2.246	82°	0.990	0.139	7.115
67°	0.921	0.391	2.356	83°	0.993	0.122	8.144
68°	0.927	0.375	2.475	84°	0.995	0.105	9.514
69°	0.934	0.358	2.605	85°	0.996	0.087	11.43
70°	0.940	0.342	2.747	86°	0.998	0.070	14.30
71°	0.946	0.326	2.904	87°	0.999	0.052	19.08
72°	0.951	0.309	3.078	88°	0.999	0.035	28.64
73°	0.956	0.292	3.271	89°	1.000	0.018	57.29
74°	0.961	0.276	3.487	90°	1.000	0.000	Infinity
75°	0.966	0.259	3.732				

Appendix B

Symbols and Abbreviations

This appendix provides a guide to common symbols and abbreviations used in the field of electronics.

Table B-1 shows common symbols.

Table B-1 Common Symbols

Item	*Symbol*
Amplifier, general	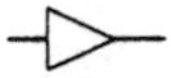
Amplifier, inverting	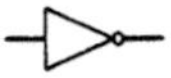
Amplifier, operational	
AND gate	
Antenna, balanced	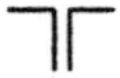
Antenna, general	
Antenna, loop	
Antenna, loop, multiturn	
Battery	
Capacitor, feed-through	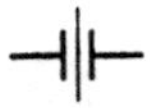
Capacitor, fixed	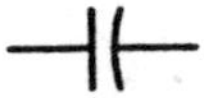
Capacitor, variable	
Cathode, electron-tube, cold	
Cathode, electron-tube, directly heated	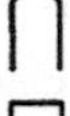
Cathode, electron-tube, indirectly heated	
Cavity resonator	

(continued)

Table B-1 *(continued)*

Cell, electrochemical	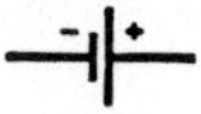
Circuit breaker	
Crystal, piezoelectric	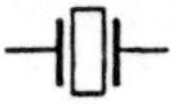
Delay line	
Diac	
Diode, field-effect	
Diode, general	
Diode, Gunn	
Diode, light-emitting	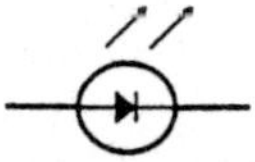
Diode, photosensitive	
Diode, PIN	
Diode, Schottky	
Diode, tunnel	
Diode, varactor	
Diode, Zener	
Exclusive OR gate	
Female contact, general	
Ferrite bead	
Filament, electron tube	
Fuse	

(continued)

Table B-1 *(continued)*

Grid, electron-tube
Ground, chassis 
Ground, Earth 
Inductor, air core 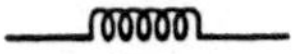
Inductor, air core, tapped 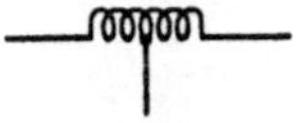
Inductor, air core, variable 
Inductor, iron core 
Inductor, iron core, tapped 
Inductor, iron core, variable 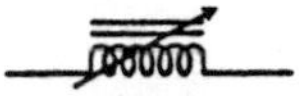
Inductor, powdered-iron core 
Inductor, powdered-iron core, tapped 
Inductor, powdered-iron core, variable 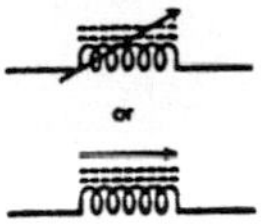
Integrated circuit, general 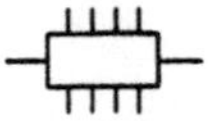
Jack, coaxial or phono 
Lamp, incandescent 
Lamp, neon 
Male contact, general 

(continued)

Table B-1 ***(continued)***

Meter, general	
Microphone	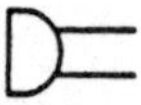
Microphone, directional	
NAND gate	
Negative voltage connection	–
NOR gate	
NOT gate	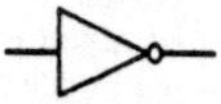
Opto-isolator	
OR gate	
Outlet, two-wire, nonpolarized	
Outlet, two-wire, polarized	
Outlet, three-wire	
Plate, electron-tube	
Plug, two-wire, nonpolarized	
Plug, two-wire, polarized	
Plug, three-wire	
Plug, coaxial or phono	
Plug, phone, two-conductor	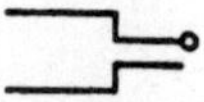
Plug, phone, three-conductor	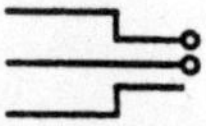

(continued)

Table B-1 ***(continued)***

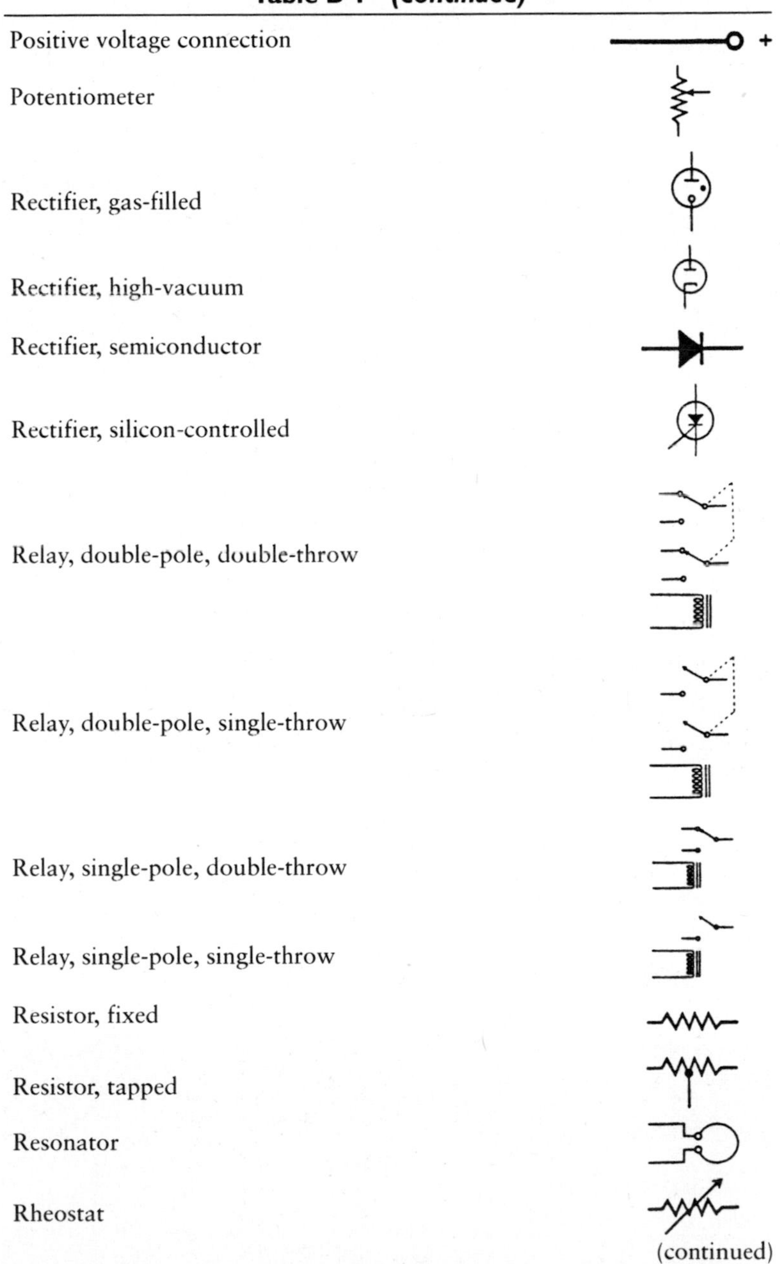

Component	Symbol
Positive voltage connection	+
Potentiometer	
Rectifier, gas-filled	
Rectifier, high-vacuum	
Rectifier, semiconductor	
Rectifier, silicon-controlled	
Relay, double-pole, double-throw	
Relay, double-pole, single-throw	
Relay, single-pole, double-throw	
Relay, single-pole, single-throw	
Resistor, fixed	
Resistor, tapped	
Resonator	
Rheostat	

(continued)

Table B-1 ***(continued)***

Saturable reactor	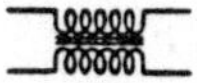
Signal generator	
Solar battery	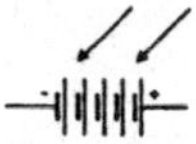
Solar cell	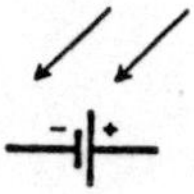
Source, constant-current	
Source, constant-voltage	
Speaker	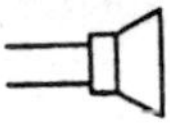
Switch, double-pole, double-throw	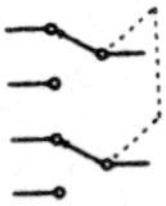
Switch, double-pole, single-throw	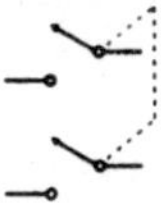
Switch, momentary-contact	
Switch, silicon-controlled	
Switch, single-pole, rotary	
Switch, single-pole, double-throw	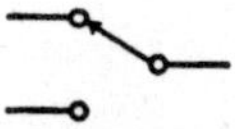
Switch, single-pole, single-throw	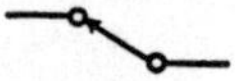
Test point	 TP
Thermocouple	
Transformer, air core	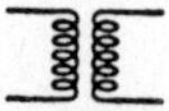

(continued)

Table B-1 (continued)

Transformer, air core, tapped primary	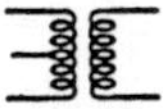
Transformer, air core, tapped secondary	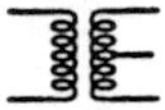
Transformer, iron core	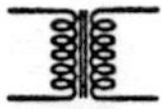
Transformer, iron core, step-down	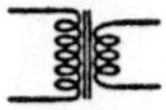
Transformer, iron core, step-up	
Transformer, iron core, tapped primary	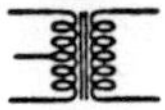
Transformer, iron core, tapped secondary	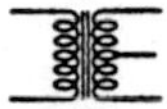
Transformer, powdered-iron core	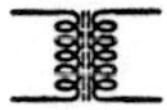
Transistor, bipolar, NPN	
Transistor, bipolar, PNP	
Transistor, field-effect, N-channel	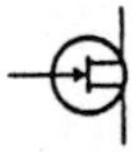
Transistor, field-effect, P-channel	
Transistor, MOS field-effect, N-channel	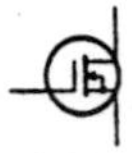
Transistor, MOS field-effect, P-channel	
Transistor, photosensitive, NPN	
Transistor, photosensitive, PNP	
Transistor, photosensitive, field-effect, N-channel	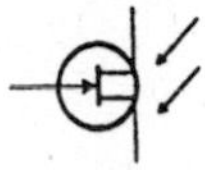

(continued)

Table B-1 *(continued)*

Transistor, photosensitive, field-effect, P-channel

Transistor, unijunction

Triac

Tube, diode

Tube, photosensitive

Tube, tetrode

Tube, triode

Waveguide, circular

Waveguide, flexible

Waveguide, rectangular

Waveguide, twisted

Wires, crossing, connected

Wires, crossing, not connected

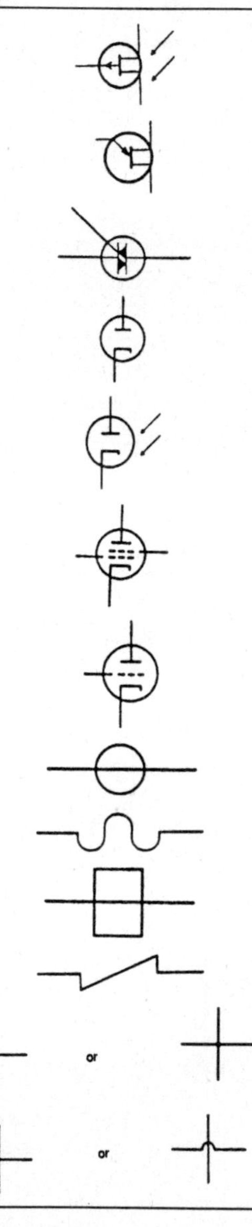

Table B-2 Common Abbreviations

Abbreviation	*Meaning*
A (amp)	Amperes
AC	Alternating current
A/D	Analog to digital
ADC	Analog-to-digital converter
AF	Audio frequency
AGC	Automatic gain control
A_I	Current gain
AM	Amplitude modulation
A_p	Power gain
A_v	Voltage gain
B	Flux density
BCD	Binary coded decimal
bfo	Beat frequency oscillator
BJT	Bipolar junction transistor
BW	Bandwidth
C	Capacitance or capacitor
c	Centi- (10^{-2})
CATV	Cable TV
cm	Centimeters
cmil	Circular mil
CPU	Central processing unit
C (Q)	Coulombs
CMOS	Complementary metal oxide semiconductor
CRT	Cathode ray tube
C_T	Total capacitance
d	Deci- (10^{-1})
D/A or D-A	Digital to analog
DC	Direct current
DIP	Dual in-line package
DPDT	Double-pole, double-throw
emf	Electromotive force
EMI	Electromagnetic interference
F	Farads
f	Frequency
FET	Field-effect transistor
FF	Flip-flop
fil	Filament
FM	Frequency modulation
f_r	Frequency at resonance
fsk	Frequency-shift keying
G	Conductance

(continued)

Table B-2 (continued)

G	Giga- (10^9)
H	Henrys
H	Magnetic field intensity
H	Magnetizing flux
hp	Horsepower
Hz	Hertz
I	Current
i	Instantaneous current
I_B	DC base current
I_C	DC collector current
IC	Integrated circuit
I_e	Total emitter current
I_{eff}	Effective current
IF	Intermediate frequency
I_{max}	Maximum current
I_{min}	Minimum current
I/O	Input/output
IR	Infrared
I_R	Resistor current
I_s	Secondary current
I_T	Total current
JFET	Junction field-effect transistor
K	Coefficient of coupling
k	Kilo- (10^3)
kHz	Kilohertz
kV	Kilovolt
kVA	Kilovolt-ampere
kW	Kilowatt
kWh	Kilowatt-hour
L	Coil, inductance
LC	Inductance-capacitance
LCD	Liquid crystal display
LCR	Inductance-capacitance-resistance
LDR	Light-dependent resistor
LED	Light-emitting diode
LF	Low-frequency
L_M	Mutual inductance
LNA	Low noise amplifier
LO	Local oscillator
LSI	Large-scale integration
L_T	Total inductance
L_W	Lambda wavelength

(continued)

Table B-2 (continued)

M	Mega- (10^6)
MC	Mutual conductance
m	Milli- (10^{-3})
MI	Mutual inductance
mA	Milliamperes
mag	Magnetron
max	Maximum
MF	Medium frequency
mH	Millihenrys
MHz	Megahertz
min	Minimum
mm	Millimeters
mmf	Magnetomotive force
MOS	Metal oxide semiconductor
MOSFET	Metal oxide semiconductor field-effect transistor
MPU	Microprocessor unit
MSI	Medium-scale integrated circuit
mV	Millivolts
mW	Milliwatts
N	Number of turns in an inductor
n	Nano- (10^{-9})
NC	Normally closed
nm	Nanometers
NO	Normally open
ns	Nanoseconds
NTSC	National Television Standards Committee
Op-amp	Operational amplifier
P	Power
p	Pico- (10^{-12})
P_{ap}	Apparent power
P_{av}	Average power
PLL	Phase-locked loop
pw	Pulse width
Q	Charge
R	Resistance
RAM	Random-access memory
RC	Resistance-capacitance
rcvr	Receiver
rect	Rectifier
ref	Reference
RF	Radio frequencies
RFI	Radio frequency interference

(continued)

Table B-2 *(continued)*

R_L	Load resistor
RLC	Resistance-capacitance-inductance
RMS, rms	Root-mean-square
ROM	Read-only memory
SNR	Signal-to-noise ratio
SPDT	Single-pole, double-throw
sq cm	Square centimeters
SSB	Single sideband
SW	Short wave
SWR	Standing-wave ratio
SYNCH, synch	Synchronous
t	Time in seconds
TC	Time constant (also temperature coefficient)
UC	Omega symbol ohms
UJT	Unijunction transistor
UV	Ultraviolet
V, v	Volts
VA	Volt-amperes
Vars	volt-amperes reactive
V_{av}	Voltage (average value)
V_{BE}	DC voltage base to emitter
V_c	Capacitive voltage
V_{CE}	DC voltage collector to emitter
V_{in}	Input voltage
V_L	Inductive voltage
V_m, V_{max}	Maximum voltage
V_{out}	Output voltage
V_p	Primary voltage
V_s	Source voltage
V_T	Total voltage
W	Watts
X_c	Capacitive reactance
X_L	Inductive reactance
Y	Admittance
Z	Impedance
Z_{in}	Input impedance
Z_o	Output impedance
Z_p	Primary impedance
Z_s	Secondary impedance
Z_T	Total impedance

Appendix C

Circuits

This appendix contains circuit diagrams for a number of interesting electronic circuits. Most of these are fairly easy to build and will provide a good start with hands-on work. Go to your local electronics store and spend some time looking around. Talk to the people who work there. Start with a simple circuit and build it carefully. It will probably feel clumsy the first time. That's normal. Work through it and try the next circuit. After a few, you'll be handling the soldering iron comfortably, and you'll start getting comfortable with components, circuit boards, and so on. Like every other skill, this one will take some time and effort to develop.

The numbers shown in some of these circuits are standard electronic part numbers, and should be good throughout the United States (and in a number of other countries, as well). Get some good electronics catalogs (usually they are free) and spend some time looking through them. Again, this will feel clumsy at first, but eventually you'll get comfortable with them. It just takes some time.

Be careful to observe wattage ratings for resistors. See what kind of current you can expect to be moving through any particular resistor and the voltage under which it operates. Ensure that the wattage rating for any resistor you buy is sufficient. If it is not, the resistor will overheat. This will either cause it to malfunction, or, in an unusual circumstance, could cause a small fire. Use caution.

Ensure that you apply capacitors at their rated voltages. If not, they will generally blow up. No, this will not be a powerful explosion, but it will take time and effort to fix.

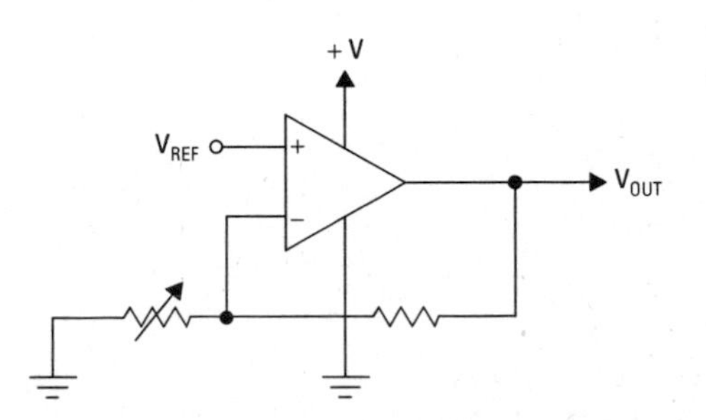

Figure C-1 Basic voltage regulation.

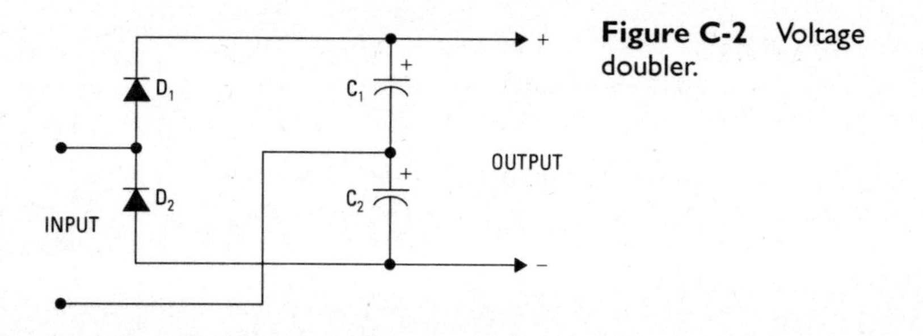

Figure C-2 Voltage doubler.

Ensure that your connections are soldered correctly, and that the entire circuit construction is mechanically strong. In the real world, circuits get bumped and dropped.

In addition to the circuits shown here, you can find thousands of circuits available at any time on the Internet.

Have fun!

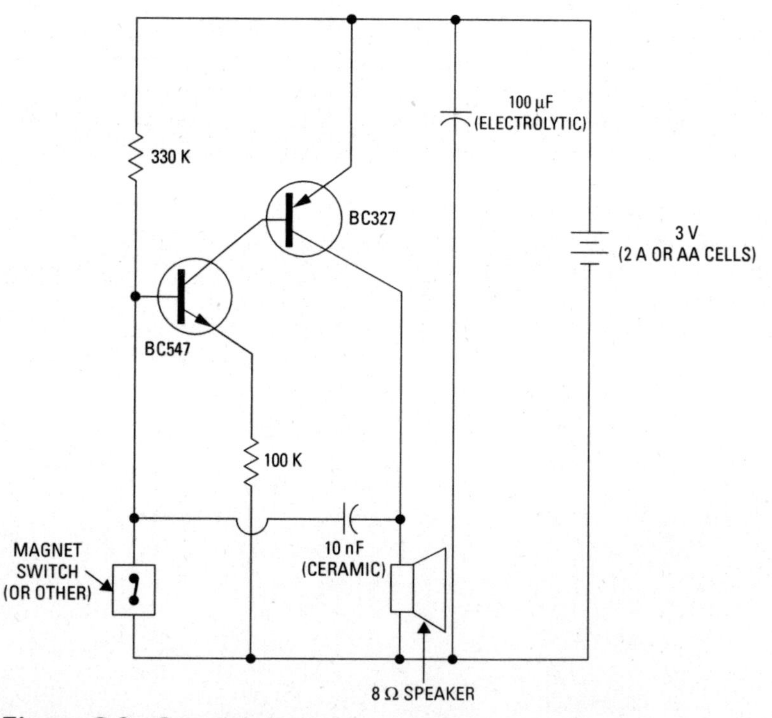

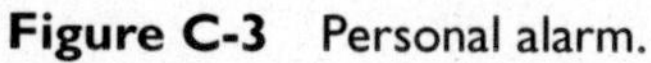
Figure C-3 Personal alarm.

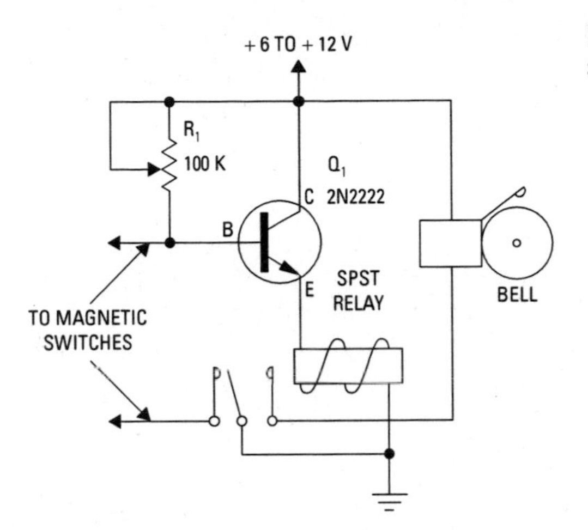

Figure C-4 Burglar alarm.

Basic Voltage Regulation

Figure C-1 shows the circuit for basic voltage regulation. Reference voltage to the noninverting input of the op-amp is amplified by the feedback and input resistors. The variable resistor modifies the output voltage.

Voltage Doubler

Figure C-2 shows the circuit for a voltage doubler. This doubles (approximately) an AC voltage and outputs it as DC. The components must be rated for the higher voltage.

Personal Alarm

Figure C-3 shows the circuit for personal alarm.

Burglar Alarm

Figure C-4 shows the circuit for a burglar alarm.

Light or Dark Detector

Figure C-5 shows the circuit for a light or dark detector. When in the L position, the speaker sounds when the photoresistor is struck with light. When in the D position, the speaker sounds when the photoresistor is not struck with light.

Light Dimmer

Figure C-6 shows the circuit for a light dimmer. The light output is determined by R_2.

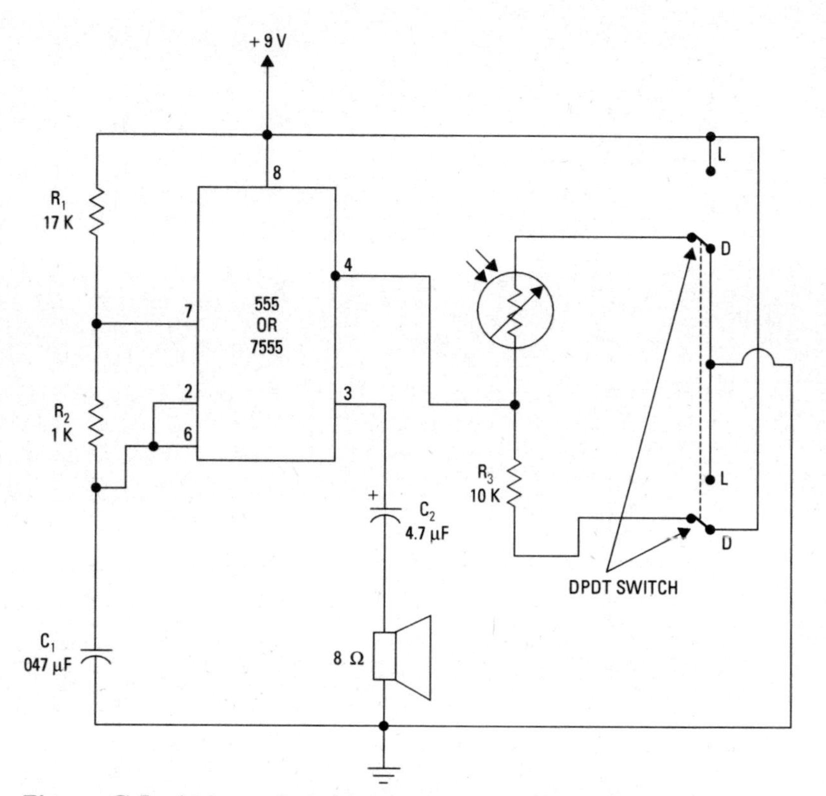

Figure C-5 Light or dark detector.

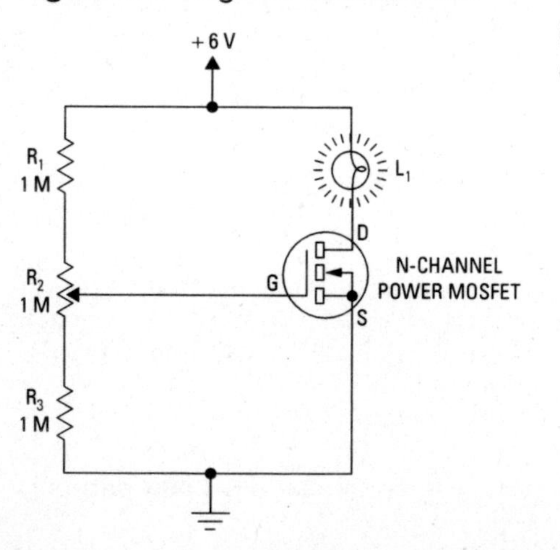

Figure C-6 Light dimmer.

Flashing Light

Figure C-7 shows the circuit for a flashing light. The rate of flash is determined by R_1.

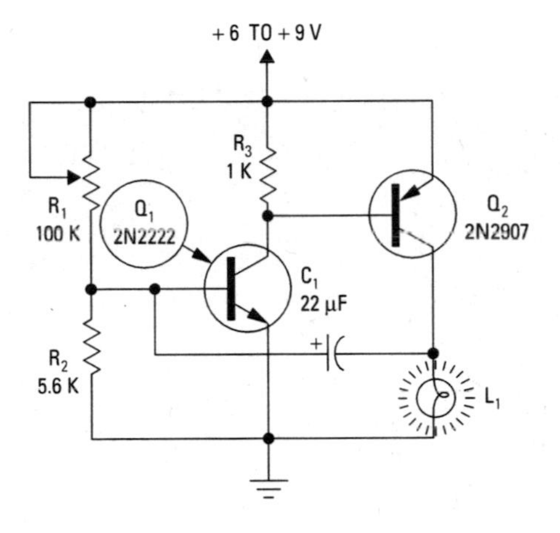

Figure C-7 Flashing light.

Digital Touch Switch

Figure C-8 shows the circuit for a digital touch switch.

Telephone Bug

Figure C-9 shows the circuit for a telephone bug. The transmitter frequency equals 88–94 megahertz FM.

Battery Charger

Figure C-10 shows the circuit for a battery charger.

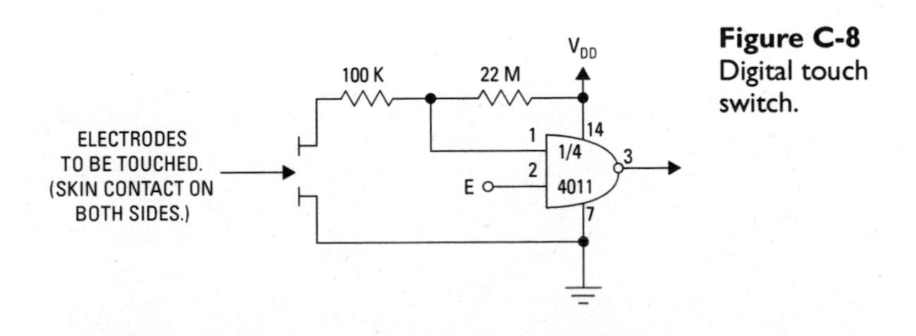

Figure C-8 Digital touch switch.

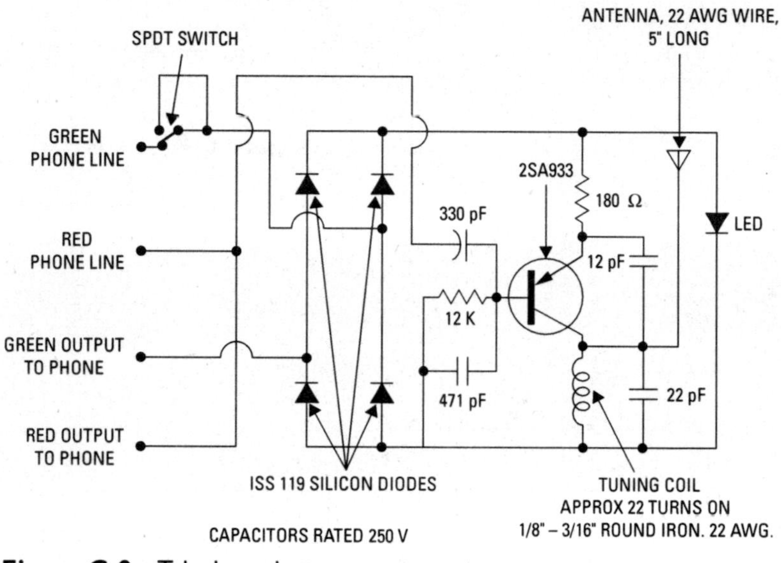

Figure C-9 Telephone bug.

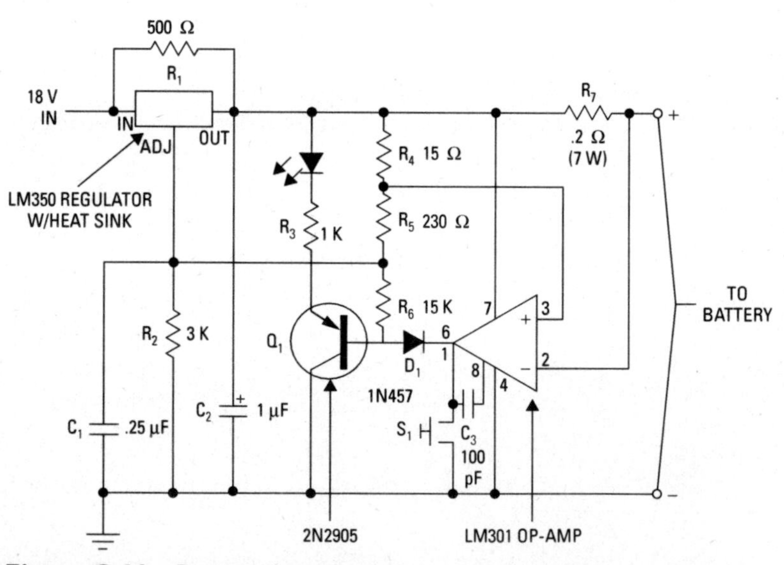

Figure C-10 Battery charger.

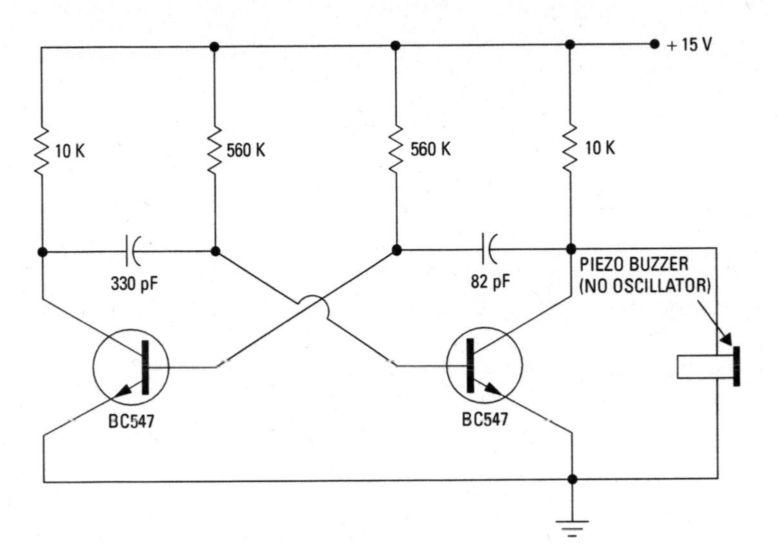

Figure C-11 Mosquito repeller.

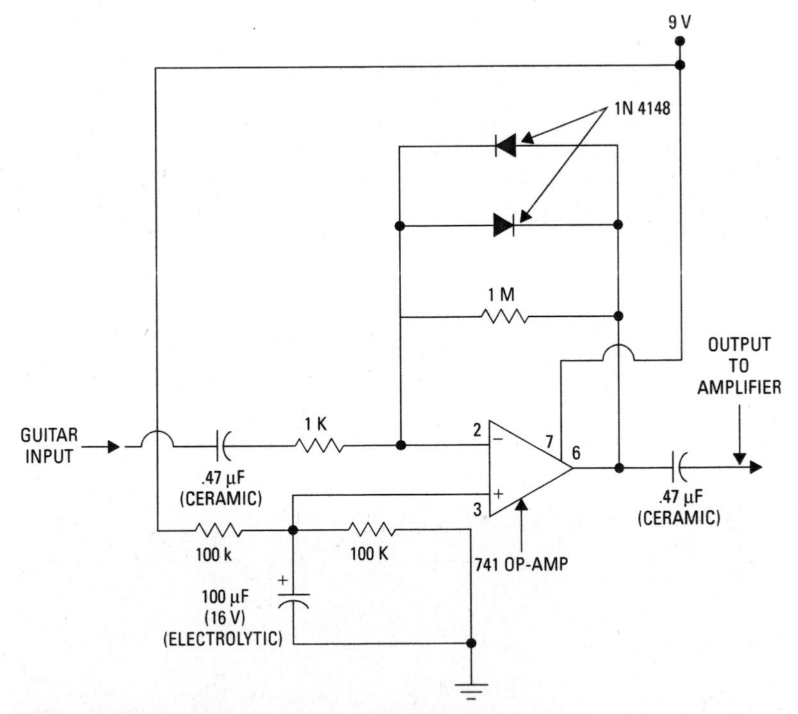

Figure C-12 Electric guitar distortion unit.

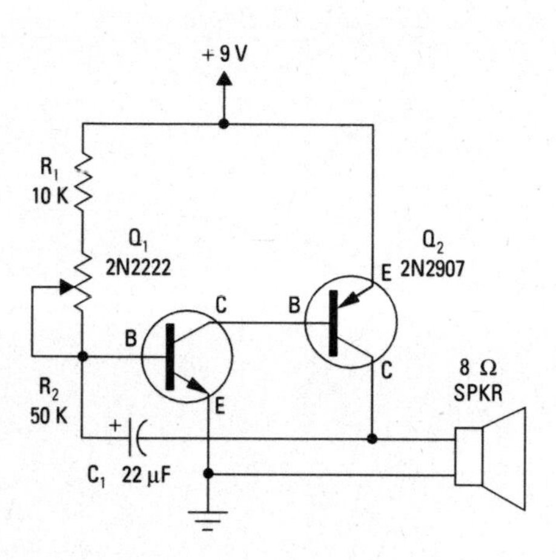

Figure C-13 Metronome.

Mosquito Repeller

Figure C-11 shows the circuit for a mosquito repeller.

Electric Guitar Distortion Unit

Figure C-12 shows the circuit for an electric guitar distortion unit.

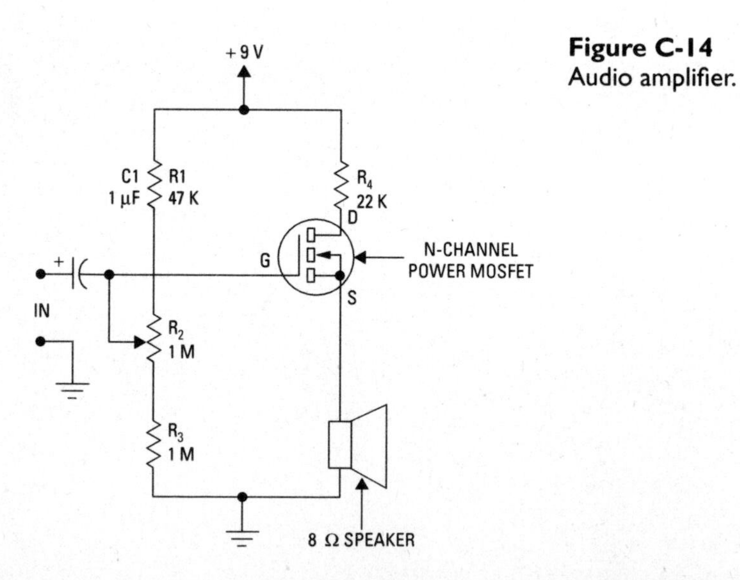

Figure C-14
Audio amplifier.

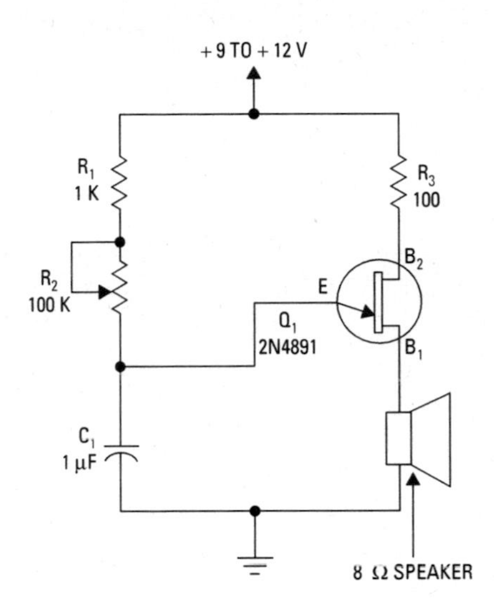

Figure C-15 Tone generator.

Metronome

Figure C-13 shows the circuit for a metronome. The rate of the beats can be modified with a variable resistor.

Audio Amplifier

Figure C-14 shows the circuit for an audio amplifier. The volume is set by R_2.

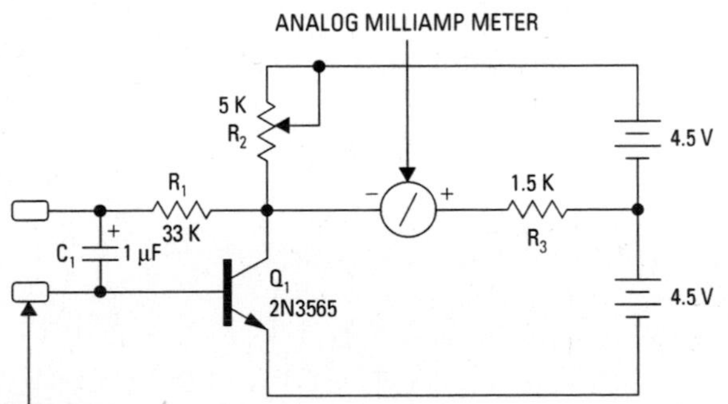

Figure C-16 Simple lie detector.

Tone Generator

Figure C-15 shows the circuit for a tone generator. The tone's pitch is set by R_2.

Simple Lie Detector

Figure C-16 shows the circuit for a simple lie detector. The electrodes attach to the back of the subject's hand, 1 inch apart. To operate, you zero the meter, and then ask questions. Skin resistance drops whenever the subject tells a lie.

Glossary

A/D converter—Analog-to-digital converter.

Admittance—The reciprocal of impedance.

Alive—Electrically connected to a source of electromotive force, or electrically charged with a potential different from that of Earth. Also, practical synonym for current-carrying or hot.

Alternating current (AC)—A periodic current, the average value of which over a period is zero.

Alternator—An alternating-current machine that changes mechanical power into electrical power.

Ambient temperature—The temperature of a surrounding cooling medium (such as gas or liquid), which comes into contact with the heated parts of an apparatus.

Ammeter—An instrument for measuring electric current.

Ampere (amp, A)—A charge flow of 1 coulomb per second. See also *Coulomb.*

Attenuation—General term used to denote the decrease in power between that transmitted and that received because of loss through equipment, lines, or other transmission devices. Usually expressed as a ratio in decibels. See also *Decibel.*

Autotransformer—A transformer in which part of the winding is common to both the primary and secondary circuits.

Bandwidth—The range of frequencies between two defined limits. Also used to express the amount of data that can be transmitted in a fixed amount of time.

Bit—A binary digit

Bus—A conductor or group of conductors that serves as a common connection.

Byte—A binary word, composed of eight bits.

Charge, electric—An inequality of positive and negative electricity in or on a body.

Choke coil—A low-resistance coil with sufficient inductance to substantially impede AC or transient currents.

Circuit, electric—A conducting path through which electric charges may flow.

Circuit, magnetic—A closed path for establishment of magnetic flux (magnetic field) that has the direction of the magnetic induction at every point. See also *Flux*.

Coaxial cable—A cable consisting of two conductors that are concentric with each other and insulated from each other.

Coercive force—The magnetizing force at which the magnetic induction is zero at a point on the hysteresis loop of a magnetic substance. See also *Hysteresis loop*.

Coil—A conductor arrangement (basically a helix or spiral) that concentrates the magnetic field produced by electric charge flow.

Conductance—A measure of permissiveness to charge flow. The reciprocal of resistance.

Conductor—A substance having free electrons or other charge carriers that permit charge flow when an electromotive force is applied across the substance. See also *Electromotive force*.

Coulomb (C)—An electric charge of 6.28×10^{18} electrons. One coulomb is transferred when a current of 1 ampere continues past a point for 1 second. See also *Electromotive force*.

Counter electromotive force (cemf)—The effective electromotive force within a system that opposes current in a specified direction. See also *Electromotive force*.

Current—The rate of charge flow. A current of 1 ampere is equal to a flow rate of 1 coulomb per second. See also *Ampere; Coulomb*.

Cycle—The complete series of values that occurs during one period of a periodic quantity. The unit of frequency (the hertz) is equal to one cycle per second. See also *Hertz*.

Dead—Functionally conducting parts of an electrical system that have no potential difference or charge (voltage of zero with respect to ground).

Decibel (dB)—A unit of measurement represented as a ratio of two voltages, currents, or powers, used to measure transmission loss or gain.

Degree, electrical—An angle equal to $^1/_{360}$ of the angle between consecutive field poles of like polarity in an electrical machine.

Diagram, connection—A drawing showing the connections and interrelations of devices employed in an electronic circuit.

Dielectric—A medium or substance in which a potential difference establishes an electric field that is subsequently recoverable as electric energy.

Digital—A system in which characters and codes are used to represent numbers in discrete steps.

Diode—Electronic component that allows current flow in one direction only.

Direct current—A unidirectional current with a constant value (defined in practice as a value that has negligible variation).

Direct emf—An electromotive force (emf) that does not change in polarity and has a constant value (one of negligible variation). Also termed direct voltage. See also *Electromotive force.*

Discharge—An energy conversion involving electrical energy. Examples include the discharge of a storage battery, discharge of a capacitor, or a lightning discharge of a thunder cloud.

Drop, voltage—An infrared voltage between two specified points in an electric circuit.

Effective value—The effective value of a sine-wave AC current or voltage is equal to 0.707 of peak. Also called the root-mean-square (rms) value, it produces the same I^2R power as an equal DC value.

Efficiency—The ratio of output power to input power, usually expressed as a percentage.

Electricity—A physical entity associated with the atomic structure of matter that occurs in polar forms (positive and negative) that are separable by expenditure of energy.

Electrode—A conducting substance through which electric current enters or leaves in devices that provide electrical control or energy conversion.

Electrolyte—A substance that provides electrical conduction when dissolved (usually in water).

Electromagnetic induction—A process of generation of electromotive force by movement of magnetic flux that cuts an electrical conductor.

Electromotive force (emf)—An energy-charge relation that results in electric pressure, which produces or tends to produce charge flow. See also *Voltage.*

Electron—The subatomic unit of negative electricity.

Electrostatics—A branch of electrical science dealing with the laws of electricity at rest.

Energy—The amount of physical work a system is capable of doing. Electrical energy is measured in watt-seconds, or the product of power and time.

Farad (F)—A unit of capacitance defined by the production of 1 volt across the capacitor terminals when a charge of 1 coulomb is stored. See also *Coulomb*.

Filament—A wire or ribbon of conducting (resistive) material that develops light and heat energy due to electric charge flow. Light radiation is also accompanied by electron emission.

Fluorescence—An electrical discharge process involving radiant energy transferred by phosphors into radiant energy that provides increased luminosity.

Flux—Electrical field energy distributed in space, in a magnetic substance, or in a dielectric. Flux is commonly represented diagramatically by means of flux lines denoting magnetic or electric forces. See also *Dielectric*.

Force—An elementary physical cause capable of modifying the motion of a mass.

Frequency—The number of periods occurring in unit time of a periodic process, such as in the flow of electric charge.

Fuse—A protective device with a fusible element that opens the circuit by melting when subjected to excessive current.

Galvanometer—An instrument for indicating or measuring comparatively small electric currents. A galvanometer usually has zero-center indication.

Gap (spark gap)—A high-voltage device with electrodes between which a disruptive discharge of electricity may pass, usually through air.

Ground—A conductor connected between a circuit and the soil. A chassis-ground is not necessary at ground potential but is taken as a zero-volt reference point. An accidental ground that occurs because of cable insulation faults, an insulator defect, and so on. Also termed earth.

Grounding electrode—A conductor buried in the earth, for connection to a circuit. The buried conductor is usually a cold-water pipe, to which connection is made with a ground clamp.

Henry (H)—The unit of inductance. It permits current increase at the rate of 1 ampere per second when 1 volt is applied across the inductor terminals.

Hertz (Hz)—International standard unit of frequency. Replaces, and is identical to, the older unit of cycles per second.

Hydrometer—An instrument for indicating the state of charge in a storage battery.

Hysteresis—The magnetic property of a substance that results from residual magnetism.

Hysteresis loop—A graph that shows the relation between magnetizing force and flux density for a cyclically magnetized substance. See also *Flux*.

Hysteresis loss—The heat loss in a magnetic substance caused by application of a cyclic magnetizing force to a magnetic substance.

Impedance—Opposition to AC current by a combination of resistance and reactance. Measured in ohms. See also *Resistance; Reactance; Ohm*.

Induced current—A current that results in a closed conductor because of cutting of lines of magnetic force.

Inductance—An electrical property of a resistanceless conductor, which may have a coil form and which exhibits inductive reactance to an AC current. All practical inductors have at least a slight amount of resistance. See also *Resistance*.

Inductor—A device such as a coil with or without a magnetic core that develops inductance, as distinguished from the inductance of a straight wire.

Interface—The junction or point of interconnection between two systems or sets of equipment having different characteristics.

Internal resistance—The effective resistance connected in series with a source of electromotive force caused by resistance of the electrolyte, winding resistance, and so on. See also *Electromotive force*.

Ion—A charged atom, or a radical. For example, a hydrogen atom that has lost an electron becomes a hydrogen ion. Sulphuric acid produces H^+ and SO_{-4} ions in water solution.

IR drop—A potential difference produced by charge flow through a resistance.

Joule—A unit of electrical energy. The transfer of 1 watt for 1 second. Also called a watt-second.

Joule's law—The rate at which electrical energy is changed into heat energy is proportional to the square of the current.

Jumper—A short length of conductor for making a connection between terminals, around a break in a circuit, or around an electrical instrument.

Junction—A point in a parallel or series-parallel circuit where current branches off into two or more paths.

Kirchhoff's law—The voltage law that states the algebraic sum of the drops around a closed circuit is equal to zero. The current law states that the algebraic sum of the currents at a junction is equal to zero.

Lenz's law—States that an induced current in a conductor is in a direction such that the applied mechanical force is opposed.

Magnet—A body that is the source of a magnetic field.

Magnetic field—The space containing distributed energy in the vicinity of a magnet, and in which magnetic forces are apparent.

Mass—Quantity of matter. The physical property that determines the acceleration of a body as the result of an applied force.

Matter—A physical entity that exhibits mass.

Meter—A unit of length equal to 39.37 inches. Also, an electrical instrument for measurement of voltage, current, power, energy, phase angle, synchronism, resistance, reactance, impedance, inductance, capacitance, and so on.

Mho—The unit of conductance defined as the reciprocal of the ohm. See also *Ohm*.

Microprocessor—A processing device in the form of a large integrated circuit.

Modulation—Alterations in the characteristics of carrier waves. Usually impressed on the amplitude and/or the frequency.

Mutual inductance—An inductance common to the primary and secondary of a transformer, resulting from primary magnetic flux that cuts the secondary winding. See also *Flux*.

Negative—A value less than zero. An electric polarity sign indicating an excess of electrons at one point with respect to another point. A current sign indicating charge flow away from a junction.

Normally closed—Denotes the automatic closure of contacts in a relay when de-energized. (Not applicable to a latching relay.)

Normally open—Denotes the automatic opening of contacts in a relay when de-energized. (Not applicable to a latching relay.)

Ohm (Ω)—The unit of resistance. A resistance of 1 ohm sustains a current of 1 ampere when 1 volt is applied across the resistance. See also *Ampere; Resistance; Volt.*

Ohmmeter—An instrument for measuring resistance values.

Ohm's law—States that current is directly proportional to applied voltage and inversely proportional to resistance, reactance, or impedance.

Period—The time required for an AC waveform to complete one cycle. See also *Waveform.*

Permanent magnet—A magnetized substance that has substantial retentivity.

Permeability—The ratio of magnetic flux density to magnetizing force. See also *Flux.*

Phase—The time of occurrence of the peak value of an AC waveform with respect to the time of occurrence of the peak value of a reference waveform. Phase is usually stated as the fractional part of a period. See also *Waveform.*

Phase angle—An angular expression of phase difference; it is commonly expressed in degrees, and is equal to the phase multiplied by 360 degrees.

Photon—The fundamental building block of electromagnetic radiation. Most generally depicted as a particle of light.

Polarity—An electrical characteristic of electromotive force that determines the direction in which current tends to flow. See also *Electromotive force.*

Pole—Area of a magnet at which its flux lines tend to converge or diverge. See also *Flux.*

Positive—A value greater than zero. An electric polarity sign denoting a deficiency of electrons at one point with respect to another point. A current sign indicating charge flow toward a junction.

Potential difference—A potential difference of 1 volt is produced when 1 unit of work is done in separating unit charges through unit distance.

Potentiometer—A resistor with a continuously variable contact arm. Electrical connections are made to both ends of the resistor and to the arm.

Power—The rate of doing work, or the rate of converting energy. When 1 volt is applied to a load and the current demand is

1 ampere, the rate of energy conversion (power) is 1 watt. See also *Ampere; Volt; Watt.*

Power, real—Developed by circuit resistance, or effective resistance.

Primary winding—The input winding of a transformer.

Protocol—A formally defined way to perform an action.

Proton—The subatomic unit of positive charge. A proton has a charge that is equal and opposite to that of an electron. See also *Electron.*

Raceway—A channel for holding wires or cables.

Rating—The rating of a device, apparatus, or machine states the limit or limits of its operating characteristics. Ratings are commonly stated in volts, amperes, watts, ohms, degrees, horsepower, and so on.

Reactance—An opposition to AC current based on the reaction of energy storage, either as a magnetic field or as an electric field. No real power is dissipated by a reactance. Reactance is measured in ohms. See also *Ohm.*

Rectifier—A device that has a high resistance in one direction and a low resistance in the other direction.

Relay—A device operated by a change in voltage or current in a circuit that actuates other devices in the same circuit or in another circuit.

Reluctance—An opposition to the establishment of magnetic flux lines when a magnetizing force is applied. Usually measured in rels. See also *Flux.*

Resistance—A physical property that opposes current and dissipates real power in the form of heat. Resistance is measured in ohms. See also *Ohm.*

Resonance—A condition in an electrical circuit, where the frequency of an externally applied force equals the natural tendency of the circuit.

Rheostat—A variable-resistive device consisting of a resistance element and a continuously adjustable contact arm.

Series circuit—A circuit that provides a complete path for current and has its components connected end-to-end.

Short-circuit—A fault path for current in a circuit that conducts excessive current. If the fault path has appreciable resistance, it is termed a leakage path.

Shunt—Denotes a parallel connection.

Sine wave—Variation in accordance with simple-harmonic motion.

Torque—Mechanical twisting force.

Transformer—A device that operates by electromagnetic induction with a tapped winding or with two or more separate windings, usually on an iron core, for the purpose of stepping voltage or current up or down, for maximum power transfer, for isolation of the primary circuit from the secondary circuit, and in special designs for automatic regulation of voltage or current.

Transient—A nonrepetitive or arbitrarily timed electrical surge.

Units—Established values of physical properties used in measurement and calculation (for example, the volt unit, the ampere unit, the ampere-turn unit, the ohms unit).

Value—The magnitude of a physical property expressed in terms of a reference unit (such as 117 volts, 60 hertz, 50 ohms, 3 henrys).

Vector—A graphical symbol for an alternating voltage or current, the length of which denotes the amplitude of the voltage or current, and the angle of which denotes the phase with respect to a reference phase.

Volt (V)—The unit of electromotive force. One volt produces a current of 1 ampere in a resistance of 1 ohm. See also *Ampere; Electromotive force; Resistance.*

Voltage—In a circuit, the greatest effective potential difference between a specified pair of circuit conductors.

Volt-amperes reactive (Vars)—The unit of apparent power (reactive power).

Watt (W)—The unit of electrical power, equal to the product of 1 volt and 1 ampere in DC values, or in root-mean-square AC values.

Wave—An electrical undulation, basically of sinusoidal form.

Waveform—Characteristic shape of an electrical current or signal.

Work—The product of force by the distance through which the force acts. Work is numerically equal to energy.

X-ray—An electromagnetic radiation with extremely short wavelength, capable of penetrating solid substances. Sometimes used in industrial plants to check the perfection of device and component fabrication (detection of flaws).

Index

S